S. BASU

The Glycoconjugates

Volume III

Glycoproteins, Glycolipids, and Proteoglycans
Part A

Contributors

LEIF C. ANDERSSON

MANJU BASU

SUBHASH BASU

JOHN W. BAYNES

LAWRENCE W. BERGMAN

THOMAS A. BEYER

MARTHA K. CATHCART

LLOYD A. CULP

R. DATEMA

CARL C. GAHMBERG

PAUL F. GOETINCK

GEIR O. GOGSTAD

LIV HELGELAND

ROBERT L. HILL

STUART KORNFELD

W. MICHAEL KUEHL

J. THOMAS LAMONT

HELLMUTH HANS RÖSSLER

HANS-JÖRG RISSE

BARRETT J. ROLLINS

R. T. SCHWARZ

BARRY D. SHUR

SUZANNE R. THORPE

KENNETH M. YAMADA

The Glycoconjugates

Volume III

Glycoproteins, Glycolipids, and Proteoglycans
Part A

EDITED BY

MARTIN I. HOROWITZ

Department of Biochemistry
New York Medical College
Valhalla, New York

1982

ACADEMIC PRESS
A Subsidiary of Harcourt Brace Jovanovich, Publishers
New York   London
Paris   San Diego   San Francisco   São Paulo   Sydney   Tokyo   Toronto

ACADEMIC PRESS, INC.
111 Fifth Avenue, New York, New York 10003

United Kingdom Edition published by
ACADEMIC PRESS, INC. (LONDON) LTD.
24/28 Oval Road, London NW1   7DX

Library of Congress Cataloging in Publication Data
Main entry under title:

Glycoproteins, glycolipids, and proteoglycans.

(The Glycoconjugates ; v. 3, pt. A)
Includes bibliographies and index.
1. Glycoproteins--Metabolism. 2. Glycolipids--
Metabolism. 3. Proteoglycans--Metabolism. I. Horowitz,
Martin I. II. Series. [DNLM: 1. Glycolipids. 2. Gly-
coproteins. 3. Proteoglycans. QU 55 G568 1977]
QP552.G59G59 vol. 3, pt. A 599.01'924s   77-4086
ISBN 0-12-356103-5   [599.01'924]   AACR2

82 83 84 85   9 8 7 6 5 4 3 2 1

Contents

Section 6

EXPRESSION OF GLYCOSPHINGOLIPID GLYCOSYLTRANSFERASES IN
DEVELOPMENT AND TRANSFORMATION
SUBHASH BASU AND MANJU BASU

3   GLYCOCONJUGATES IN CELLULAR ADHESION AND AGGREGATION

Section 1

FIBRONECTIN–PROTEOGLYCAN BINDING AS THE MOLECULAR BASIS
FOR FIBROBLAST ADHESION TO EXTRACELLULAR MATRICES
BARRETT J. ROLLINS, MARTHA K. CATHCART, AND LLOYD A. CULP

Section 2

BIOCHEMISTRY OF FIBRONECTIN
KENNETH M. YAMADA

Contents

List of Contributors

Numbers in parentheses indicate the pages on which the authors' contributions begin.

LEIF C. ANDERSSON (231), Transplantation Laboratory, Department of Surgery, University of Helsinki, 00290 Helsinki 29, Finland, and Department of Immunology, Biomedicum, Uppsala University, Uppsala, Sweden

MANJU BASU (265), Department of Chemistry, Biochemistry and Biophysics Program, University of Notre Dame, Notre Dame, Indiana 46556

SUBHASH BASU (265), Department of Chemistry, Biochemistry and Biophysics Program, University of Notre Dame, Notre Dame, Indiana 46556

JOHN W. BAYNES (113), Department of Chemistry, School of Medicine, University of South Carolina, Columbia, South Carolina 29208

LAWRENCE W. BERGMAN[1] (81), Department of Microbiology, University of Virginia, Charlottesville, Virginia 22908

THOMAS A. BEYER (25), The Department of Biochemistry, Duke University Medical Center, Durham, North Carolina 27710

MARTHA K. CATHCART[2] (289), Department of Microbiology, School of Medicine, Case Western Reserve University, Cleveland, Ohio 44106

LLOYD A. CULP (289), Department of Microbiology, School of Medicine, Case Western Reserve University, Cleveland, Ohio 44106

R. DATEMA (47), Institut für Virologie, Justus-Liebig-Universität Giessen, D-6300 Giessen, Federal Republic of Germany

CARL C. GAHMBERG[3] (231), Department of Bacteriology and Immunology, University of Helsinki, 00290 Helsinki 29, Finland

---

[1] Present address: Laboratory of Nutrition and Endocrinology, Department of Health and Human Services, National Institute of Arthritis, Diabetes, and Digestive and Kidney Diseases, National Institutes of Health, Bethesda, Maryland 20205.

[2] Present address: Department of Immunology, Cleveland Clinic Foundation, Cleveland, Ohio 44106

[3] Present address: Department of Biochemistry, University of Helsinki, 00170 Helsinki 17, Finland

PAUL F. GOETINCK (197), Department of Animal Genetics and Department of Biological Science, University of Connecticut, Storrs, Connecticut 06268

GEIR O. GOGSTAD (99), Research Institute for Internal Medicine, Section on Hemostasis and Thrombosis, University of Oslo, Rikshospitalet, Oslo 1, Norway

LIV HELGELAND (99), Department of Biochemistry, University of Oslo, Blindern, Oslo 3, Norway

ROBERT L. HILL (25), Department of Biochemistry, Duke University Medical Center, Durham, North Carolina 27710

STUART KORNFELD (3), Division of Hematology–Oncology, Washington University School of Medicine, St. Louis, Missouri 63110

W. MICHAEL KUEHL (81), Department of Microbiology, School of Medicine, University of Virginia, Charlottesville, Virginia 29908

J. THOMAS LAMONT[4] (187), Division of Gastroenterology, Peter Bent Brigham Hospital, Harvard Medical School, Boston, Massachusetts 02115

HELLMUTH HANS RÖSSLER (135), Institut für Veterinär–Biochemie, Freie Universität Berlin, Berlin, Federal Republic of Germany

HANS-JÖRG RISSE (135), Institut für Veterinär–Biochemie, Freie Universität Berlin, Berlin, Federal Republic of Germany

BARRETT J. ROLLINS[5] (289), Department of Microbiology, Case Western Reserve University, Cleveland, Ohio 44106

R. T. SCHWARZ (47), Institut für Virologie, Justus-Liebig-Universität Giessen, D-6300 Giessen, Federal Republic of Germany

BARRY D. SHUR (145), Department of Anatomy, University of Connecticut Health Center, Farmington, Connecticut 06032

SUZANNE R. THORPE (113), Department of Chemistry, University of South Carolina, Columbia, South Carolina 29208

KENNETH M. YAMADA (331), Membrane Biochemistry Section, Laboratory of Molecular Biology, National Cancer Institute, Bethesda, Maryland 20205

[4]Present address: Section of Gastroenterology, University Hospital, Boston, Massachusetts 02118

[5]Present address: Department of Medicine, Beth Israel Hospital, Boston, Massachusetts 02146

Preface

Since the publication of Volume II of "The Glycoconjugates" we have witnessed a quickening of the pace of research. As a result sufficient new trends and concepts have emerged to warrant preparation of this new volume.

One of the most rapidly growing areas has been that of glycosylation (Chapter 1, Volume III) with particular emphasis on lipid-linked sugar intermediates. Progress has been made in elucidating the structure, site of synthesis, and addition of oligosaccharides to their protein acceptors. The processing of oligosaccharides, a procedure which in some respects resembles the processing of proteins and nucleic acids in that larger structures are synthesized first and then are reduced in size or "trimmed" and then modified enzymatically, is the youngest subject in this area. We are fortunate to have for our opening contribution a chapter on processing by an investigator who has pioneered in the development of this subject. Following are a series of essays which discuss glycosylation, the transition between translation and glycosylation, and the intracellular transport of glycoproteins. It should be possible for the individual working in any of these areas to obtain a good review of his specific area and an overview of related areas.

The subject of nonenzymatic glycosylation of proteins also has been included even though this is a separate and distinct topic from the others discussed above. This subject, which originated as a relatively narrow one concerning the glycosylation of hemoglobin and the application of concentration levels of the glycosylated hemoglobins for monitoring the status of diabetics, has emerged as a general one with nonenzymatic glycosylation of many proteins being recognized as occurring in both normal and diseased states. The parameters of nonenzymatic glycosylation and how to monitor the process are clearly discussed in Chapter 1.6, and the reader is then prepared to appreciate and interpret future developments in this area.

Another area of considerable interest is the relationship of glycosylation and of glycosyltransferases to embryogenesis, differentiation, and development (Chapter 2, Volume III). Chapters were solicited from investigators who have studied those developmental processes in which glycosylation figures prominently. The authors have been instructed to assume that the reader is not an embryologist, but is interested in learning about the embryo (or in some instances a particular developing tissue) as a model system in which glycosylation may be a crucial event and glycosylated molecules important markers that affect the sequence of events and possibly provide signals for important qualitative developmental changes. Thus the authors have provided the background and principles relating to the specific system which they are discussing prior to introducing recent developments in the area. Chapter 2.2 on Fertilization and Early Embryogenesis is particularly recommended since it will provide orientation for the novice to the general subject of embryogenesis.

Membrane glycoconjugates, their renewal, shedding, and participation in cell–cell interactions are discussed in Chapter 3, Volume III and Chapters 1 and 2, Volume IV. Though membrane glycoconjugates were discussed in earlier volumes of "The Glycoconjugates," they were considered incidentally as constituents of tissue systems with the emphasis placed on reviews of methods for isolation, analysis, detection, and the role of these substances in the health and disease of specific systems, i.e., pulmonary, nervous, gastrointestinal, and genitourinary. In this volume, emphasis is placed on the glycoconjugates as constituents of membranes as such and on their participation in membrane dynamics. Our understanding in these areas has been facilitated by the development and application of methods for labeling and modifying these structures so that the dynamics of their insertion, transport, and degradation could be studied.

In reviewing their topics on membrane related phenomena, the authors have endeavored to transmit to the reader the rationale for the approaches used, logic behind the inferences drawn, and the advantages and possible pitfalls of the procedure employed.

Viral glycoconjugates were not reviewed in our earlier volumes that emphasized mammalian glycoconjugates. Since information is now available on the structures of several viral glycoconjugates, their biosynthesis, and interactions with host cells, considerable space is allocated for these topics in Chapter 3 of Volume IV. A certain amount of overlap between material in this chapter and that in Chapters 1.3–1.5, Volume III and Chapters 1 and 2, Volume IV was unavoidable because certain viral glycoproteins, most notably the G protein of VSV, have been used as model systems in studies of biosynthesis and transport of glycoproteins and the process of membrane fusion. The contributing authors were informed of the intersection of their areas of coverage, and as a result the duplication of coverage was kept to the minimum consonant with the need for properly developing a line of argument or rounding out a discussion. Liberal use was made of cross references to avoid excessive duplication and to guide the reader to related areas of interest.

In the final chapter of Volume IV, the reader will find reviews of glycosyltransferases and glycoconjugates as they relate to tumorigenesis, atherosclerosis, and arthritis. Chapter 4 is a well balanced essay which could have been placed equally well in Chapter 1, Volume III but was placed in Chapter 4 because of the incisive discussion of the uses and design of membrane sugar analogs to tumor chemotherapy. Chapter 4.3 on Glycosyltransferases in Cancer was solicited since information was accumulating on differences between normal individuals and cancer patients in the varieties and amounts of certain glycosyltransferases present in their respective sera, and there was a need to summarize and evaluate these findings. It is anticipated that in the future one will encounter an increasing amount of research on this topic and on the related topic of tumor antigens (see Chapter 7.2 in Volume I). The subject of Glycosaminoglycans in Atherosclerosis (Chapter 4.2) was included to call attention to the important role that glycosaminoglycans may play in the development of the atheromatous lesion. Much of the focus on the development of this disease has been on cholesterol and lipoproteins. Yet the interaction of lipoproteins with the arterial wall and their retention there appears to occur via glycosaminoglycans. The authors of this essay discuss the biology of this disease and review the pertinent research on glycosaminoglycans without being dogmatic about the role of these compounds in the pathogenesis of atherosclerosis. The final chapter on proteoglycans in aging and osteoarthritis (Chapter 4.4) is a logical sequel to Chapter 1 in Volume II of "The Glycoconjugates" which discusses basic aspects of glycosaminoglycan research and of Chapter 2.4 of Volume III on Proteoglycans in Developing Embryonic Cartilage. Owing to improvements in the methodology for fractionation of proteoglycans, it has become possible to study changes occurring in proteoglycans during aging and during the degradation of articular cartilage.

With the completion of the fourth volume in this series, I believe that we have been able to bring to the reader an up-to-date survey of the major areas of glycoconjugate research and to convey a number of the outstanding problems that remain to be attacked together with some of the approaches that may be used to that end. We intend to issue future volumes in this series at suitable intervals to help the reader keep abreast of the progress being made in these areas and to evaluate critically concepts and trends which may be emerging. The editor welcomes critical comments from the reader and suggestions for future reviews.

The editor gratefully acknowledges the cooperation of the authors in complying with the guidelines and timetable set for them. Thanks also are extended to the staff of Academic Press who worked hard to bring about the publication of this volume in a timely fashion.

Martin I. Horowitz

Contents of Previous Volumes

Volume I

1

Glycosylation of Proteins

SECTION 1

Oligosaccharide Processing during Glycoprotein Biosynthesis

STUART KORNFELD

I. INTRODUCTION

Many soluble and membrane-bound glycoproteins contain asparagine-linked oligosaccharide units, which can be divided into three main categories, termed high-mannose (or simple), hybrid, and complex (Kornfeld and Kornfeld, 1980). Typical examples of these oligosaccharides are shown in Figure 1. They all share a common core structure consisting of Manα1→3(Manα1→6)Manβ1→4-GlcNAcβ1→4GlcNAc→Asn but differ in their outer branches.* The sequence of the outer chains of complex-type units is most often SA→Gal→GlcNAc, as shown in the figure, but numerous variations on this sequence have been described (Kornfeld and Kornfeld, 1980). Some complex-type oligosaccharides

---

*Abbreviations: SA, sialic acid; GlcNAc, N-acetylglucosamine; Gal, galactose; Man, mannose; Fuc, fucose.

THE GLYCOCONJUGATES, VOL. III

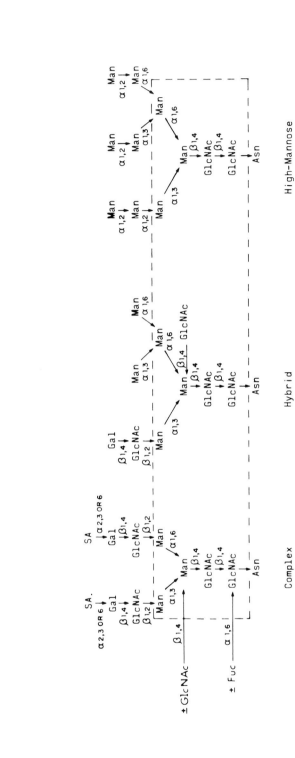

Figure 1 Examples of the major classes of asparagine-linked oligosaccharides. The dashed box shows the common core structure of all three classes of molecules.

also have an N-acetylglucosamine residue linked $\beta 1 \rightarrow 4$ to the β-linked mannose and a fucose residue linked $\alpha 1 \rightarrow 6$ to the innermost N-acetylglucosamine. High-mannose oligosaccharides typically have two to six additional mannose residues linked to the pentasaccharide core. Hybrid molecules are so named because they have features of both high-mannose and complex-type units. Recent experiments from several laboratories have demonstrated that all three classes of asparagine-linked oligosaccharides are derived from a single lipid-linked oligosaccharide precursor (Robbins *et al.*, 1977; Tabas *et al.*, 1978; Hunt *et al.*, 1978; Harpaz and Schachter, 1980b). This oligosaccharide is transferred to the nascent polypeptide and then processed to give rise to these various structures. This finding explains why the core regions of the different classes of molecules have identical structures. The purpose of this section is to describe the steps involved in the processing of asparagine-linked oligosaccharides and to discuss some of the possible reasons why this pathway developed.

II. SEQUENCE OF OLIGOSACCHARIDE PROCESSING

Starting with the work of Behrens *et al.* (1973), it has been well established that the biosynthesis of asparagine-linked oligosaccharide units is initiated by the *en bloc* transfer of a preformed oligosaccharide from an oligosaccharide pyrophosphoryldolichol intermediate to an asparagine residue of the nascent polypeptide chain [for a recent review, see Parodi and Leloir (1979)]. The structure of this molecule is shown in Figure 2. This oligosaccharide has three branches, one of which contains three glucose residues in the sequence Glc1$\rightarrow$2Glc1$\rightarrow$3-Glc1$\rightarrow$3 (Li *et al.*, 1978; Liu *et al.*, 1979). The glucoses appear to be in α linkage (Spiro *et al.*, 1979; Ugalde *et al.*, 1981). These residues function as a signal for transfer of the oligosaccharide to the nascent protein (Turco *et al.*, 1977). After this precursor molecule is transferred to the nascent polypeptide, it is extensively processed.

The sequence of the oligosaccharide processing has been determined in tissue culture cells infected with vesicular stomatitis virus (Kornfeld *et al.*, 1978; Hubbard and Robbins, 1979). This virus contains a single glycoprotein, termed G, which has two complex-type oligosaccharides. The experimental approach was to incubate virus-infected cells for 1–5 minutes with [2-^{3}H]mannose, [^{3}H]-glucosamine, or [^{3}H]galactose to label the mannose, N-acetylglucosamine, and glucose residues, respectively, of the lipid-linked oligosaccharide. The cells were then suspended in fresh medium without radioactive sugars, and aliquots were harvested at various times for up to 2 hours. Glycopeptides were prepared from the cells after various chase intervals and treated with either *Clostridium perfringens* endo-β-N-acetylglucosaminidase C_{II} or *Streptomyces plicatus* endo-β-N-acetylglucosaminidase H. These enzymes cleave the di-N-acetylchitobiose

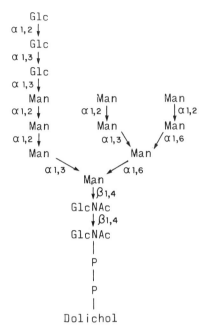

Figure 2 Proposed structure of the lipid-linked oligosaccharide precursor of asparagine-linked oligosaccharides.

unit of glycopeptides with the structure $Man_{>3}GlcNAc_2 Asn$ but do not act on $Man_3 GlcNAc_2 Asn$ or glycopeptides with complex-type oligosaccharides (Kobata, 1979). Consequently, they can be used to distinguish these two classes of glycopeptides. The released oligosaccharides were then separated by paper chromatography or gel filtration and their structures determined (Kornfeld *et al.*, 1978; Hubbard and Robbins, 1979). With this information the sequence of sugar removal could be deduced. The results of these studies are summarized in Figure 3. Shortly after transfer to the protein, processing is initiated by the removal of the glucose residues. In chick embryo fibroblasts the first glucose residue is removed very rapidly, with a half-time for the reaction of less than 2 minutes. The second glucose is removed more slowly (half-time of about 5 minutes), and the last glucose is removed even more slowly (Hubbard and Robbins, 1979). At approximately 15–30 minutes the last glucose residue is removed, and in normal cells processing then proceeds rapidly with the ultimate removal of six of the nine mannose residues and the addition of the outer branch sugars.

Studies with a mutant line of Chinese hamster ovary cells that lack UDP-GlcNAc:α-D-mannoside N-acetylglucosaminyltransferase I (GlcNAc-transferase I) have improved our understanding of the events involved in the later stages of processing. The finding (Tabas *et al.*, 1978; Li and Kornfeld, 1978; Robertson

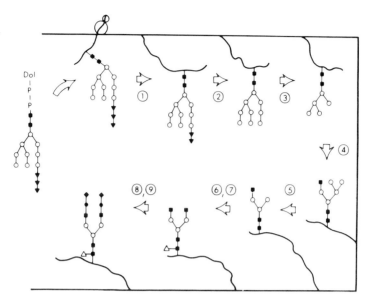

Figure 3 Proposed sequence for the processing of asparagine-linked oligosaccharides. The enzymes involved are (1) glucosidase I, (2) glucosidase II, (3) α-mannosidase I, (4) UDP-GlcNAc : α-D-mannoside $\beta 1 \rightarrow 2$ N-acetylglucosaminyltransferase I, (5) α-mannosidase II, (6) UDP-GlcNAc : α-D-mannoside $\beta 1 \rightarrow 2$ N-acetylglucosaminyltransferase II, (7) GDP-fucose : β-D-N-acetylglucos- aminide fucosyltransferse, (8) UDP-Gal : glycoprotein galactosyltransferase, (9) CMP-SA : glyco- protein sialyltransferase. Key to symbols: ■, N-acetylglucosamine residues; ○, mannose residues; ▼, glucose residues; ●, galactose residues; ◆, sialic acid residues; and △, fucose residues.

et al., 1978) that this cell line accumulates molecules with the structure Man_5- $GlcNAc_2$ was the clue that processing occurs in the following sequence. First, the four $\alpha 1,2$-linked mannoses are removed to give the $Man_5 GlcNAc_2$ structure, which appears to be the physiological substrate for GlcNAc-transferase I. This enzyme transfers an N-acetylglucosamine residue to the mannose residue that is linked $\alpha 1,3$ to the β-linked core mannose. The resulting oligosaccharide acts as a substrate for a second, highly specific α-mannosidase, which cleaves the two mannose residues linked to the mannose that is linked $\alpha 1,6$ to the β-linked mannose. A second N-acetylglucosaminyltransferase (GlcNAc-transferase II) then adds another outer N-acetylglucosamine residue to the mannose residue exposed by the removal of two mannose residues. The synthesis of the complex- type oligosaccharide is then completed by the addition of galactose, fucose, and sialic acid residues. Since high-mannose type oligosaccharides are derived from the same lipid-linked oligosaccharide precursor, they must also be processed by the removal of the glucose residues and, in some instances, by the excision of some of the $\alpha 1,2$-linked mannose residues.

These early processing steps occur in yeast as well as in animal cells. Several investigators have established that the synthesis of yeast mannoproteins is initiated by the transfer of a lipid-linked oligosaccharide with the composition $Glc_3Man_9GlcNAc_2$ to nascent protein (Parodi, 1979; Lehle, 1980; Trimble *et al.*, 1980). Subsequently, this oligosaccharide is processed by the removal of the three glucose residues and possibly some of the mannose residues (Parodi, 1979; Lehle, 1980) before the residual oligosaccharide is extended by the addition of more mannose residues. Although the processing of the glucose residues seems to be a general phenomenon, one exception has already been described. An established cell line of the mosquito *Aedes aegypti* contains a plasma membrane glycoprotein with an asparagine-linked oligosaccharide having the composition $Glc_3Man_9GlcNAc_2$ (Butters and Hughes, 1981). One explanation is that these cells lack the processing glucosidases.

Recently, a dolichyl diphosphate oligosaccharide with the composition $Glc_3Man_9GlcNAc_2$ has been identified in a number of plants (Staneloni *et al.*, 1980; Lehle, 1981), indicating that plants also use this pathway to glycosylate asparagine residues.

III. PROCESSING ENZYMES

A. Glucosidases I and II

Glucose processing is mediated by two distinct glucosidases, termed glucosidase I and glucosidase II. Glucosidase I removes the outermost glucose, which is linked $\alpha 1 \rightarrow 2$, whereas glucosidase II removes the two inner glucoses, both of which are linked $\alpha 1 \rightarrow 3$. The enzymes have been detected in hen oviduct (Chen and Lennarz, 1978), rat liver (Ugalde *et al.*, 1978; Grinna and Robbins, 1979), calf brain (Scher and Waechter, 1979), calf thyroid (Spiro *et al.*, 1979), and calf liver (Michael and Kornfeld, 1980). They are located in the microsomes (Ugalde *et al.*, 1978; Grinna and Robbins, 1979; Elting *et al.*, 1980), which is consistent with the evidence that glycosylation occurs on nascent polypeptides (Kiely *et al.*, 1976) and that glucose removal begins immediately after glycosylation. The two enzymes can be differentially solubilized from microsomal preparations and are physically separable by gel filtration or ion-exchange chromatography (Grinna and Robbins, 1979; Michael and Kornfeld, 1980; Ugalde *et al.*, 1979). Glucosidase I has been purified 1600-fold from hen oviduct, in which it is localized in the rough endoplasmic reticulum (Elting *et al.*, 1980). Glucosidase II has been purified from both rat liver (Grinna and Robbins, 1979; Ugalde *et al.*, 1981) and calf liver (Michael and Kornfeld, 1980). The enzymes can also be differentiated by their response to various inhibitors. Ugalde *et al.* (1980) have shown that kojibiose ($Glc\alpha 1 \rightarrow 2Glc$) is a specific inhibitor of glucosidase I, whereas nigerose (Glcα1→3Glc) is a specific inhibitor of glucosidase II. Maltose (Glcα1→4Glc) also inhibits glucosidase II but to a lesser extent than nigerose. None of the disaccharides with β-linked glucose is inhibitory. Both enzymes are highly specific for the intact glucosylated high-mannose oligosaccharide substrate. Thus, removal of one or more of the mannose residues from the 6'-pentamannosyl branch of the substrate (see Fig. 2) results in a significant decrease in the rate of glucose removal (Spiro *et al.*, 1979; Michael and Kornfeld, 1980; Ginna and Robbins, 1980). However, the enzymes do display some activity toward species containing only four and five mannose residues. This is of interest since a mutant line of mouse lymphoma cells is known to transfer a truncated oligosaccharide with the composition Glc$_3$Man$_5$GlcNAc$_2$ to protein and then to process this oligosaccharide to complex-type units (Kornfeld *et al.*, 1979). Several investigators have shown that the glucosidases can act on glucosylated lipid-linked oligosaccharides as well as protein-bound and free oligosaccharides. However, since kinetic studies in intact cells have shown that there is virtually no turnover of lipid-linked Glc$_3$Man$_9$GlcNAc$_2$ other than that which can be accounted for by transfer to protein (Hubbard and Robbins, 1980), it seems unlikely that the glucosidases act on the lipid-linked oligosaccharides *in vivo*.

A glucosidase that cleaves the terminal glucose from Glc$_3$Man$_9$GlcNAc$_2$ has been partially purified from *Saccharomyces cerevisiae* X-2180 (Kilker *et al.*, 1981). This enzyme is equally distributed between the particulate and soluble fractions of a cell homogenate. The pH optimum of the enzyme is 6.8, and there is no requirement for divalent cations. Since cell-free preparations are capable of removing all three glucose residues, the cells presumably contain glucosidase II as well as glucosidase I.

B. (α1,2)-Specific Mannosidase (Mannosidase I)

The next step in the conversion of the deglucosylated Man$_9$GlcNAc$_2$ oligosaccharide to a complex-type unit is the removal of the four (α1,2)-linked mannose residues. This reaction appears to be catalyzed by an (α1,2)-specific mannosidase, which is localized in the Golgi region of the cell (Tabas and Kornfeld, 1979). The enzyme has been solubilized with the nonionic detergent Triton X-100 and purified 6000-fold from rat liver (Tabas and Kornfeld, 1979). The purified enzyme is capable of removing all four (α1,2)-linked mannoses. The Man$_5$GlcNAc$_2$ intermediate, which lacks (α1,2)-linked mannose residues, is a very poor substrate for the enzyme. The enzyme has a pH optimum between 6.0 and 6.5. It has an apparent K_m for a Man$_8$GlcNAc oligosaccharide substrate of 100 μM. When tested with Man$_8$GlcNAc$_2$-Asn-peptide, a nonlinear double reciprocal plot was generated so that the only parameter that could be derived was the substrate concentration required for half-maximal velocity, which was 17

μM. Kinetic studies of mannose removal are consistent with a mechanism whereby the enzyme sequentially attaches to a processing intermediate, removes one mannose residue, detaches from this intermediate, and then begins the cycle again on another intermediate.

C. UDP-GlcNAc : α-D-Mannoside $\beta1\rightarrow2$ N-Acetylglucosaminyltransferase I

GlcNAc-Transferase I has been highly purified from bovine colostrum and rabbit liver by affinity chromatography on UDP-hexanolamine–agarose (Harpaz and Schachter, 1980a; Oppenheimer and Hill, 1981). As previously mentioned, the physiological substrate for the enzyme is the $Man_5GlcNAc_2$ processing intermediate (Fig. 3). This is reflected by the finding that the purified enzyme has the lowest K_m value for this substrate, although it is also quite active toward $Man\alpha1\rightarrow3(Man\alpha1\rightarrow6)Man\beta1\rightarrow4GlcNAc$ and $Man\alpha1\rightarrow3Man\beta1\rightarrow4GlcNAc$. The K_m values of these two acceptors are approximately 2 and 10 times that of the $Man_5GlcNAc$ acceptor. The most purified enzyme does not transfer N-acetylglucosamine residues to the mannose linked $\alpha1\rightarrow6$ to the β-linked mannose. The finding that $Man\alpha1\rightarrow3(Man\alpha1\rightarrow6)Man\beta1\rightarrow4GlcNAc$ is a good acceptor is in agreement with its proposed role in the alternate processing pathway described in Section IV.

D. ($\alpha1,3$)- and ($\alpha1,6$)-Specific Mannosidase (Mannosidase II)

Evidence for the existence of an α-mannosidase that is dependent on the prior action of GlcNAc-transferase I was first obtained by demonstrating that membrane preparations of Chinese hamster ovary cells are unable to release mannose from $[^3H]Man_5GlcNAc_2$-peptide unless UDP-GlcNAc is included in the reaction mixture (Tabas and Kornfeld, 1978). Furthermore, the presence of this mannosidase could not be demonstrated in membrane extracts from clone 15B cells, which are deficient in GlcNAc-transferase I activity. However, extracts of both Chinese hamster ovary and 15B cells could cleave mannose from the proposed intermediate $GlcNAcMan_5GlcNAc_2$-peptide. The structure of the major product formed in the reactions, which contained UDP-GlcNAc, was shown to be $GlcNAc\beta1\rightarrow2Man\alpha1\rightarrow3(GlcNAc\beta1\rightarrow2Man\alpha1\rightarrow6)Man\beta1\rightarrow4GlcNAc\beta1\rightarrow4GlcNAc$-peptide. These data indicate that the late-stage processing reactions occur as shown in Figure 4. Subsequent studies by Harpaz and Schachter localized this mannosidase to the Golgi membranes of rat liver (Harpaz and Schachter, 1980b). Starting with $GlcNAcMan_5GlcNAc_2$-Asn as substrate, these workers were able to isolate two $GlcNAcMan_4GlcNAc_2$-Asn intermediates. Although there was not enough material to perform a complete structural

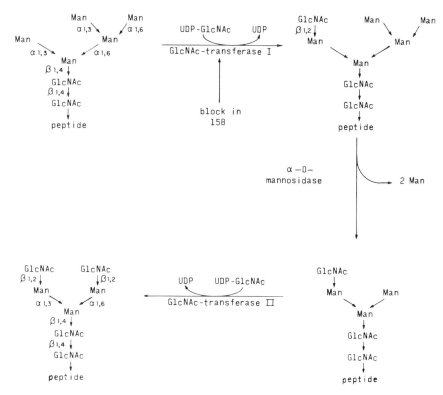

Figure 4 Sequence of reactions in the late stages of processing of complex oligosaccharide units. The α- D- mannosidase is mannosidase II.

analysis, the data were consistent with the major intermediate having the structure GlcNAcβ1→ 2Manα1→ 3(Manα1→3Manα1→6) Manβ1→ 4GlcNAcβ1→ 4GlcNAc-Asn. This suggests that mannosidase II acts preferentially on the Manα1→6 terminal residue.

Harpaz and Schachter also demonstrated that if the GlcNAcMan₅ GlcNAc₂- Asn substrate is substituted with another GlcNAc linked β1→4 to the β-linked mannose residue, the resulting compound is no longer a substrate for mannosidase II. This finding provides an explanation for the generation of "hybrid" structures, which have been described in ovalbumin (Tai *et al.*, 1977; Yamashita *et al.*, 1978). These molecules have *N*-acetylglucosamine or *N*-acetyllactosamine groups attached to the mannose residue that is linked α1,3 to the β-linked mannose and one or two mannose residues attached to the mannose that is linked α1,6 to the β-linked mannose. As shown in Figure 5, the hybrid molecules with five mannose residues could arise if a third GlcNAc-transferase (GlcNAc-transferase III) acts before mannosidase II. If the GlcNAc-transferase

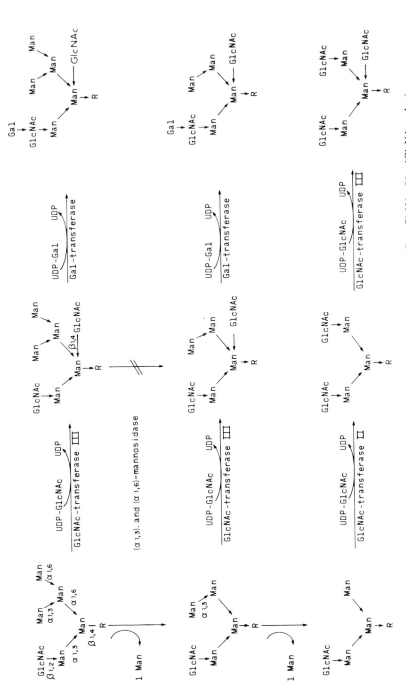

Figure 5 Proposed mechanism for the generation of hybrid-type oligosaccharides (R = GlcNAc-β1→4GlcNAc→Asn).

III acts after the α-mannosidase has removed a single mannose residue, then a hybrid molecule with four mannoses will result. Since a number of complex-type molecules have also been shown to have an N-acetylglucosamine residue linked $\beta1\rightarrow4$ to the β-linked mannose (Kornfeld and Kornfeld, 1980), one must postulate that this residue can also be transferred to the oligosaccharide after the α-mannosidase has removed both of the susceptible mannoses (see Fig. 5). It is not known whether one or more GlcNAc-transferases participate in the transfer of the N-acetylglucosamine residues to the β-linked mannose of these various acceptors.

Recent evidence (Tulsiani *et al.*, 1981) indicates that mannosidase II is identical to the previously described rat liver Golgi enzyme which is active toward p-nitrophenyl-α-D-mannopyranoside with a pH optimum of 5.5 (Opheim and Touster, 1978).

E. UDP-GlcNAc : α-D-Mannoside $\beta1\rightarrow2$ N-Acetylglucosaminyltransferase II and GDP-Fucose : β-D-N-Acetylglucosaminide Fucosyltransferase

The compound formed by the action of mannosidase II serves as a substrate for GlcNAc-transferase II (Narasimhan *et al.*, 1977) (see Fig. 4) and for a fucosyltransferase which transfers a fucose residue $\alpha1\rightarrow6$ to the GlcNAc residue that is linked to the asparagine (Wilson *et al.*, 1976). The assembly of biantennary complex-type oligosaccharides is then completed by the stepwise addition of galactose and sialic acid residues. The formation of the triantennary and tetraantennary molecules requires the addition of more N-acetylglucosamine residues to the α-linked mannose residues. (See Chapter 1, Section 2, of this volume for a detailed discussion of the glycosyltransferases.)

IV. ALTERNATE PROCESSING PATHWAY

Although the processing pathway illustrated in Figure 3 is the major pathway utilized by cells, an alternate processing pathway occurs in a mutant cell line that has a block in the synthesis of the $Glc_3Man_9GlcNAc_2$ lipid-linked oligosaccharide (Trowbridge and Hyman, 1979; Chapman *et al.*, 1979). Thy-1$^-$ mutant mouse lymphoma cells of the class E complementation group are unable to synthesize dolichol-P-mannose, the mannose donor for four of the nine mannose residues of the lipid-linked oligosaccharide (Chapman *et al.*, 1980). Consequently, these cells synthesize a truncated lipid-linked oligosaccharide with the structure $Glc1\rightarrow2Glc1\rightarrow3Glc1\rightarrow3Man\alpha1\rightarrow2Man\alpha1\rightarrow2Man\alpha1\rightarrow3(Man\alpha1\rightarrow6)$-$Man\beta1\rightarrow4GlcNAc\beta1\rightarrow4GlcNAc$. This oligosaccharide is transferred to protein and processed to give typical complex-type oligosaccharides (Kornfeld *et al.*, 1979). The sequence of processing is shown in Figure 6. The truncated oligosaccharide is transferred *en bloc* from the lipid carrier to newly synthesized protein. Within minutes the first glucose residue is excised, and then the next two glucose residues are removed. Subsequently, the two mannose residues linked $\alpha 1 \rightarrow 2$ are removed, presumably by the processing $(\alpha 1,2)$-specific mannosidase, since this enzyme acts on the $Man_5GlcNAc_2$ species *in vitro*. The resultant $Man_3GlcNAc_2$ species serves as a substrate for GlcNAc-transferase I, which transfers an *N*-acetylglucosamine residue to the mannose residue linked $\alpha 1 \rightarrow 3$ to the β-linked mannose. The assembly of the final complex-type oligosaccharide is then completed by the addition of one or more *N*-acetylglucosamine residues as well as the addition of galactose, fucose, and sialic acid residues.

This alternate pathway for complex-type oligosaccharide synthesis also functions in some normal tissue culture cells under certain experimental conditions. For example, glucose starvation of Chinese hamster ovary cells leads to a rapid cessation of the synthesis of the $Glc_3Man_9GlcNAc_2$ lipid-linked oligosaccharide and the accumulation of the $Man_5GlcNAc_2$ lipid-linked species. The

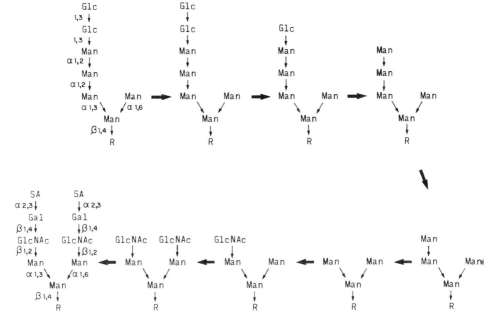

Figure 6 Steps in oligosaccharide processing by the alternate pathway (R = GlcNAc-$\beta 1 \rightarrow 4$GlcNAc$\rightarrow$Asn).

latter compound is glucosylated and transferred to protein, where it is subsequently processed (Rearick *et al.*, 1981). A similar phenomenon probably occurs in glucose-starved BHK cells that are infected with vesicular stomatitis virus (Turco, 1980) and possibly in other virus-infected cells that are deprived of glucose (Kaluza, 1975; Sefton, 1977).

Another variation in oligosaccharide processing has been described in a double mutant cell line (CHO-Pha^R Con A^R) selected sequentially for resistance to both phytohemagglutinin and concanavalin A (Robertson *et al.*, 1978). This cell line has a block in the synthesis of the lipid-linked oligosaccharide precursor, which leads to the formation of structures containing seven instead of nine mannose residues (Hunt, 1980). When this species is transferred to protein, it is processed to an oligosaccharide with the composition $Man_4GlcNAc_2$ rather than the usual $Man_5GlcNAc_2$. Since the mutant cell line is missing GlcNAc-transferase I, processing ceases at this point. It seems likely that a defect in the mutant cell line is the failure to add the last two mannose residues to the 6′-pentamannosyl portion of the lipid-linked oligosaccharide.

V. ASSEMBLY OF PHOSPHORYLATED HIGH-MANNOSE OLIGOSACCHARIDES

It is now established that 6-phosphomannosyl residues present on high-mannose oligosaccharide units of newly synthesized acid hydrolases serve as an essential component of the recognition marker that mediates enzyme uptake by various cell types and targeting to lysosomes (Kaplan *et al.*, 1977) [see Neufeld and Ashwell (1980) and Chapter 1, Section 1, Volume IV, for reviews]. Since phosphorylation occurs on the protein-bound oligosaccharide rather than on the lipid-linked oligosaccharide precursor (Goldberg and Kornfeld, 1981), this pathway represents a specialized form of oligosaccharide processing, which appears to be unique to acid hydrolases. The clue to the mechanism of phosphorylation was the finding that the biosynthetic intermediates of the acid hydrolases contain phosphate groups in diester linkage between mannose residues of the underlying oligosaccharide and outer α-linked N-acetylglucosamine residues (Tabas and Kornfeld, 1980; Hasilik *et al.*, 1980). The structures of these intermediates have been determined (Varki and Kornfeld, 1980b). These studies have shown that phosphates may be linked to five different mannose residues on the oligosaccharide, and individual molecules may contain one, two, and perhaps even three phosphates. A composite picture is shown in Figure 7. Since the phosphate is present in the mature enzyme as a phosphomonoester moiety, it was proposed that the mechanism of phosphorylation of acid hydrolases involves the transfer of α-N-acetylglucosamine 1-phosphate from UDP-GlcNAc to mannose residues of the high-mannose oligosaccharide followed by the selective removal

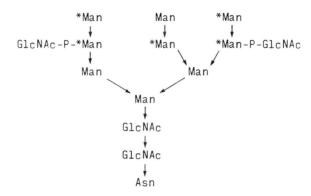

Figure 7 Structure of a phosphorylated high-mannose oligosaccharide containing two phos-phodiester units. The asterisk identifies the mannose residues that are phosphorylated in various isomers (Varki and Kornfeld, 1980b).

of the N-acetylglucosamine residue to expose a phosphomannosyl group (Tabas and Kornfeld, 1980; Hasilik et al., 1980).

As shown in Figure 8, this pathway requires two enzymes: an N-acetyl-glucosaminylphosphotransferase and an α-N-acetylglucosaminyl phosphodies-terase. Assays for both of these enzymes have been developed and the enzyme activities detected in a number of tissues. UDP-GlcNAc : glycoprotein N-acetyl-glucosaminyl-1-phosphotransferase activity has been detected in homogenates of Chinese hamster ovary cells, human diploid fibroblasts, and rat liver (Reitman and Kornfeld, 1981; Hasilik et al., 1981). Using $[\beta$-^{32}P]UDP-$[^3$H]GlcNAc as donor, it was shown that the enzyme transfers N-acetylglucosamine 1-phosphate to the 6-hydroxyl of mannose residues present in high-mannose oligosaccharides of endogenous glycoproteins and of exogenous thyroglobulin glycopeptides. The membrane-associated transferase is not inhibited by tunicamycin or stimulated by dolichol phosphate, indicating that the reaction does not proceed via a dolichol–pyrophosphoryl-N-acetylglucosamine intermediate. Fibroblasts from patients with the lysosomal enzyme storage diseases, I-cell disease (mucolipidosis II), and pseudo-Hurler polydystrophy (mucolipidosis III) are severely deficient in this enzyme activity (Hasilik et al., 1981; Reitman et al., 1981). These diseases are characterized by a generalized failure to target acid hydrolases to lysosomes in spite of normal rates of acid hydrolase synthesis (Hickman and Neufeld, 1972) and by the failure to incorporate $[^{32}$P]phosphate into newly synthesized acid hydrolases (Hasilik and Neufeld, 1980; Bach et al., 1979). The absence of the GlcNAc-phosphotransferase explains the lack of phosphorylation of the newly synthesized acid hydrolases. This failure to generate the phosphomannosyl rec-ognition signal prevents the receptor-mediated targeting of the newly synthesized acid hydrolases to lysosomes, and consequently the enzymes are secreted into the extracellular milieu.

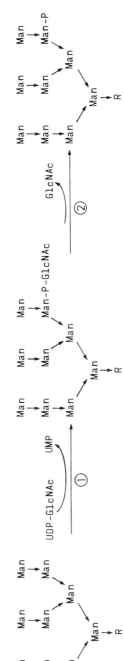

Figure 8 Proposed mechanism for the phosphorylation of the high-mannose units of acid hydrolases. The enzymes involved are (1) UDP-GlcNAc : glycoprotein N-acetylglucosaminyl-1-phosphotransferase and (2) α-N-acetylglucosaminyl phosphodiesterase (R = GlcNAcβ1→4GlcNAc→Asn).

An α-N-acetylglucosaminyl phosphodiesterase capable of removing the "covering" N-acetylglucosamine residues has been purified from rat liver (Varki and Kornfeld, 1981) and human placenta (Waheed *et al.*, 1981a). The enzyme is localized to Golgi-enriched smooth membranes and is distinct from the lysosomal α-N-acetylglucosaminidase, which can also remove covering N-acetylglucosamine residues (Varki and Kornfeld, 1980a; Waheed *et al.*, 1981b). It is highly specific for phosphorylated high-mannose oligosaccharides but shows some activity toward other molecules such as UDP-GlcNAc which contain N-acetylglucosamine 1-phosphate residues.

VI. REGULATION OF OLIGOSACCHARIDE PROCESSING

Since there is a common protein-bound oligosaccharide precursor which can be processed to form either a high-mannose or a complex-type chain, there is the potential of having either type of oligosaccharide chain at any particular asparagine that is glycosylated. Yet this type of heterogeneity is rarely seen. It does occur in bovine pancreatic RNase, where there is either a high-mannose oligosaccharide (as in RNase B) or a complex-type oligosaccharide (as in RNase C and D) at Asn-34 (Plummer, 1968). Presumably, some of the RNase molecules are partially processed, whereas others are completely processed. The typical situation, however, is exemplified by IgM immunoglobulin, which has five glycosylation sites. At three of the sites complex-type molecules are found, whereas at the other two high-mannose units are present (Putnam *et al.*, 1973). The factors governing the degree of processing are not well understood. The finding that a single polypeptide such as the IgM heavy chain may have both high-mannose and complex-type oligosaccharides rules out subcellular compartmentation as a mechanism for controlling processing. It seems most likely that the conformation of the protein and the location of the oligosaccharide along the polypeptide influence the extent of processing. The simplest case would be that high-mannose units result when the oligosaccharide is not accessible to the processing enzymes, and complex-type units are generated when the oligosaccharide precursor is located in an "exposed" position. The problem with this argument is that most high-mannose chains are processed to some extent; e.g., they lose all three glucoses and frequently one to four mannoses. For instance, in human IgM, the major high-mannose oligosaccharides at Asn-402 have the composition $Man_6GlcNAc_2$ and $Man_5GlcNAc_2$, whereas at Asn-563 the major oligosaccharides are $Man_8GlcNAc_2$ and $Man_6GlcNAc_2$ (Chapman and Kornfeld, 1979a,b; Cohen and Ballou, 1980). Although it is possible that this heterogeneity could arise from degradation of the oligosaccharides as the IgM circulates, this seems unlikely due to the low levels of plasma α-mannosidase activity at the physiological pH of 7.4 and the specific differences in the isomers at the two locations on the protein. The reasonable conclusion is that the extent and even the sequence of processing at any individual glycosylation site are influenced by the conformation of the protein at that site.

Hakimi and Atkinson presented evidence that there are also growth-dependent alterations in oligosaccharide processing (Hakimi and Atkinson, 1980). These investigators infected rapidly growing and resting chicken embryo fibroblasts with Sindbis virus and then analyzed the oligosaccharides present on the viral glycoproteins. They found that virus derived from the rapidly growing cells had greater quantities of larger high-mannose type of units than did virus from non-growing cells. Since the viral polypeptides are identical in both instances, these data show that the metabolic state of the host cell can also influence processing. However, the differences observed reflect the extent of mannose processing of high-mannose units and not the ratio of high-mannose to complex-type units. A similar growth-dependent alteration in mannose processing has been reported in human diploid cells (Muramatsu *et al.*, 1976).

VII. WHY PROCESSING?

Having described the complicated sequence of events that occur during oligosaccharide processing, it might be useful to speculate about the function of this pathway. One explanation for the development of processing on the polypeptide is that it represents a mechanism for the generation of diversity of oligosaccharide structures. This is readily apparent from Figure 9, which summarizes the types of oligosaccharides that are known to be derived from the precursor lipid-linked oligosaccharide. As shown in the figure, five different classes of oligosaccharides are formed from the single precursor, and within each class there are many subclasses. Therefore, the processing pathway allows the cell to generate a great variety of oligosaccharide structures. We can speculate that, as the complexity of organisms increased, there was a need to generate a wider repertoire of oligosaccharide structures to participate in more and varied biological functions. The processing pathway served this need without the necessity for developing a second lipid-linked oligosaccharide pathway. It is of interest that many primitive organisms such as yeast and fungi appear to contain only high-mannose oligosaccharides, whereas complex-type oligosaccharides appear to have developed at a later time in evolution. This observation suggests that high-mannose oligosaccharides are the ancestors of complex-type oligosaccharides and is consistent with the theory that processing developed in order to increase the diversity of oligosaccharide structures.

Another possible advantage of oligosaccharide processing is that it allows these molecules to have a dual function. In order to explain this we must first review some of the recent work on the role of asparagine-linked oligosaccharides

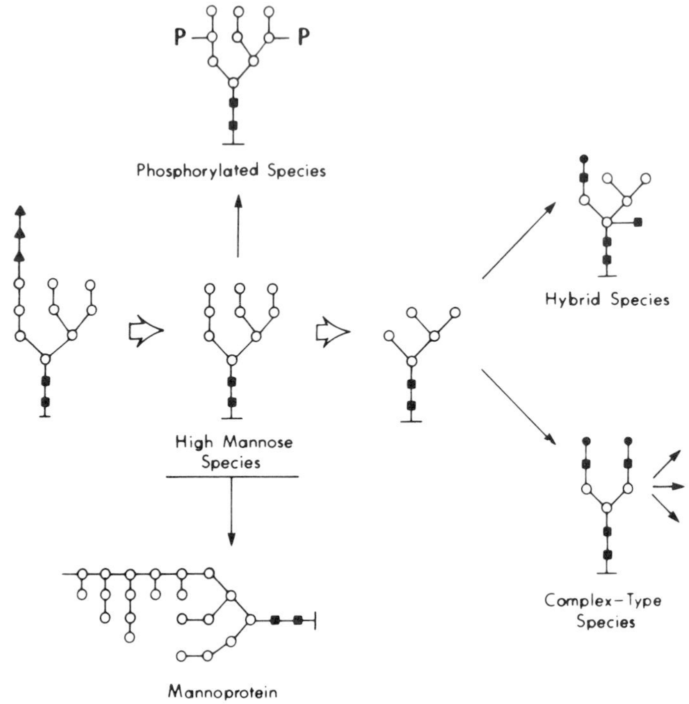

Figure 9 Summary of the types of oligosaccharides known to be generated from the common lipid-linked oligosaccharide precursors.

in stabilizing protein structure [for a more extensive review see Gibson *et al.* (1980)]. The availability of agents such as tunicamycin that are capable of blocking glycosylation has provided a means of examining how the absence of sugars affects those proteins normally destined to become glycoproteins. The results have been diverse, ranging from no effect to complete inactivation of the polypeptide (Gibson *et al.*, 1980). These differences in the requirement for glycosylation can be best understood if it is assumed that different glycoproteins have different requirements for glycosylation. In several of the systems evidence has been obtained that the oligosaccharide units of the glycoprotein have a major effect on the physical properties of the protein. Schwarz *et al.* (1976) used tunicamycin to block the glycosylation of influenza virus hemagglutinin and found that the carbohydrate-free hemagglutinin precursor was degraded intracellularly by proteases. This suggests that the oligosaccharide units of the hemagglutinin protect it against nonspecific cleavage and possibly have a positive role in identifying the normal sites of cleavage. Leavitt *et al.* (1977) demonstrated that the nonglycosylated glycoproteins of Sindbis virus and the San Juan strain of vesicular stomatitis virus are stable in the infected cell but are unable to migrate to the cell surface. The nonglycosylated glycoproteins of both viruses are insoluble in nonionic detergents owing to aggregation, whereas their glycosylated counterparts are soluble (Leavitt *et al.*, 1977; Gibson *et al.*, 1979). It is not only the nonglycosylated form of the vesicular stomatitis virus G protein that tends to aggregate. Aggregation can also be observed when the G protein lacks the full complement of carbohydrate at the time of polypeptide folding (Gibson *et al.*, 1981). The importance of glycosylation for proper protein foldings has also been demonstrated by Chu *et al.* (1978) in the case of yeast invertase. These workers were able to detect differences in the physical properties of the normal and carbohydrate-depleted forms of the enzyme only after denaturation and subsequent renaturation.

These data indicate that some glycoproteins require the presence of the large $Glc_3Man_9GlcNAc_2$ oligosaccharide at the time of folding. Since high-mannose chains appear to be the most primitive forms of oligosaccharides, these glycoproteins may have evolved with the large high-mannose unit assuming an important role during the initial folding. The retention of this large oligosaccharide would then be essential for these glycoproteins to achieve their correct tertiary structure. However, once the initial folding has occurred, the large oligosaccharide may no longer be necessary, and the sugar chain may be processed so that it can serve a different function. In this manner the oligosaccharide unit at a specific glycosylation site may serve a dual function.

REFERENCES

Bach, G., Bargal, R., and Cantz, M. (1979). *Biochem. Biophys. Res. Commun.* **91,** 976-981.
Behrens, N. H., Carminatti, H., Staneloni, R. J., Leloir, L. F., and Cantarella, A. I. (1973). *Proc. Natl. Acad. Sci. U.S.A.* **70,** 3390-3394.
Butters, T. D., and Hughes, R. C. (1981). *Biochim. Biophys. Acta* **640,** 655-671.
Chapman, A., and Kornfeld, R. (1979a). *J. Biol. Chem.* **254,** 816-823.
Chapman, A., and Kornfeld, R. (1979b). *J. Biol. Chem.* **254,** 824-828.
Chapman, A., Trowbridge, I. S., Hyman, R., and Kornfeld, S. (1979). *Cell* **17,** 509-515.
Chapman, A., Fujimoto, K., and Kornfeld, S. (1980). *J. Biol. Chem.* **255,** 4441-4445.
Chen, W. W., and Lennarz, W. J. (1978). *J. Biol. Chem.* **253,** 5780-5785.
Chu, F. K., Trimble, R. B., and Maley, F. (1978). *J. Biol. Chem.* **253,** 8691-8693.
Cohen, R. E., and Ballou, C. E. (1980). *Biochemistry* **19,** 4345-4358.
Elting, J. J., Chan, W. W., and Lennarz, W. J. (1980). *J. Biol. Chem.* **255,** 2325-2331.
Gibson, R., Schlesinger, S., and Kornfeld, S. (1979). *J. Biol. Chem.* **254,** 3600-3607.
Gibson, R., Kornfeld, S., and Schlesinger, S. (1980). *TIBS* **5,** 290-293.
Gibson, R., Kornfeld, S., and Schlesinger, S. (1981). *J. Biol. Chem.* **256,** 456-462.
Goldberg, D., and Kornfeld, S. (1981). *J. Biol. Chem.* **256,** 13060-13067.
Grinna, L. S., and Robbins, P. W. (1979). *J. Biol. Chem.* **254,** 8814-8818.
Grinna, L. S., and Robbins, P. W. (1980). *J. Biol. Chem.* **255,** 2255-2258.
Hakimi, J., and Atkinson, P. H. (1980). *Biochemistry* **19,** 5619-5624.

Harpaz, N., and Schachter, H. (1980a). *J. Biol. Chem.* **255**, 4885–4893.

Harpaz, N., and Schachter, H. (1980b). *J. Biol. Chem.* **255**, 4894–4902.

Hasilik, A., and Neufeld, E. F. (1980). *J. Biol. Chem.* **255**, 4946–4950.

Hasilik, A., Klein, U., Waheed, A., Strecker, G., and von Figura, K. (1980). *Proc. Natl. Acad. Sci. U.S.A.* **77**, 7074–7078.

Hasilik, A., Waheed, A., and von Figura, K. (1981). *Biochem. Biophys. Res. Commun.* **98**, 761–767.

Hickman, S., and Neufeld, E. F. (1972). *Biochem. Biophys. Res. Commun.* **49**, 992–999.

Hubbard, S. C., and Robbins, P. W. (1979). *J. Biol. Chem.* **254**, 4568–4576.

Hubbard, S. C., and Robbins, P. W. (1980). *J. Biol. Chem.* **255**, 11782–11793.

Hunt, L. A. (1980). *Cell* **21**, 407–415.

Hunt, L. A., Etchison, J. R., and Summers, D. F. (1978). *Proc. Natl. Acad. Sci. U.S.A.* **75**, 754–758.

Kaluza, G. (1975). *J. Virol.* **16**, 602–612.

Kaplan, A., Achord, D. T., and Sly, W. S. (1977). *Proc. Natl. Acad. Sci. U.S.A.* **74**, 2026–2030.

Kiely, M. L., McKnight, G. S., and Schimke, R. T. (1976). *J. Biol. Chem.* **251**, 5490–5495.

Kilker, R. D., Saunier, B., Tkacz, J. S., and Herscovics, A. (1981). *J. Biol. Chem.* **256**, 5299–5303.

Kobata, A. (1979). *Anal. Biochem.* **100**, 1–14.

Kornfeld, R., and Kornfeld, S. (1980). *In* "The Biochemistry of Glycoproteins and Proteoglycans" (W. J. Lennarz, ed.), pp. 1–34. Plenum, New York.

Kornfeld, S., Li, E., and Tabas, I. (1978). *J. Biol. Chem.* **253**, 7771–7778.

Kornfeld, S., Gregory, W., and Chapman, A. (1979). *J. Biol. Chem.* **254**, 11649–11654.

Leavitt, R., Schlesinger, S., and Kornfeld, S. (1977). *J. Biol. Chem.* **252**, 9018–9023.

Lehle, L. (1980). *Eur. J. Biochem.* **109**, 589–601.

Lehle, L. (1981). *FEBS Lett.* **123**, 63–66.

Li, E., and Kornfeld, S. (1978). *J. Biol. Chem.* **253**, 6426–6431.

Li, E., Tabas, I., and Kornfeld, S. (1978). *J. Biol. Chem.* **253**, 7762–7770.

Liu, T., Stetson, B., Turco, S. J., Hubbard, S. C., and Robbins, P. W. (1979). *J. Biol. Chem.* **254**, 4554–4559.

Michael, J. M., and Kornfeld, S. (1980). *Arch. Biochem. Biophys.* **199**, 249–258.

Muramatsu, T., Koide, N., Ceccarini, C., and Atkinson, P. H. (1976). *J. Biol. Chem.* **251**, 4673–4679.

Narasimhan, S., Stanley, P., and Schachter, H. (1977). *J. Biol. Chem.* **252**, 3926–3933.

Neufeld, E. F., and Ashwell, G. (1980). *In* "The Biochemistry of Glycoproteins and Proteoglycans" (W. J. Lennarz, ed.), pp. 252–257. Plenum, New York.

Opheim, D. J., and Touster, O. (1978). *J. Biol. Chem.* **253**, 1017–1023.

Oppenheimer, C. L., and Hill, R. L. (1981). *J. Biol. Chem.* **256**, 799–804.

Parodi, A. J. (1979). *J. Biol. Chem.* **254**, 10051–10060.

Parodi, A. J., and Leloir, L. (1979). *Biochim. Biophys. Acta* **559**, 1–37.

Plummer, T. H. (1968). *J. Biol. Chem.* **243**, 5961–5966.

Putnam, F. W., Florent, G., Paul, C., Shinoda, T., and Shimizu, A. (1973). *Science* **182**, 287–290.

Rearick, J., Chapman, A., and Kornfeld, S. (1981). *J. Biol. Chem.* **256**, 6255–6261.

Reitman, M. L., and Kornfeld, S. (1981). *J. Biol. Chem.* **256**, 4275–4281.

Reitman, M. L., Varki, A., and Kornfeld, S. (1981). *J. Clin. Invest.* **62**, 1574–1579.

Robbins, P. W., Hubbard, S. C., Turco, S. J., and Wirth, D. F. (1977). *Cell* **12**, 893–900.

Robertson, M. A., Etchison, J. R., Robertson, J. S., Summers, D. F., and Stanley, P. (1978). *Cell* **13**, 515–526.

Scher, M. G., and Waechter, C. J. (1979). *J. Biol. Chem.* **254**, 2630–2637.

Schwarz, R. T., Rohrschneider, J. M., and Schmidt, M. F. G. (1976). *J. Virol.* **19**, 782–791.

Sefton, B. M. (1977). *Cell* **10,** 654–668.

Spiro, R. G., Spiro, M. J., and Bhoyroo, V. D. (1979). *J. Biol. Chem.* **254,** 7659–7667.

Staneloni, R. J., Tolmasky, M. E., Petriella, C., Ugalde, R. A., and Leloir, L. (1980). *Biochem. J.* **191,** 257–260.

Tabas, I., and Kornfeld, S. (1978). *J. Biol. Chem.* **253,** 7779–7786.

Tabas, I., and Kornfeld, S. (1979). *J. Biol. Chem.* **254,** 11655–11663.

Tabas, I., and Kornfeld, S. (1980). *J. Biol. Chem.* **255,** 6633–6639.

Tabas, I., Schlesinger, S., and Kornfeld, S. (1978). *J. Biol. Chem.* **253,** 716–722.

Tai, T., Yamashita, K., and Kobata, A. (1977). *Biochem. Biophys. Res. Commun.* **78,** 434–491.

Trimble, R. B., Maley, F., and Tarentino, A. L. (1980). *J. Biol. Chem.* **255,** 10232–10238.

Trowbridge, I. S., and Hyman, R. (1979). *Cell* **17,** 503–508.

Tulsiani, D. R. P., Hubbard, S. C., Robbins, P. W., and Touster, O. (1981). *Fed. Proc., Fed. Am. Soc Exp. Biol.* **10,** 1883.

Turco, S. J. (1980). *Arch. Biochem. Biophys.* **205,** 330–339.

Turco, S. J., Stetson, B., and Robbins, P. W. (1977). *Proc. Natl. Acad. Sci. U.S.A.* **74,** 4411–4414.

Ugalde, R. A., Staneloni, R. J., and Leloir, L. F. (1978). *FEBS Lett.* **91,** 209–212.

Ugalde, R. A., Staneloni, R. J., and Leloir, L. F. (1979). *Biochem. Biophys. Res. Commun.* **91,** 1174–1181.

Ugalde, R. A., Staneloni, R. J., and Leloir, L. F. (1980). *Eur. J. Biochem.* **113,** 97–103.

Varki, A., and Kornfeld, S. (1980a). *J. Biol. Chem.* **255,** 8398–8401.

Varki, A., and Kornfeld, S. (1980b). *J. Biol. Chem.* **255,** 10847–10858.

Varki, A., and Kornfeld, S. (1981). *J. Biol. Chem.* **256,** 9937–9943.

Waheed, A., Hasilik, A. and von Figura, K. (1981a). *J. Biol. Chem.* **256,** 5717–5721.

Waheed, A., Pohlmann, R., Hasilik, A., and von Figura, K. (1981b). *J. Biol. Chem.* **256,** 4150–4152.

Wilson, J. R., Williams, D., and Schachter, H. (1976). *Biochem. Biophys. Res. Commun.* **72,** 909–916.

Yamashita, K., Tachibana, Y., and Kobata, A. (1978). *J. Biol. Chem.* **253,** 3862–3869.

SECTION 2

Glycosylation Pathways in the Biosynthesis of Nonreducing Terminal Sequences in Oligosaccharides of Glycoproteins

THOMAS A. BEYER AND ROBERT L. HILL

I. INTRODUCTION

Oligosaccharide groups in glycoproteins are synthesized by the sequential actions of a series of glycosyltransferases, with each transferase extending the length of the oligosaccharide by one monosaccharide residue. A specific transferase is thought to be required for the synthesis of each of the nearly 100 different disaccharide sequences that are known to be present in glycoconjugates (Dawson, 1978), but less than one-half of the predicted transferase activities have been identified in *in vitro* systems, and only about a dozen of these have been purified and enzymatically characterized. Thus, with the increasingly large number of new oligosaccharide structures that are reported each year (Montreuil, 1980; Kornfeld and Kornfeld, 1980), the number of glycosyltransferases that

THE GLYCOCONJUGATES, VOL. III
Copyright © 1982 by Academic Press, Inc.
All rights of reproduction in any form reserved.
ISBN 0-12-356103-5

must be identified to account for oligosaccharide biosynthesis is also becoming increasingly large.

Most of the glycosyltransferases that have been purified and characterized at present are those associated with the synthesis of nonreducing terminal sequences in glycoproteins (Beyer *et al.*, 1981). Although many kinds of experimental approaches are required to reveal the details of the pathways for oligosaccharide biosynthesis, it has been possible with the aid of these transferases to deduce certain aspects of oligosaccharide synthesis that were unknown until recently. Thus, this section focuses on our current understanding of the biosynthesis of the nonreducing terminal sequences in N- and O-linked oligosaccharides of glycoproteins as revealed by studies with purified glycosyltransferases and their structurally well defined acceptor substrates. The biosynthesis of dolichylphosphoryl oligosaccharides, their transfer to protein, and their processing are reviewed elsewhere in this series (Chapter 2, Volume II; Chapter 1, Section 1, this volume), as is the biosynthesis of proteoglycans (Chapter 1, Section 1, Volume II) and glycolipids (Chapter 4, Volume II).

II. GLYCOSYLTRANSFERASES

Sialic acid, fucose, galactose, *N*-acetylgalactosamine, *N*-acetylglucosamine, and mannose are found most frequently at nonreducing terminal positions in oligosaccharides of glycoproteins. More than one monosaccharide may be terminal since most oligosaccharides are branched structures. Moreover, because of structural microheterogeneity, the oligosaccharide groups in pure glycoproteins may differ somewhat from molecule to molecule. Oligosaccharides that are attached to a specific amino acid side chain may have the same carbohydrate–protein linkage and identical core structures but may differ in the number of branches or in the sequence of terminal residues on a given branch (Fournet *et al.*, 1978). It is presently unclear how such variation in oligosaccharide structure is related to the functional roles of the oligosaccharide groups. In addition, very little is known about mechanisms that aid in the regulation of oligosaccharide synthesis except that the strict substrate acceptor specificities as well as the subcellular localization of glycosyltransferases contribute to the ultimate structure of an oligosaccharide in a glycoconjugate. Nevertheless, before considering what is currently known about the sequential and concerted actions of glycosyltransferases (Section III), it is valuable to consider first certain properties of the glycosyltransferases that form the terminal nonreducing structures in oligosaccharides of glycoproteins. The methods for assay and purification of these transferases and their molecular properties have been reviewed recently (Sadler *et al.*, 1981). Each class of transferase will be considered in turn.

A. Sialyltransferases

Table I lists the types of sialic acid (Sia) linkages known to occur in mammalian glycoproteins. Each of the linkages listed is thought to be synthesized by a different transferase, although only five of the nine expected transferase activities have been shown to exist in *in vitro* systems, and only three of the five have been purified extensively. Each of the known transferases catalyzes a reaction of the following form

$$\text{CMP-SIA} + \text{acceptor} \rightarrow \text{Sia}\alpha2 \rightarrow n\text{-acceptor} + \text{CMP}$$

where n refers to the substituted hydroxyl group in the acceptor.

The most common sequences containing sialic acid are Sia$\alpha2\rightarrow$6Gal and Sia$\alpha2\rightarrow$3Gal, which are found not only in a variety of soluble glycoproteins such as those of blood plasma, but also in membrane-bound glycoproteins on cell surfaces or on intracellular membranes. The transferase ($\beta1\rightarrow4$ galactoside $\alpha2\rightarrow6$ sialyltransferase) that synthesizes the Sia$\alpha2\rightarrow$6Gal sequence was purified to homogeneity from bovine colostrum (Paulson *et al.*, 1977a) and its substrate specificity examined (Paulson *et al.*, 1977b). Substrates with the nonreducing terminal sequence Gal$\beta1\rightarrow$4GlcNAc were the only effective acceptors; those with the terminal sequence Gal$\beta1\rightarrow$3GalNAc were less than 1% as effective as acceptors. This high degree of acceptor substrate specificity is observed with most glycosyltransferases and accounts for the fact that there are two galactoside $\alpha2\rightarrow3$ sialyltransferases: one that forms the Sia$\alpha2\rightarrow$3Gal$\beta1\rightarrow$4GlcNAc sequence in N-linked oligosaccharides, and another that forms the Sia$\alpha2\rightarrow$3Gal-

TABLE I

Types of Sialic Acid Linkage in Glycoproteins

Sialic acid–galactose
  Sia$\alpha2$,6Gal$\beta1$,4GlcNAc-[a]
  Sia$\alpha2$,3Gal$\beta1$,4GlcNAc-[b]
  Sia$\alpha2$,3Gal$\beta1$,3GalNAc-[a]
  Sia$\alpha2$,4Gal$\beta1$,4GlcNAc-
  Sia$\alpha2$,4Gal$\beta1$,3GlcNAc-
Sialic acid–N-acetylgalactosamine
  Sia$\alpha2$,6GalNAcα-[a]
Sialic acid–N-acetylglucosamine
  Sia$\alpha2$,4GlcNAc-
  Sia$\alpha2$,6GlcNAc-
Sialic acid–sialic acid
  Sia$\alpha2$,8Sia$\alpha2$[b]

[a] Sialytransferases purified.
[b] Sialytransferases known.

$\beta1{\rightarrow}3$GalNAc sequence in O-linked oligosaccharides. The latter enzyme ($\beta1{\rightarrow}3$ galactoside $\alpha2{\rightarrow}3$ sialyltransferase) was purified to homogeneity from porcine submaxillary glands (Sadler *et al.*, 1979a), and examination of its substrate specificity showed that the only effective acceptors are those containing Gal-$\beta1{\rightarrow}3$GalNAc terminal sequences (Rearick *et al.*, 1979). The other $\alpha2{\rightarrow}3$ sialyltransferase ($\beta1{\rightarrow}4$ galactoside $\alpha2{\rightarrow}3$ sialyltransferase) has not been purified extensively, but characterization of its substrate structural requirements will most likely provide considerable insight into aspects of oligosaccharide synthesis. It would appear to require the same acceptors as the $\alpha2{\rightarrow}6$ sialyltransferase; however, this may not be strictly the case since, as discussed in Section III, structural differences that extend well beyond the Gal$\beta1{\rightarrow}4$GlcNAc sequence may determine the rate at which each transferase acts and thus the ultimate structure of an oligosaccharide.

The sialyltransferases that synthesize the other types of sialic acid linkages (Table I) have not been identified or are poorly characterized, with the exception of the transferase that forms the Sia$\alpha2{\rightarrow}6$GalNAc sequence (N-acetylgalactosaminide $\alpha2{\rightarrow}6$ sialyltransferase), which was purified to homogeneity from porcine submaxillary glands and characterized enzymatically (Sadler *et al.*, 1979a,b). Two acceptor substrates were found for this transferase: GalNAcα-O-Ser/Thr and Gal$\beta1{\rightarrow}3$GalNAcα-O-Ser/Thr in glycoproteins; the latter, however, is a much better substrate than the former.

B. Galactosyltransferases

The biosynthesis of only three of the nine galactose linkages found in terminal nonreducing sequences in glycoproteins (Table II) has been examined. The most common sequence, Gal$\beta1{\rightarrow}4$GlcNAc, which occurs in glycolipids and proteoglycans as well as in the O- and N-linked glycoprotein oligosaccharides, is formed

TABLE II
Types of Galactose Linkage in Glycoproteins

Galactose–N-acetylglucosamine	Galactose–N-acetylgalactosamine
Gal$\beta1$,4GlcNAc-[a]	Gal$\beta1$,3GalNAc-[b]
Gal$\beta1$,3GlcNAc-	Gal$\beta1$,6GalNAc-
Gal$\beta1$,6GlcNAc-	Gal$\alpha1$,3GalNAc-
Galactose–galactose	
Gal$\alpha1$,3Gal-[a]	
Gal$\beta1$,3Gal-	
Gal$\beta1$,6Gal-	

[a] Highly purified and characterized.
[b] Activity identified in tissues.

by the N-acetylglucosaminide $\beta1\rightarrow4$ galactosyltransferase. The enzyme has been purified to homogeneity from bovine and human milk, where it interacts with α-lactalbumin to form lactose synthase, and from a variety of nonmammary tissues and fluids, where it appears to function in glycoprotein biosynthesis (Beyer *et al.*, 1981). Although N-acetylglucosamine is utilized as a substrate,

$$\text{UDP-Gal} + \text{GlcNAc} \rightarrow \text{Gal}\beta1\rightarrow4\text{GlcNAc} + \text{UDP}$$

structures with the N-acetylglucosamine in β linkage to other sugar residues are much better acceptors (Schanbacher and Ebner, 1970). Thus, the apparent K_m values for GlcNAc, GlcNAc$\beta1\rightarrow4$GlcNAc, and ovalbumin (with terminal GlcNAc$\beta1\rightarrow2$Man residues) are 8.3, 0.6, and 1.7 mM, respectively.

The B blood group galactosyltransferase (fucosyl $\alpha1\rightarrow2$ galactoside $\alpha1\rightarrow3$ galactosyltransferase) has been purified to apparent homogeneity from human serum (Nagai *et al.*, 1978b). The enzyme catalyzes the reaction

$$\text{UDP-Gal} + \text{Fuc}\alpha1\rightarrow2\text{Gal} \rightarrow \begin{array}{c} \text{Gal}\alpha1\rightarrow3\diagdown \\ \qquad\qquad \text{Gal}\beta\text{-} + \text{UDP} \\ \text{Fuc}\alpha1\rightarrow2\diagup \end{array}$$

The Fuc$\alpha1\rightarrow2$Gal sequence is an essential structure in acceptor substrates, since galactosides free of $(\alpha1\rightarrow2)$-linked fucose are not acceptors.

C. Fucosyltransferases

The types of fucose linkages found in nonreducing terminal sequences in glycoproteins are listed in Table III. Four of the seven linkage types have been shown to be synthesized by glycosyltransferase activities in *in vitro* systems, and three of the enzymes have been highly purified. The transferase (β-galactoside $\alpha1\rightarrow2$ fucosyltransferase) that forms the H blood group structure, Fuc$\alpha1\rightarrow2$Gal, has been purified to homogeneity from porcine submaxillary glands (Beyer *et*

TABLE III
Types of Fucose Linkage in Glycoproteins

Fucose–galactose	Fucose–N-acetylglucosamine
Fucα1,2Galβ-[a]	Fucα1,4(Galβ1,3)GlcNAcβ-[a]
Fucα1,6Galβ-	Fucα1,3(Galβ1,4)GlcNAcβ-[a]
Fucα1,3Galβ-	Fucα1,6(R-GlcNAcβ1,4)GlcNAcβ-[b]
Fucose–fucose	
Fucα1,3Fuc	

[a] Fucosyltransferase purified extensively.
[b] Fucosyltransferase activity identified in tissues.

al., 1980) and characterized enzymatically (Beyer and Hill, 1980). It catalyzes the reaction

$$\text{GDP-Fuc} + \text{Gal}\beta\text{-R} \rightarrow \text{Fuc}\alpha1{\rightarrow}2\text{Gal}\beta\text{-R} + \text{GDP}$$

where R is any of a number of structures. Thus, unlike many transferases, its acceptor specificity is rather broad. A fucosyltransferase that forms fucosides with *N*-acetylglucosamine has also been purified extensively from human milk and characterized (Prieels *et al.*, 1981). Surprisingly, a preparation purified about 5×10^5-fold had a constant ratio of two different activities throughout its purification and catalyzed each of the following reactions:

$$\text{GDP-Fuc} + \text{Gal}\beta1{\rightarrow}4\text{GlcNAc} \rightarrow \begin{array}{c}\text{Gal}\beta1{\rightarrow}4\\ \\ \text{Fuc}\alpha1{\rightarrow}3\end{array}\!\!\!\!\!\! \diagdown\!\!\!\!\diagup \text{GlcNAc} + \text{GDP}$$

$$\text{GDP-Fuc} + \text{Gal}\beta1{\rightarrow}3\text{GlcNAc} \rightarrow \begin{array}{c}\text{Gal}\beta1{\rightarrow}3\\ \\ \text{Fuc}\alpha1{\rightarrow}4\end{array}\!\!\!\!\!\! \diagdown\!\!\!\!\diagup \text{GlcNAc} + \text{GDP}$$

Thus, this transferase appears to form both $\alpha1{\rightarrow}3$ and $\alpha1{\rightarrow}4$ linkages and is an exception to the concept that one specific transferase is required for the synthesis of each type of disaccharide sequence in glycoconjugates. The fucosyltransferase that forms the Fuc$\alpha1{\rightarrow}6$GlcNAc sequence with the *N*-acetylglucosamine residue linked to asparagine in complex-type N-linked oligosaccharides appears to be widely distributed, since many glycoproteins contain this structure. Although the enzyme has not been extensively purified, characterization of its substrate specificity indicates that it may play a role in regulation of N-linked oligosaccharide synthesis, as discussed in Section III (Wilson *et al.*, 1976; Longmore and Schachter, 1980).

D. *N*-Acetylglucosaminyl- and *N*-Acetylgalactosaminyltransferases

N-Acetylglucosamine is known to occur in 14 different types of glycosidic linkages (Beyer *et al.*, 1981) in glycoproteins from higher organisms, a far greater number than that for any other monosaccharide. Many of these, however, are not in terminal nonreducing sequences. The most frequently encountered terminal *N*-acetylglucosamine residues are those that form branches with mannose in di-, tri- and tetraantennary N-linked oligosaccharides, as shown in the following hypothetical sequence:

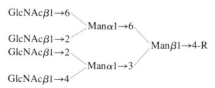

Only two of the N-acetylglucosaminyltransferases that form such sequences, however, have been purified and characterized. One catalyzes the formation of the GlcNAcβ1→2Manα1→3Man sequence, and the other the GlcNAcβ1→2Manα1→6Man sequence. As discussed in Section III, the former may aid in the regulation of N-linked oligosaccharide synthesis.

The only N-acetylgalactosaminyltransferase that forms terminal sequences catalyzes the reaction

$$\text{UDP-GalNAc} + \text{Fuc}\alpha1{\rightarrow}2\text{Gal}\beta\text{-R} \rightarrow \begin{array}{c} \text{Fuc}\alpha1{\rightarrow}2 \\ \diagdown \\ \text{GalNAc}\alpha1{\rightarrow}3 \diagup \end{array} \text{Gal}\beta\text{R-} + \text{UDP}$$

and forms the A$^+$ blood group structure with a variety of acceptors (Schwyzer and Hill, 1977b), each of which requires the Fucα1→2Gal sequence. It has been purified from human serum and porcine submaxillary glands (Nagai *et al.*, 1978a; Schwyzer and Hill, 1977a).

E. Mannosyltransferases

Most of the known mannosyltransferases are involved in the biosynthesis of the ''high-mannose'' dolichylphosphoryl oligosaccharide intermediates in N-linked oligosaccharides. They employ either GDP-Man or dolichylphosphorylmannose as donor substrates in reactions of the following form,

$$\text{GDP-Man} + \text{R} \rightarrow \text{Man}\alpha/\beta\text{-R} + \text{GDP}$$
$$\text{Dol-PO}_4\text{-Man} + \text{R} \rightarrow \text{Man}\alpha\text{-R} + \text{Dol-PO}_4$$

where R is any one of a number of acceptors (Rearick *et al.*, 1981). At present, no one of these transferases has been purified and characterized, and thus little is known of their enzymatic properties except what can be deduced from the identification of their substrates and products.

III. BIOSYNTHESIS OF N-LINKED OLIGOSACCHARIDES

The biosynthesis of N-linked oligosaccharides of glycoproteins proceeds in four steps: (a) synthesis of an oligosaccharide rich in mannose and glucose on a phosphoryldolichol intermediate; (b) transfer of the oligosaccharide from phosphoryldolichol to a protein acceptor; (c) processing of the protein-bound oligosaccharide by glycosidases that remove specific, terminal glucose and mannose residues; and (d) elongation of the oligosaccharide by the sequential actions of specific glycosyltransferases (Chapter 2, Volume II). The sequence of reactions in steps (a), (b), and (c) has been established, although very little is known about the glycosyltransferases involved. In contrast, several of the transferases that participate in elongation [step (d)] have been purified and well characterized (Section II), although the exact sequence in which they act to form the final nonreducing terminal sequences remains poorly understood. Since most N-linked oligosaccharides differ primarily in structure from one glycoprotein to another in their nonreducing terminal sequences, our knowledge of the reactions in step (d) that lead to the synthesis of terminal regions is essential for a complete understanding of oligosaccharide biosynthesis.

A. Elongation

Typical complex-type oligosaccharides contain from two to four branches, each of which consists of the sequence Galβ1→4GlcNAc attached to the α-linked mannose residues of the core oligosaccharide.

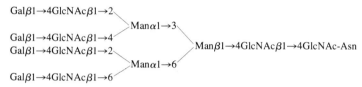

These, in turn, may be substituted with sialic acid or fucose in a variety of linkages. In principle, the formation of each of these branches might be completely independent of the elongation of the other branches such that a great number of possible reaction sequences could lead to the formation of the same completed oligosaccharide chain. However, it appears that the acceptor substrate specificities of the glycosyltransferases involved effectively prohibit many of these potentially possible pathways and, in some instances, may restrict the synthesis of a particular structure to a specific sequence of reactions.

The first step unique to the synthesis of the complex-type chains is the transfer of N-acetylglucosamine in β1→2 linkage to the α-linked mannose residues in the core oligosaccharide. In a mutant Chinese hamster ovary cell line defective in glycoprotein glycosylation (Gottlieb et al., 1974, 1975), Narasimhan et al. (1977) demonstrated that two different N-acetylglucosaminyltransferases were responsible for synthesis of the GlcNAcβ1→2Man linkages. Wild-type cells could transfer N-acetylglucosamine to both structures I and II below,

Manα1→3
          ⟩Manβ1→4GlcNAcβ1→4GlcNAc-Asn   (I)
Manα1→6

GlcNAcβ1→2Manα1→3
                ⟩Manβ1→rGlcNAcβ1→4GlcNAc-Asn   (II)
          Manα1→6

whereas the mutant cells could transfer only to structures of type II. The mutant cells were proposed to lack the enzyme that transfers N-acetylglucosamine to structure I, designated N-acetylglucosaminyltransferase I. These conclusions have been substantiated by the resolution and purification of the two transferases. Harpaz and Schachter (1980) reported the 11,000-fold purification of transferase I from bovine colostrum. The enzyme transferred to both structures I and II, but the K_m for the former (0.2 mM) was 50 times lower than the K_m for the latter (10 mM). Oppenheimer and Hill (1981) purified transferase I to near homogeneity from rabbit liver. In contrast to the colostrum enzyme, this transferase was totally inactive toward structure II.

N-Acetylglucosaminyltransferase II has been partially purified from bovine colostrum (Harpaz and Schachter, 1980) and porcine trachea (Reddy *et al.*, 1979). The former was inactive with structure I as an acceptor, whereas the latter used both acceptors, with structure II 20 times more effective than structure I.

On the basis of the oligosaccharide structures from the mutant CHO cell line which lacks transferase I, Li and Kornfeld (1978) proposed that physiological acceptors for this enzyme would have structure III.

$$\text{Man}\alpha1\rightarrow3 \diagdown$$
$$\qquad\qquad \text{Man}\alpha1\rightarrow6 \diagdown$$
$$\text{Man}\alpha1\rightarrow6 \diagup \qquad\qquad \text{Man}\beta1\rightarrow4\text{GlcNAc}\beta1\rightarrow4\text{GlcNAc-Asn} \qquad\qquad (\text{III})$$
$$\qquad\qquad \text{Man}\alpha1\rightarrow3 \diagup$$

This is in accord with the substrate specificity of the enzyme since acceptors with this structure have the lowest K_m values of all acceptors tested (Harpaz and Schachter, 1980; Oppenheimer and Hill, 1981). Confirmation of the proposal came from studies by Tabas and Kornfeld (1978) on the α-mannosidase that removes the two most distal mannose residues on structure III. This oligosaccharide served as a substrate for the mannosidase only in wild-type CHO cells and only in the presence of UDP-GlcNAc. In contrast, structure IV

$$\text{Man}\alpha1\rightarrow3 \diagdown$$
$$\qquad\qquad \text{Man}\alpha1\rightarrow6 \diagdown$$
$$\text{Man}\alpha1\rightarrow6 \diagup \qquad\qquad \text{Man}\beta1\rightarrow4\text{GlcNAc}\beta1\rightarrow4\text{GlcNAc-Asn} \qquad\qquad (\text{IV})$$
$$\text{GlcNAc}\beta1\rightarrow2\text{Man}\alpha1\rightarrow3 \diagup$$

served as a substrate for the mannosidase in both the wild-type and mutant cells and showed no requirement for UDP-GlcNAc. Thus, on the basis of the substrate specificities of the two N-acetylglucosaminyltransferases and the processing α-mannosidase, the following pathway for transfer of these N-acetylglucosamine residues was proposed:

$$\text{Man}\alpha1\rightarrow3 \diagdown \qquad\qquad\qquad\qquad \text{Man}\alpha1\rightarrow3 \diagdown$$
$$\qquad\qquad \text{Man}\alpha1\rightarrow3 \diagdown \qquad\qquad\qquad\qquad \text{Man}\alpha1\rightarrow6 \diagdown$$
$$\text{Man}\alpha1\rightarrow6 \diagup \qquad\qquad \text{Man-R}\rightarrow \text{Man}\alpha1\rightarrow6 \diagup \qquad\qquad \text{Man-R}$$
$$\qquad\qquad \text{Man}\alpha1\rightarrow6 \diagup \qquad\qquad \text{GlcNAc}\beta1\rightarrow2\text{Man}\alpha1\rightarrow3 \diagup$$
$$\downarrow$$
$$\text{GlcNAc}\beta1\rightarrow2\text{Man}\alpha1\rightarrow6 \diagdown \qquad\qquad\qquad\qquad \text{Man}\alpha1\rightarrow6 \diagdown$$
$$\qquad\qquad\qquad\qquad \text{Man-R} \leftarrow \qquad\qquad\qquad\qquad \text{Man-R}$$
$$\text{GlcNAc}\beta1\rightarrow2\text{Man}\alpha1\rightarrow3 \diagup \qquad\qquad \text{GlcNAc}\beta1\rightarrow2\text{Man}\alpha1\rightarrow3 \diagup$$

N-Acetylglucosaminyltransferases that form other N-acetylglucosamine–mannose linkages have not been described; thus, the pathways for the formation of tri- and tetraantennary oligosaccharide chains are not known.

A fucosyltransferase activity catalyzing the reaction

$$R\text{-Man}\alpha 1 \rightarrow 6$$
$$\text{Man}\beta 1 \rightarrow 4\text{GlcNAc}\beta 1 \rightarrow 4\text{GlcNAc}\beta\text{-Asn} + \text{GDP-fucose} \rightarrow$$
$$\text{GlcNAc}\beta 1 \rightarrow 2\text{Man}\alpha 1 \rightarrow 3$$

$$R\text{-Man}\alpha 1 \rightarrow 6$$
$$\text{Man}\beta 1 \rightarrow 4\text{GlcNAc}\beta 1 \rightarrow 4$$
$$\text{GlcNAc}\beta 1 \rightarrow 2\text{Man}\beta 1 \rightarrow 3 \text{GlcNAc}\beta\text{-Asn} + \text{GDP}$$
$$\text{Fuc}\alpha 1 \rightarrow 6$$

has been described by Schachter and co-workers (Wilson *et al.*, 1976; Longmore and Schachter, 1980) in rat liver microsomes. The enzyme displays an absolute specificity for acceptors with an unsubstituted GlcNAc$\beta 1 \rightarrow 2$Man$\alpha 1 \rightarrow 3$Man branch but shows less specificity for the nature of the R substituent on the Man$\alpha 1 \rightarrow 6$Man branch. Thus, the fucosyltransferase is constrained to act after the transfer of N-acetylglucosamine by transferase I but before the addition of galactose to the Man$\alpha 1 \rightarrow 3$Man branch.

The incorporation of galactose into these oligosaccharide chains is catalyzed by the N-acetylglucosaminide $\beta 1 \rightarrow 4$ galactosyltransferase. Despite the fact that this enzyme has been purified to homogeneity from a number of sources and its enzymatic properties have been extensively studied, surprisingly little is known about its reaction with glycoprotein acceptors. In an investigation of the enzymatic basis for the apparent incomplete chains that occur in the oligosaccharide chains of IgG, Rao and Mendicino (1978) tested the influence of glycopeptide structure on the activity of the galactosyltransferase. Both human (Baenziger and Kornfeld, 1974) and porcine (Rao *et al.*, 1976) IgG contain monosialylated, monogalactosylated biantennary chains in which one branch terminates with an N-acetylglucosamine residue. The following glycopeptides prepared from porcine IgG were tested as acceptors for the porcine lymph node galactosyltransferase:

$$\text{GlcNAc}\beta 1 \rightarrow 2\text{Man}\alpha 1 \rightarrow 6$$
$$\text{Man}\beta 1 \rightarrow 4\text{GlcNAc}\beta 1 \rightarrow 4\text{GlcNAc-Asn}$$
$$\text{GlcNAc}\beta 1 \rightarrow 2\text{Man}\alpha 1 \rightarrow 3$$
$$(\text{V})$$

$$\text{Gal}\beta 1 \rightarrow 4\text{GlcNAc}\beta 1 \rightarrow 2\text{Man}\alpha 1 \rightarrow 6$$
$$\text{Man}\beta 1 \rightarrow 4\text{GlcNAc}\beta 1 \rightarrow 4\text{GlcNAc-Asn}$$
$$\text{GlcNAc}\beta 1 \rightarrow 2\text{Man}\alpha 1 \rightarrow 3$$
$$(\text{VI})$$

$$\text{NeuAc}\alpha 2 \rightarrow 6\text{Gal}\beta 1 \rightarrow 4\text{GlcNAc}\beta 1 \rightarrow 2\text{Man}\alpha 1 \rightarrow 6$$
$$\text{Man}\beta 1 \rightarrow 4\text{GlcNAc}\beta 1 \rightarrow 4\text{GlcNAc-Asn}$$
$$\text{GlcNAc}\beta 1 \rightarrow 2\text{Man}\alpha 1 \rightarrow 3$$
$$(\text{VII})$$

Each of the glycopeptides was an acceptor for the transferase, but the apparent K_m values increased from 0.25 mM for glycopeptide V, to 2 mM for VI, and to 10 mM for VII. Thus, the sialylation of one branch might effectively prohibit completion of the second branch. The existence of these "incomplete" structures on the IgG was proposed to arise from a higher ratio of sialyltransferase to galactosyltransferase at the site of synthesis of the IgG, which would result in rapid sialylation of the monogalactosylated oligosaccharide. Structural analysis of the monogalactosylated product synthesized *in vitro* was not performed, so it is not known whether the incorporation of galactose occurs randomly on the two branches or preferentially on one branch. However, the occurrence of monogalactosylated chains in nature in which the galactose occurs exclusively on one branch suggests that the latter possibility is probably the case.

The final step in the synthesis of the complex-type chains is the addition of one or more residues of α-linked sialic acid or fucose. These reactions are considered in more detail in Section III,B, but it should be pointed out here that, like the N-acetylglucosaminyltransferases and the galactosyltransferase, at least some of these enzymes show preferential glycosylation of one branch of a complex-type chain. Paulson *et al.* (1978) reported that the incorporation of sialic acid into asialotransferrin and asialo-α_1-acid glycoprotein by the bovine colostrum β-galactoside $\alpha2\rightarrow6$ sialyltransferase was biphasic, with incorporation into the first 50% of the sites being 10^2–10^3 times faster than into the second 50%. The authors suggested that this might reflect a preferential sialylation of one branch of a complex-type chain over the other branches. This proposal was substantiated by van den Eijnden *et al.* (1980), who examined the structures of bi- and triantennary α_1-acid glycoprotein glycopeptides that had been partially sialylated *in vitro* by the $\beta1\rightarrow4$ galactoside $\alpha2\rightarrow6$ sialyltransferase. With the following biantennary structure as substrate,

Galβ1→4GlcNAcβ1→2Manα1→3
$\qquad\qquad\qquad\qquad\qquad\qquad$ Manβ1→4GlcNAcβ1→4GlcNAc-Asn
Galβ1→4GlcNAcβ1→2Manα1→6

greater than 90% of the sialic acid was incorporated into the Manα1→3Man branch. The GlcNAcβ1→2Manα1→3Man branch in the triantennary glycopeptide,

Galβ1→4GlcNAcβ1→4
$\qquad\qquad\qquad\qquad$ Manα1→3
Galβ1→4GlcNAcβ1→2 $\qquad\qquad\qquad$ Manβ1→4GlcNAcβ1→4GlcNAc-Asn
$\quad$ Galβ1→4GlcNAcβ1→2Manα1→6

was also preferentially sialylated, with incorporation into the GlcNAcβ1→4Manα1→3Man branch being intermediate between the other two.

Differential incorporation has also been observed with several other glycosyltransferases that terminate complex-type oligosaccharide synthesis (Beyer *et al.*,

1979; Beyer and Hill, 1980), but interpretation of these findings must await structural characterization of the partially glycosylated products.

A schematic pathway summarizing the possible biosynthetic steps in the formation of the peripheral portion of a typical biantennary oligosaccharide chain is shown in Figure 1. Reactions designated by a solid arrow have been demonstrated *in vitro*, whereas those designated by an open arrow have not been tested. Reactions that are prohibited based on the acceptor substrate specificities of the enzymes are not shown. Sugar residues on the Manα1→3Man branch are denoted by solid symbols to distinguish between the two branches.

As previously discussed, the initial steps in complex-type oligosaccharide synthesis are catalyzed by the two β1→2 N-acetylglucosaminyltransferases and the processing α-mannosidase, as shown in reactions 1, 2, and 6. However, the products of each of these reactions is an acceptor substrate for the N-acetylglucosaminide α1→6 fucosyltransferase (reactions 14, 15, and 16) (Longmore and Schachter, 1980). Thus, the transfer of fucose may precede removal of the mannose residues and transfer of the second N-acetylglucosamine residue. The products of reactions 14 and 15 are acceptor substrates for the N-acetylglucosaminide β1→4 galactosyltransferase (reactions 18 and 19) (C. L. Oppenheimer and R. L. Hill, unpublished observations), but these products have not been tested as acceptors for the α2→6 sialyltransferase (reactions 22 and 23). The occurrence of hybrid-type oligosaccharide structures in nature (Yamashita *et al.,* 1978) with a lactosamine moiety typical of complex-type chains on the Manα1→3Man branch, and α-mannosyl residues typical of high-mannose type of chains on the Manα1→6Man branch, suggests that reactions 14 and 18 occur *in vivo*. However, the products of reactions 14 and 18 have not been tested as substrates for the α-mannosidase (reactions 3 and 4), nor has the product of reaction 19 been tested as an acceptor for the β1→2 N-acetylglucosaminyl-transferase II (reaction 8). Thus, it is not known whether flux through these pathways can lead to the formation of the completed oligosaccharide chain.

At present it is unclear whether the synthesis of the digalactosylated product occurs preferentially via the pathway involving reactions 11 and 21 or the pathway involving reactions 20 and 12, or whether synthesis via both pathways is allowed. The occurrence of a monogalactosylated oligosaccharide in human IgG with the same structure as the product of reaction 20 (Baenziger and Kornfeld, 1974) suggests that the latter pathway might be preferred. However, this preferential galactosylation of the Manα1→3Man branch could also be due to transfer of galactose to this branch before the transfer of the N-acetylglucosamine residue to the Manα1→6Man branch (reactions 19 and 8).

Of course, such a scheme cannot take into account such factors as the compartmentalization of the enzymes within the endoplasmic reticulum and Golgi apparatus or the availability of nucleotide sugar donor substrates at the site of synthesis. Thus, the actual sequence of reactions *in vivo* may be even more

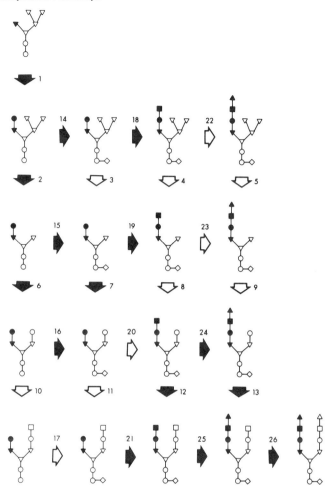

Figure 1 Possible biosynthetic pathways in the elongation of a biantennary complex-type oligosaccharide. Reactions that have been demonstrated *in vitro* are designated by a solid arrow. Sugar residues on the Manα1→3Man branch are shown in solid symbols to distinguish between the two branches. Reactions are catalyzed by the following enzymes: reaction 1, α1→3 mannoside β1→2 N-acetylglucosaminyltransferase (Harpaz and Schachter, 1980; Oppenheimer and Hill, 1981); reactions 2, 3, 4, and 5, processing α-mannosidase (Tabas and Kornfeld, 1978); reactions 6, 7, 8, and 9, α1→6 mannoside β1→2 N-acetylglucosaminyltransferase (Harpaz and Schachter, 1980); reactions 10, 11, 12, 13, 18, 19, 20, and 21, N-acetylglucosaminide β1→4 galactosyltransferase (Rao and Mendicino, 1978; C. L. Oppenheimer and R. L. Hill, unpublished observations); reactions 14, 15, 16, and 17, N-acetylglucosaminide α1→6 fucosyltransferase (Longmore and Schachter, 1980); reactions 22, 23, 24, 25, and 26, β-galactoside α2→6 sialyltransferase (Paulson *et al.*, 1978; van den Eijnden *et al.*, 1980). Key: ○, N-acetylglucosamine; ▽, mannose; □, galactose; ◇, fucose; △, sialic acid.

specific than shown in Figure 1. Also, the construction of such a scheme assumes that the acceptor substrate specificities of the glycosyltransferases that catalyze the same reaction in different organisms will be identical. This is a reasonable assumption, but its validity awaits experimental confirmation.

B. Chain Termination

As mentioned above, the final step in the synthesis of the complex-type oligosaccharide chains is often the transfer of one or more α-linked sugar residues, typically sialic acid and fucose and, less commonly, galactose, N-acetylgalactosamine, N-acetylglucosamine or mannose, to nonreducing terminal positions on the oligosaccharide. This process is referred to as chain termination because attachment of these residues usually blocks further elongation of the oligosaccharide structure. It is unclear why the addition of specific sugars results in chain termination or why the addition of different sugars is required for termination in different glycoproteins, except perhaps for lysosomal hydrolases containing phosphomannose residues (Varki and Kornfeld, 1980; see Chapter 4, Section 1, Volume IV). Nevertheless, since more than one residue is often attached to a single chain, producing a branched structure, the sequence of addition of the terminal residues cannot be deduced simply from the structure of the product. However, certain aspects of the biosynthesis of some of these structures can be predicted from the acceptor substrate specificities of the glycosyltransferases that form these linkages.

In recent years, a number of the glycosyltransferases involved in chain termination of complex-type oligosaccharides have been obtained in pure or highly purified form (Beyer *et al.*, 1981). Beyer *et al.* (1979) used four of these enzymes, a β-galactoside $\alpha2{\rightarrow}6$ sialyltransferase, a β-galactoside $\alpha1{\rightarrow}2$ fucosyltransferase, a β-N-acetylglucosaminide $\alpha1{\rightarrow}3$ fucosyltransferase, and a fucosyl $\alpha1{\rightarrow}2$ galactoside $\alpha1{\rightarrow}3$ N-acetylgalactosaminyltransferase, to study all of the possible sequential reactions with human asialotransferrin as a model substrate containing complex-type oligosaccharide chains. The acceptor substrate specificities of these enzymes are outlined in Section II. On the basis of the structures that served as acceptors for each enzyme and the relative rates of these reactions, the biosynthetic pathway shown in Figure 2 was proposed.

Asialotransferrin, which contains oligosaccharides with the nonreducing terminal sequence Gal$\beta1{\rightarrow}4$GlcNAc (structure j), is an acceptor for the sialyltransferase and both fucosyltransferases, giving structures k, l, and m in close to theoretical yield under appropriate conditions. As reported by Paulson *et al.* (1978), structure k is not an acceptor for the $\alpha1{\rightarrow}3$ fucosyltransferase and structure l is not an acceptor for the sialyltransferase; thus, product n cannot be synthesized. Since structure k is not an acceptor for the $\alpha1{\rightarrow}2$ fucosyltransferase, the other monosialylated, monofucosylated product (structure o) can

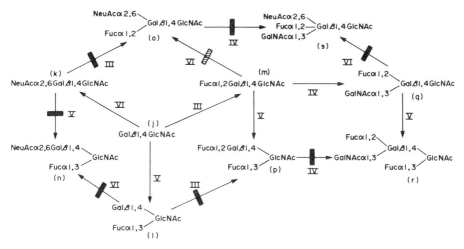

Figure 2 Proposed pathways for the biosynthesis of the nonreducing terminal sequences of complex-type oligosaccharide chains. The solid bars indicate that a reaction does not occur, and the hatched bars indicate that the reaction occurs very slowly. III, β-Galactoside α1,2 fucosyltransferase; IV, α1,2 fucosylgalactoside α1,3 N-acetylgalactosaminyltransferase; V, N-acetylglucosaminide α1,3 fucosyltransferase; VI, β-galactoside α2,6 sialyltransferase. From Beyer *et al.* (1979).

be formed only by the action of the sialyltransferase on structure m. However, the rate of this reaction is extremely slow, suggesting that it may not be significant *in vivo*. In accord with these specificities, no naturally occurring products corresponding to structures n or o have been reported. Oligosaccharide chains similar to structure n, in which the sialic acid is in α2→3 linkage to the galactose, have been described (Krusius and Finne, 1978). The sialyltransferase that forms this linkage has not been characterized; therefore, the pathway for synthesis of this structure cannot be predicted.

The difucosylated product, structure p, can be formed only if the fucose α1→2 galactose linkage is formed first, since structure *l* is not an acceptor for the α1→2 fucosyltransferase. However, since structure m is actually a better acceptor for the α1→3 fucosyltransferase than is structure j, the difucosylated product is easily formed. This structure occurs in both glycoproteins (Lloyd *et al.*, 1966) and glycolipids (McKibbin, 1978).

The fucosyl α1→2 galactoside α1→3 N-acetylgalactosaminyltransferase is specific for acceptors with the nonreducing terminal sequence Fucα1→2Gal (Schwyzer and Hill, 1977b). Therefore, asialotransferrin (structure j) is not an acceptor, but the product of the α1→2 fucosyltransferase (structure m) is an acceptor. As reported by Kobata *et al.* (1968), the difucosylated product, structure p, is not an acceptor for the N-acetylgalactosaminyltransferase. Therefore, synthesis of structure r can occur only via the pathway j→m→q→r. Synthesis of

structure s is forbidden since the actions of the sialyltransferase and the
N-acetylgalactosaminyltransferase are mutually exclusive; that is, structure o is
not an acceptor for the N-acetylgalactosaminyltransferase, and structure q is not
an acceptor for the sialyltransferase.

The pathways proposed in Figure 2 are based on the substrate specificities of
glycosyltransferases from three different organisms. Since these enzymes have
not been purified from other sources for comparison of the substrate specificities,
the general validity of the scheme cannot be tested directly. However, the excellent agreement between the structures that can be synthesized *in vitro* and those
that have been reported in naturally occurring oligosaccharides suggests that the
pathways shown in Figure 2 accurately reflect oligosaccharide biosynthesis in
many organisms.

IV. BIOSYNTHESIS OF O-LINKED OLIGOSACCHARIDES

The Thr/Ser-linked oligosaccharide chains typical of epithelial mucins display
a much greater heterogeneity than do the N-linked oligosaccharides, ranging in
size from as few as 1 to as many as 20 monosaccharide residues. This results in
part from the fact that these chains do not arise from a common lipid-linked
precursor, but rather are synthesized one residue at a time, directly on the
protein. These oligosaccharides can be grouped into two major classes: those that
contain a core consisting only of the disaccharide sequence Galβ1→3GalNAc,
and those that have in addition one or more branches composed of a repeating
galactose–N-acetylglucosamine sequence (Kornfeld and Kornfeld, 1980).

The discussion here focuses primarily on the former class, since most of the
glycosyltransferases that have been studied to date are involved in the synthesis
of these oligosaccharide chains. However, it should be emphasized that the
nonreducing terminal residues on the core regions of the latter class of oligosaccharides often contain the sequence Galβ1→4GlcNAc. Thus, the glycosylation
pathways outlined in Section III, B for chain termination of complex-type
N-linked chains may well apply to the chain termination of these structures as
well.

A. Glycosylation Pathways

Most of our information about the biosynthesis of these structures is obtained
from studies on the glycosyltransferases involved in the formation of ovine and
porcine submaxillary mucins. Roseman and co-workers (McGuire, 1970) reported solubilized preparations of all of the transferases required for synthesis of
the most complex oligosaccharide chains on these proteins. Subsequently, four
of these enzymes were purified to homogeneity and the acceptor substrate spe-

cificities extensively characterized (Section II). In experiments similar to those described in Section III, B, Beyer *et al.* (1979) studied all of the possible sequential reactions of the four enzymes with antifreeze glycoprotein, a model acceptor containing oligosaccharides with the structure Galβ1→3GalNAc. The results of these studies are summarized in Figure 3.

Antifreeze glycoprotein (structure a) can be completely glycosylated with the α-N-acetylgalactosaminide α2→6 sialyltransferase (structure b), the β-galactoside α2→3 sialyltransferase (structure c), and the β-galactoside α1→2 fucosyltransferase (structure d). The disialylated product (structure e), which is found on many serum and membrane glycoproteins, can be synthesized by either pathway a→b→e or a→c→e, but comparison of the rates of these reactions suggests that the latter is preferred since structure b is a relatively poor acceptor for the α2→3 sialyltransferase. Similarly, structure g, which occurs as one of the oligosaccharides on porcine submaxillary mucin (Carlson, 1968), can be formed by pathway a→b→g or a→d→g, but since structure d is a poor acceptor for the sialyltransferase, the former pathway is probably preferred. The other monosialyl monofucosyl product, structure f, cannot be formed since structure c is not an acceptor for the fucosyltransferase and structure d is not an acceptor for the α2→3 sialyltransferase.

Both structures d and g are acceptors for the A blood group N-acetylgalactosaminyltransferase, yielding structures h and i, oligosaccharides found on A$^+$ porcine submaxillary mucin (Carlson, 1968). However, structure h is not an acceptor for the α2→6 sialyltransferase. Therefore, structure i can be synthesized only

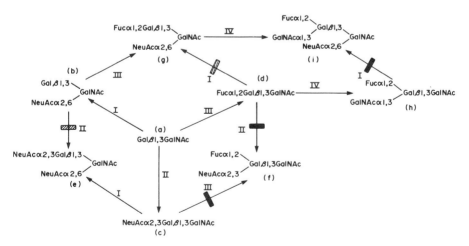

Figure 3 Proposed pathways for the biosynthesis of mucin-type oligosaccharide chains. The solid bars indicate that the reaction does not occur, and the hatched bars indicate that the reaction occurs very slowly. I, N-Acetylgalactosaminide α2,6 sialyltransferase; II, β-galactoside α2,3 sialyltransferase; III, β-galactoside α1,2 fucosyltransferase; IV, α1,2 fucosylgalactoside α1,3 N-acetylgalactosaminyltransferase. From Beyer *et al.* (1979).

from structure g by the action of the N-acetylgalactosaminyltransferase. Takasaki *et al.* (1978) reported that the A blood group $\alpha 1 \rightarrow 3$ N-acetylgalactosaminyltransferase from human milk was unable to transfer to erythrocyte oligosaccharides identical to structure g. Thus, there may be a difference in the acceptor substrate specificities of the two enzymes.

Williams and Schachter (1980) have described an N-acetylglucosaminyltransferase in canine submaxillary glands which catalyzes the following reaction:

$$Gal\beta 1 \rightarrow 3GalNAc\alpha \rightarrow Thr/Ser + UDP\text{-}GlcNAc \rightarrow$$

$$\begin{matrix} Gal\beta 1 \rightarrow 3 \searrow \\ \qquad\qquad\qquad GalNAc\alpha \rightarrow Thr/Ser + UDP \\ GlcNAc\beta 1 \rightarrow 6 \nearrow \end{matrix}$$

Preliminary analysis of the acceptor substrate specificity of the enzyme suggests that it may play an important role in determining the ultimate structure of O-linked oligosaccharide chains. Obviously, the actions of the N-acetylglucosaminyltransferase and the $\alpha 2 \rightarrow 6$ sialyltransferase are mutually exclusive. Thus, as proposed by the authors, this transferase may be involved in determing the acidic character of the oligosaccharide chain. In this regard, it will be of interest to see what effect transfer of the N-acetylglucosamine has on the activity of the β-galactoside $\alpha 2 \rightarrow 3$ sialyltransferase, and vice versa. The transferase also appears to be inactive toward substrates with the structure $Fuc\alpha 1 \rightarrow 2Gal\beta 1 \rightarrow 3GalNAc$, whereas the product of the N-acetylglucosaminyltransferase is a good acceptor for the fucosyltransferase. Thus, the preferred pathway for synthesis of the structure $Fuc\alpha 1 \rightarrow 2Gal\beta 1 \rightarrow 3(GlcNAc\beta 1 \rightarrow 6)GalNAc$ is almost certainly via the initial transfer of the N-acetylglucosamine followed by the addition of the fucose.

As the N-acetylglucosaminyltransferases and galactosyltransferases involved in synthesis of the larger O-linked oligosaccharides become available, it should be possible to examine the sequential actions of these enzymes and those currently available so that the pathways outlined in Figure 3 can be expanded to include the biosynthesis of these more complex structures.

B. Biosynthesis of Porcine Submaxillary Mucin

Of the five glycosyltransferases involved in the synthesis of the most complex chains of porcine submaxillary A^+ blood group mucin, three have been purified to homogeneity (Section II) and the other two have been fairly well characterized. Thus, probably more is known about the biosynthesis of this structure than any other glycoprotein oligosaccharide. The proposed biosynthetic pathway, based on the acceptor substrate specificities of the transferases, is shown in Figure 4.

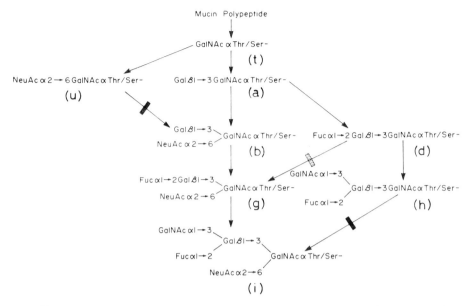

Figure 4 Proposed pathway for the biosynthesis of the oligosaccharide chains of blood group A⁺ porcine submaxillary mucin. From Beyer *et al.* (1979).

The mucin polypeptide α-N-acetylgalactosaminyltransferase that forms the oligosaccharide–protein linkage has been partially purified and characterized (McGuire, 1970; Hill *et al.*, 1977). The enzyme transfers to peptides prepared from deglycosylated ovine submaxillary mucin, but the acceptor activity decreases with decreasing peptide size. Sequence analysis of these peptides, which allowed comparison of the sequences adjacent to 28 serine and threonine residues normally glycosylated in the mucin, suggested that there is no unique amino acid sequence that acts as a signal for glycosylation. Although there are indeed minimal structural requirements necessary for glycosylation (Young *et al.*, 1979), the primary prerequisite appears to be that the potential serine or threonine acceptor site be accessible to the enzyme, probably in regions with little secondary or tertiary structure.

The next step in the synthesis of the pentasaccharide must be the transfer of galactose by the N-acetylgalactosaminide $\beta1\rightarrow3$ galactosyltransferase. Schachter *et al.* (1971) demonstrated that the galactosyltransferase is inactive with acceptors having the sequence NeuAc$\alpha2\rightarrow6$GalNAc$\alpha\rightarrow$Thr/Ser. Thus, transfer of sialic acid to structure t yields a "dead-end" product (structure u) that cannot be elongated. Structure u is the major oligosaccharide chain on ovine submaxillary mucin and occurs on the porcine mucin as well (Chapter 3, Section 2, Volume I). It has been proposed that the relative concentrations of the galactosyltransferase and the $\alpha2\rightarrow6$ sialytransferase in these tissues dictates whether the oligosaccharides will terminate at the disaccharide as in ovine submaxillary mucin or will proceed to more complex chains as in porcine submaxillary mucin (Schachter *et al.*, 1971).

As shown in Figure 3, structure a is an acceptor for the β-galactoside $\alpha1\rightarrow2$ fucosyltransferase and the N-acetylgalactosaminide $\alpha2\rightarrow6$ sialyltransferase, yielding structures d and b, respectively. Since structure d is a fairly poor acceptor for the sialyltransferase, the preferred pathway for synthesis of structure g is probably a→b→g. Synthesis of the pentasaccharide product must proceed via structure g by the action of the $\alpha1\rightarrow3$ N-acetylgalactosaminyltransferase because the alternative pathway results in the accumulation of structure h. This is another "dead-end" product since it is not an acceptor for the $\alpha2\rightarrow6$ sialyltransferase.

Thus, the synthesis of this structure, the pentasaccharide chain of porcine submaxillary A^+ blood group mucin, requires that the glycosyltransferases act in a very specific order. However, the occurrence of "dead-end" products such as structures u and h in the native mucin (Carlson, 1968) indicates that the transferases are not constrained to act in one particular sequence. Rather, it appears that the heterogeneity observed in the oligosaccharide chains of porcine submaxillary mucin results, at least in part, from the competition of glycosyltransferases with overlapping acceptor substrate specificities for common acceptor sites. Presently, it is unknown whether the system is controlled simply by the relative abundance of the various transferases or whether other factors such as availability of nucleotide sugar donor substrates or modulation of enzyme activity are involved.

REFERENCES

Baenziger, J., and Kornfeld, S. (1974). *J. Biol. Chem.* **249,** 7270–7281.

Beyer, T. A., and Hill, R. L. (1980). *J. Biol. Chem.* **255,** 5373–5379.

Beyer, T. A., Rearick, J. I., Paulson, J. C., Prieels, J. P., Sadler, J. E., and Hill, R. L. (1979). *J. Biol. Chem.* **254,** 12531–12541.

Beyer, T. A., Sadler, J. E., and Hill, R. L. (1980). *J. Biol. Chem.* **255,** 5364–5372.

Beyer, T. A., Sadler, J. E., Rearick, J. I., Paulson, J. C., and Hill, R. L. (1981). *Adv. Enzymol.* **52,** 23–175.

Carlson, D. M. (1968). *J. Biol. Chem.* **243,** 616–626.

Dawson, G. (1978). *In* "Methods in Enzymology" (V. Ginsburg, ed.), Vol. 50, p. 272. Academic Press, New York.

Fournet, B., Montreuil, J., Strecker, G., Dorland, L., Haverkamp, J., Vliegenthart, J. F. G., Binette, J. P., and Schmid, K. (1978). *Biochemistry* **17,** 5206–5214.

Gottlieb, C., Skinner, A. M., and Kornfeld, S. (1974). *Proc. Natl. Acad. Sci. U.S.A.* **71,** 1078–1082.

Gottlieb, C., Baenziger, J., and Kornfeld, S. (1975). *J. Biol. Chem.* **250,** 3303–3309.

Harpaz, N., and Schachter, H. (1980). *J. Biol. Chem.* **255,** 4885–4893.

Hill, H. D., Schwyzer, M., Steinman, H. M., and Hill, R. L. (1977). *J. Biol. Chem.* **252**, 3799–3804.

Kobata, A. Grollman, E. F., and Ginsburg, V. (1968). *Arch. Biochem. Biophys.* **124**, 609–612.

Kornfeld, R., and Kornfeld, S. (1980). In "The Biochemistry of Glycoproteins and Proteoglycans" (W. J. Lennarz, ed.), pp. 1–34. Plenum, New York.

Krusuis, T., and Finne, J. (1978). *Eur. J. Biochem.* **84**, 395–403.

Li, E., and Kornfeld, S. (1978). *J. Biol. Chem.* **253**, 6426–6431.

Lloyd, K. O., Kabat, E. A., Layug, E. J., and Gruezo, F. (1966). *Biochemistry* **5**, 1489–1501.

Longmore, G., and Schachter, H. (1980). *Fed. Proc., Fed. Am. Soc. Exp. Biol.* **39**, 2002.

McGuire, E. J. (1970). *In* "Blood and Tissue Antigens" (D. Aminoff, ed.), pp. 461–474. Academic Press, New York.

McKibbin, J. M. (1978). *J. Lipid Res.* **19**, 131–147.

Montreuil, J. (1980). *Adv. Carbohydr. Chem. Biochem.* **37**, 157–223.

Nagai, M., Davé, V., Kaplan, B. E., and Yoshida, A. (1978a). *J. Biol. Chem.* **253**, 377–379.

Nagai, M., Davé, V., Muensch, H., and Yoshida, A. (1978b). *J. Biol. Chem.* **253**, 380–381.

Narasimhan, S., Stanley, P., and Schachter, H. (1977). *J. Biol. Chem.* **252**, 3926–3933.

Oppenheimer, C. L., and Hill, R. L. (1981). *J. Biol. Chem.* **256**, 799–804.

Paulson, J. C., Beranek, W. E., and Hill, R. L. (1977a). *J. Biol. Chem.* **252**, 2356–2362.

Paulson, J. C., Rearick, J. I., and Hill, R. L. (1977b). *J. Biol. Chem.* **252**, 2363–2371.

Paulson, J. C., Prieels, J. P., Glasgow, L. R., and Hill, R. L. (1978). *J. Biol. Chem.* **253**, 5617–5624.

Prieels, J. P., Monnom, D., Dolmans, M., Beyer, T. A., and Hill, R. L. (1981). *J. Biol. Chem.* **256**, 10456–10463.

Rao, A. K., and Mendicino, J. (1978). *Biochemistry* **17**, 5632–5638.

Rao, A. K., Garver, F., and Mendicino, J. (1976). *Biochemistry* **15**, 5001–5009.

Rearick, J. I., Sadler, J. E., Paulson, J. C., and Hill, R. L. (1979). *J. Biol. Chem.* **254**, 4444–4451.

Rearick, J. I., Fujimoto, K., and Kornfeld, S. (1981). *J. Biol. Chem.* **256**, 3762–3769.

Reddy, S., Davilla, M., Winters, W. C., and Mendicino, J. (1979). *Fed. Proc., Fed. Am. Soc. Exp. Biol.* **38**, 631.

Sadler, J. E., Rearick, J. I., Paulson, J. C., and Hill, R. L. (1979a). *J. Biol. Chem.* **254**, 4434–4443.

Sadler, J. E., Rearick, J. I., and Hill, R. L. (1979b). *J. Biol. Chem.* **254**, 5934–5941.

Sadler, J. E., Beyer, T. A., Oppenheimer, C. L., Paulson, J. C., Prieels, J. P., Rearick, J. I., and Hill, R. L. (1981). *In* "Methods in Enzymology" (V. Ginsburg, ed.) Vol. 83 (In press). Academic Press. New York.

Schachter, H., McGuire, E. J., and Roseman, S. (1971). *J. Biol. Chem.* **246**, 5321–5328.

Schanbacher, F. L., and Ebner, K. E. (1970). *J. Biol. Chem.* **245**, 5057–5061.

Schwyzer, M., and Hill, R. L. (1977a). *J. Biol. Chem.* **252**, 2338–2345.

Schwyzer, M., and Hill, R. L. (1977b). *J. Biol. Chem.* **252**, 2346–2355.

Tabas, I., and Kornfeld, S. (1978). *J. Biol. Chem.* **252**, 7779–7786.

Takasaki, S., Yamashita, K., and Kobata, A. (1978). *J. Biol. Chem.* **253**, 6086–6091.

van den Eijnden, D. H., Joziasse, D. H., Dorland, L., van Halbeek, H., Vliegenthart, J. F. G., and Schmid, K. (1980). *Biochem. Biophys. Res. Commun.* **92**, 839–845.

Varki, A., and Kornfeld, S. (1980). *J. Biol. Chem.* **255**, 10847–10858.

Williams, D., and Schachter, H. (1980). *J. Biol. Chem.* **255**, 11247–11252.

Wilson, J. R., Williams, D., and Schachter, H. (1976). *Biochem. Biophys. Res. Commun.* **72**, 909–916.

Yamashita, K., Tachibana, Y., and Kobata, A. (1978). *J. Biol. Chem.* **253**, 3862–3869.

Young, J. D., Tsuchiya, D., Sandlin, D. E., and Holroyde, M. J. (1979). *Biochemistry* **18**, 4444–4448.

SECTION 3

Inhibition of Lipid-Dependent Glycosylation

R. T. SCHWARZ AND R. DATEMA

I. INTRODUCTION

Many proteins in eukaryotes are glycosylated (Sharon and Lis, 1981). The glycosylation machinery is complex and has survived evolution from "yeast to man" (Sharon and Lis, 1981). The effort of a cell to glycosylate proteins is substantial, as shown in the various sections of Chapter 1 of this volume, yet little is known about the biological significance of the carbohydrate chains of glycoproteins. Interfering with glycosylation, however, is one approach to learning something about the roles of the carbohydrate side chains (Schwarz and Datema, 1982b).

As described in Sections 1 and 2, Chapter 1, of this volume and in recent reviews (Kornfeld and Kornfeld, 1980; Montreuil, 1980), protein-linked oligosaccharides can be divided into two groups: oligosaccharides linked N-glycosidically to asparaginyl residues and those linked O-glycosidically to seryl or threonyl residues. Very few inhibitors specifically interfering with the formation of the O-glycosidically linked chains are known, and we have there-

THE GLYCOCONJUGATES, VOL. III

47

fore restricted our discussion to inhibitors of the biosynthesis of N-glycosidically linked oligosaccharides.

Three phases can be discerned in the biosynthesis of those oligosaccharides. First, an oligosaccharide is assembled on a lipid, dolichol diphosphate (see Chapter 1, Section 1, of this volume). Second, this oligosaccharide is transferred to a protein, usually a nascent polypeptide chain (Chapter 1, Sections 4 and 5). It has been suggested that transfer of a preassembled oligosaccharide to a nascent polypeptide chain is advantageous because it allows glycosylation to keep pace with the rate of translation (Hanover *et al.*, 1980). However, this argument may not be valid because N-glycosylation of completed chains may also occur (Klenk *et al.*, 1972; Bergman and Kuehl, 1978), and furthermore the rate of oligosaccharide assembly is high and proportional to the rate of protein synthesis (Hubbard and Robbins, 1980). Third, the protein-bound oligosaccharide is "processed" to its mature form by the concerted action of a series of glycosidases and, to generate the so-called complex oligosaccharides, glycosyltransferases (see Chapter 1, Section 1). Thus, there are several possibilities for interfering with protein N-glycosylation, and among the compounds used for this purpose the inhibitors of assembly of dolichol-linked oligosaccharides are best known (Schwarz and Datema, 1982b). Inhibitors of transfer to protein are not known (see, however, Hortin and Boime, 1980), and only a few inhibitors of the processing reactions have been described (Schwarz and Datema, 1982b). Because the different steps of glycosylation occur in different cell compartments, inhibitors of intracellular transport (see Tartakoff, 1980, for a review) indirectly interfere with processing. In addition, cell mutants with a defect in one of the many glycosyltransferases (Stanley, 1980) or viral glycoproteins with defects in intracellular migration (Lohmeyer and Klenk, 1979; Pesonen *et al.*, 1981) or propagation of viruses in different cell types (Keegstra *et al.*, 1975) can be employed as tools to provide glycoproteins with different carbohydrate chains.

Here we restrict our discussion to inhibitors of the lipid pathway of protein glycosylation. For a more comprehensive treatise, see Schwarz and Datema (1982b). After a brief outline of the pathway of lipid-dependent glycosylation of proteins (Section II), the inhibitors and their mode of action are presented (Section III). The effects of the inhibitors on biosynthesis and the properties of individual (unglycosylated) proteins are discussed in Section IV.

II. THE DOLICHOL PATHWAY OF PROTEIN GLYCOSYLATION

Two different lipid-linked oligosaccharides have so far been recognized as donors for protein-linked oligosaccharides (Fig. 1). The tetradecasaccharide-lipid seems to occur in most eukaryotes (Hubbard and Ivatt, 1981) and is a

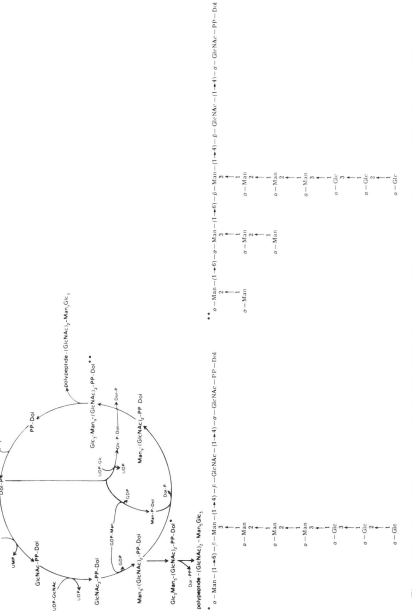

Figure 1 A pathway of the biosynthesis of dolichol-linked oligosaccharide precursors of asparagine-linked oligosaccharides and the structures for the glycan units of lipid-linked decasaccharide and tetradecasaccharide precursors (Kornfeld and Kornfeld, 1980; Chapman *et al.*, 1979b).

precursor for both complex-type and high-mannose-type oligosaccharides (Schachter and Roseman, 1980). The biosynthesis of the tetradecasaccharide-lipid requires Dol-P, UDP-GlcNAc, GDP-Man, Man-P-Dol, and Glc-P-Dol (Fig. 1). The pathway of assembly of $Man_9(GlcNAc)_2$-PP-Dol shown in Figure 1 is based on a limited number of studies (Chapman *et al.*, 1979a,b, 1980; Li and Kornfeld, 1979), and a different pattern of assembly is possible (Vijay and Perdew, 1980). The decasaccharide–lipid has only recently been described in cells in which the biosynthesis of Man-P-Dol is inhibited (see below). The pathway of protein glycosylation via the decasaccharide–lipid will be called the alternate pathway.

The enzymes involved in the assembly of the lipid-linked oligosaccharides are membrane bound and appear to occur mainly in the rough endoplasmic reticulum (for a discussion, see Schwarz and Datema, 1982b). However, other subcellular locations have also been noted, especially the outer membrane of mitochondria (Gateau *et al.*, 1978) and the plasma membrane (Cacan *et al.*, 1980; Welten-Versteegen *et al.*, 1980; Hoflack *et al.*, 1980). Only some of the enzymes have been obtained in a soluble and partially purified form—for example, the enzyme transferring the oligosaccharide from Dol-PP to a peptide acceptor (Das and Heath, 1980). Structural requirements in the tetradecasaccharide for effective transfer to protein are the di-*N*-acetylchitobiosyl residue at the reducing end (Reuvers *et al.*, 1977) and the glucosyl residues, especially the glucose group, at the nonreducing end (Turco *et al.*, 1977; Staneloni *et al.*, 1980). The peripheral α-mannosyl residues and groups can be removed from the tetradecasaccharide-lipid without hampering transfer of the oligosaccharide to protein (Spiro *et al.*, 1979).

It has been suggested (Staneloni *et al.*, 1980) that the glucosyl residues are introduced by way of Glc-P-Dol in the tetradecasaccharide–lipid, whereas their origin in the decasaccharide–lipid (Chapman *et al.*, 1979b; Kornfeld *et al.*, 1979) has not yet been investigated. It is reasonable to assume that at least two glucosyltransferases participate in the assembly, because two different glycosidic linkages are formed. However, the two enzyme activities have not yet been separated or purified.

Both GDP-Man and Man-P-Dol are donors of mannosyl residues in the assembly of tetradecasaccharide–lipid. As shown in Figure 1, a heptasaccharide–lipid can be formed in the absence of Man-P-Dol in cell-free systems (Chambers *et al.*, 1977; Schutzbach *et al.*, 1980; Datema *et al.*, 1980a; Spencer and Elbein, 1980) and in intact cells (Chapman *et al.*, 1979a,b; Datema *et al.*, 1980b). This endoglucosaminidase H-resistant oligosaccharide is an acceptor of glucosyl residues (see above) but can also be elongated further by stepwise additions of mannosyl residues (Hubbard and Robbins, 1980) from Man-P-Dol (Chapman *et al.*, 1979a) and is therefore at the bifurcation of the normal and alternate pathway of protein glycosylation. Thus, when no Man-P-Dol is present (Chapman *et al.*,

1979b, 1980) or when its formation is inhibited (Datema *et al.*, 1980b; Datema and Schwarz, 1981), endoglucosaminidase H-resistant oligosaccharide–lipids up to the heptasaccharide– or even decasaccharide–lipid can be isolated from intact cells. However, this does not imply that under normal conditions, when Man-P-Dol is present, only this pathway of assembly is followed. Alternative pathways have become evident from studies in cell-free systems (Vijay and Perdew, 1980; Jensen *et al.*, 1980). Therefore, the assembly of the Man$_9$(GlcNAc)$_2$-PP-Dol need not proceed via only one pathway of ordered additions of mannosyl residues (Chapman *et al.*, 1979a).

The reactions leading to the trisaccharide–lipid βMan(GlcNAc)$_2$-PP-Dol and to the several monosaccharide–lipids have been amply verified (Parodi and Leloir, 1979; Struck and Lennarz, 1980; Hubbard and Ivatt, 1981). Four different monosaccharide–lipids can be formed from Dol-P, and three different pathways of formation of Dol-P have been found (Fig. 2). These reactions do not necessarily occur in all cells. For example, Glc-PP-Dol has been described as a precursor for cellulose formation in the alga *Prototheca zopfii* (Hopp *et al.*, 1978a) and in peas (Pont Lezica *et al.*, 1978). The formation of Dol-P from Man$_9$(GlcNAc)$_2$-PP-Dol has been described only in lymphocytes as a presumptive first step in the catabolism of the oligosaccharide–lipid (Cacan *et al.*, 1980).

The formation *de novo* of Dol-P occurs via 3-hydroxy-3-methylglutaryl-coenzyme A (HMG-CoA) (Schwarz and Datema, 1982b), a precursor not only of dolichol, but also of cholesterol and ubiquinone. The available evidence suggests that phosphatase action on 2,3-dehydrodolichol diphosphate precedes saturation, as shown in Figure 2 (Wellner and Lucas, 1979, 1980; Grange and Adair, 1977;

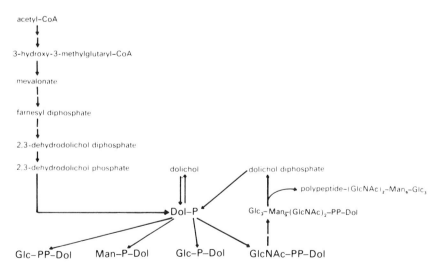

Figure 2 Biosynthesis and metabolic roles of dolichol phosphate.

Adair and Trepanier, 1980). A dolichol-diphosphate phosphatase has been described in human lymphocytes (Wedgwood and Strominger, 1980) and in rat liver (Kato *et al.*, 1980), underlining the role of Dol-P as a reutilizable substrate or coenzyme (Hemming, 1977). On the other hand, the finding that exogenous dolichol can be phosphorylated (Chojnacki *et al.*, 1980) indicates that the reserves of dolichol can be mobilized *in vivo* to give active Dol-P. Dol-P synthesized via formation *de novo,* cleavage of Dol-PP, or phosphorylation of dolichol can participate directly in glycosylation reactions (Martin and Thorne, 1974; Daleo and Pont Lezica, 1977; Daleo *et al.*, 1977; Burton *et al.*, 1979; Kato *et al.*, 1980), yet evidence, though indirect so far, is accumulating that recycling of Dol-P via Dol-PP after transfer of the oligosaccharide to protein may be rate limiting in protein glycosylation (Hubbard and Robbins, 1980; Hemming, 1977; Hubbard and Ivatt, 1981).

III. INHIBITORS OF PROTEIN GLYCOSYLATION

In Tables I–III are listed compounds for which inhibitory effects on the lipid pathway of protein glycosylation have been demonstrated. Procedures for measuring the biosynthesis of lipid-linked oligosaccharides and monosaccharides and for evaluating the inhibition of biosynthesis in intact cells and in cell-free systems have been worked out (Behrens and Tábora, 1978; Schwarz and Datema, 1982a). Thus, it has been possible to determine for many drugs which step(s) they inhibit. These results are summarized in the tables. Enzymes of the lipid cycles, however, have not yet been purified to an extent that would allow accurate determination of inhibitory effects at the molecular level. We have distinguished two categories of inhibitors:

1. Analogs of simple sugars (Table I). Analysis of the antiviral effects of these sugars has revealed that they are inhibitors of the dolichol cycle of protein glycosylation.
2. Antibiotics (Table II; Fig. 3). With the exception of showdomycin, these substances also interfere with the lipd cycle of peptidoglycan synthesis.

A few other compounds directly or indirectly interfere with lipid-dependent protein glycosylation and are shown in Table III. Among these compounds are the inhibitors of HMG-CoA reductase, an enzyme catalyzing the conversion of HMG-CoA to mevalonate. To date no specific inhibitors of *de novo* synthesis of dolichol phosphate have been recognized.

A. Sugar Analogs

The sugar analogs 2-deoxy-D-glucose (deoxyglucose), 2-deoxy-2-fluoro-D-glucose (fluoroglucose), and 2-deoxy-2-fluoro-D-mannose (fluoromannose), but

TABLE I

Inhibitors of Lipid-Dependent Protein Glycosylation: Sugar Analogs

Inhibitor	Metabolites detected under inhibitory conditions	Inhibitory agents	Reaction of the dolichol cycle inhibited	Effects on pool size of nucleotide sugars in chick embryo cells[a]		
				UDP-Glc	GDP-Man	UDP-HexNac
Deoxyglucose (dGlc) (2-deoxy-D-arabino-hexose)	dGlc-1-P dGlc-6-P Deoxygluconic acid-6-P UDP-dGlc GDP-dGlc dGlc-P-Dol	UDP-dGlc GDP-dGlc dGlc-P-Dol	By UDP-dGlc 1. Formation of Glc-P-Dol By GDP-dGlc 1. Formation of dolichol-linked monosaccharides by trapping of Dol-P as dGlc-P-Dol 2. Assembly of the lipid-linked oligosaccharide by incorporation of dGlc instead of Man By dGlc-P-Dol 1. Glucosylation of Man_9-$(GlcNAc)_2$-PP-Dol by incorporation of dGlc instead of Glc	−	0	+
Fluoroglucose (FGlc) (2-Deoxy-2-fluoro-D-glucose)	FGlc-6-P UDP-FGlc GDP-FGlc	Unknown	1. Formation of Man-P-Dol 2. Formation of Glc-P-Dol	−	0	+
Fluoromannose (FMan) (2-Deoxy-2-fluoro-D-mannose)	FMan-6-P UDP-FMan GDP-FMan	Unknown	Unknown	Unknown	Unknown	Unknown
Glucosamine (GlcN) (2-Deoxy-2-amino-D-glucose)	GlcNAc GlcNAc-6-P UDP-GlcNAc	GlcN	Unknown	−	+	+

[a] Key: −, pool size decreases; +, pool size increases; 0, no effect on pool size.

TABLE II

Inhibitors on Lipid-Dependent Protein Glycosylation: Antibiotics

Inhibitor	Category and origin	Related antibiotics	Structure shown in	Reactions of the dolichol cycle inhibited	Mechanism of inhibition
Tunicamycin	Nucleoside antibiotic from *Streptomyces lysosuperificus*	Mycospocidins, streptovirudin, A24010	Fig. 3A	1. Formation of GlcNAc-PP-Dol 2. At high concentration, the formation of Glc-P-Dol	Tunicamycin is a tight-binding, competitive inhibitor of UDP-GlcNAc: dolichol-phosphate GlcNAc-1-P-transferase
Amphomycin	Polypeptide antibiotic from *Streptomyces canus*	Tsushimycin, asparticin, laspartomycin	Fig. 3B	Formation of dolichol-linked monosaccharides	Amphomycin inactivates glycosyltransferases
Bacitracin	Polypeptide antibiotic from *Bacillus licheniformis*		Fig. 3C	Formation of dolichol phosphate from dolichol diphosphate or dolichol diphosphate-linked oligosaccharides	The antibiotic forms a complex with polyprenol diphosphates and a metal ion
Showdomycin	Nucleoside antibiotic from *Streptomyces showdoensis*		Fig. 3D	Formation of Glc-P-Dol; in solubilized enzyme preparations, the formation of all dolichol-linked monosaccharides	Showdomycin is a non-competitive inhibitor of Glc-P-Dol formation
Diumycin	Phosphoglycolipid antibiotic from *Streptomyces umbrinus*	Moenomycin, macarbomycin		1. Formation of (GlcNAc)$_2$-PP-Dol from UDP-GlcNAc and GlcNAc-PP-Dol 2. At higher concentrations, the formation of Man-P-Dol, GlcNAc-PP-Dol	

TABLE III
Other Inhibitors of Lipid-Dependent Protein Glycosylation[a]

Inhibitor	Structure	Compounds that have or may have similar effects	Mechanism of inhibition	System(s) in which inhibition of glycosylation was shown
25-Hydroxy-cholesterol	Systematic name. cholest-5-ene-3β, 25-diol	7-Ketocholesterol, diosgenin (Ref. 1)	Decreases the activity of HMG-CoA reductase (Ref. 2)	Aortic, smooth muscle cells in culture (Ref. 1)
Compactin	Shown in Fig. 3E	Monacolin K (Ref. 3)	Competitive inhibitor of HMG-CoA reductase with respect to HMG-CoA (Ref. 4)	Sea urchin embryos (Ref. 5)
Cycloheximide		Other inhibitors of protein synthesis	Unknown	Several (Refs. 6,7,8)
m-Chlorocarbonyl-cyanide phenyl-hydrazone (CCCP)		Other respiratory inhibitors	Inhibition of formation of Man-P-Dol (Ref. 9)	Influenza-virus-infected chick embryo cells
Coumarin	Systematic name. 2H-1-benzopyran-2-one		Inhibition of transfer of dolichol-linked cello-oligosaccharide to a protein acceptor (Ref. 10)	Crude membrane fraction from the alga *Prototheca zopfii*
Warfarin	3-(α-Acetonylbenzyl)-4-hydroxycoumarin		Inhibition of incorporation of GlcN into lipid-linked intermediates (Ref. 11)	Glycosylation of prothrombin in rat liver
Interferon			Membranes from interferon-treated cells have a decreased capacity to form Dol-PP-GlcNAc from Dol-P and UDP-GlcNAc (Ref. 12)	L cells infected with vesicular stomatitis virus

[a] Key to references: 1: Mills and Adamany (1978); 2: Kandutsch und Chen (1974); 3: Endo (1981); 4: Tanzawa and Endo (1979); 5: Carson and Lennarz (1979); 6: Schmitt and Elbein (1979); 7: Hubbard and Robbins (1980); 8: White and Speake (1980); 9: Datema and Schwarz (1981); 10: Hopp *et al.* (1978); 11: Meeks and Couri (1980); 12: Maheshwari *et al.* (1980).

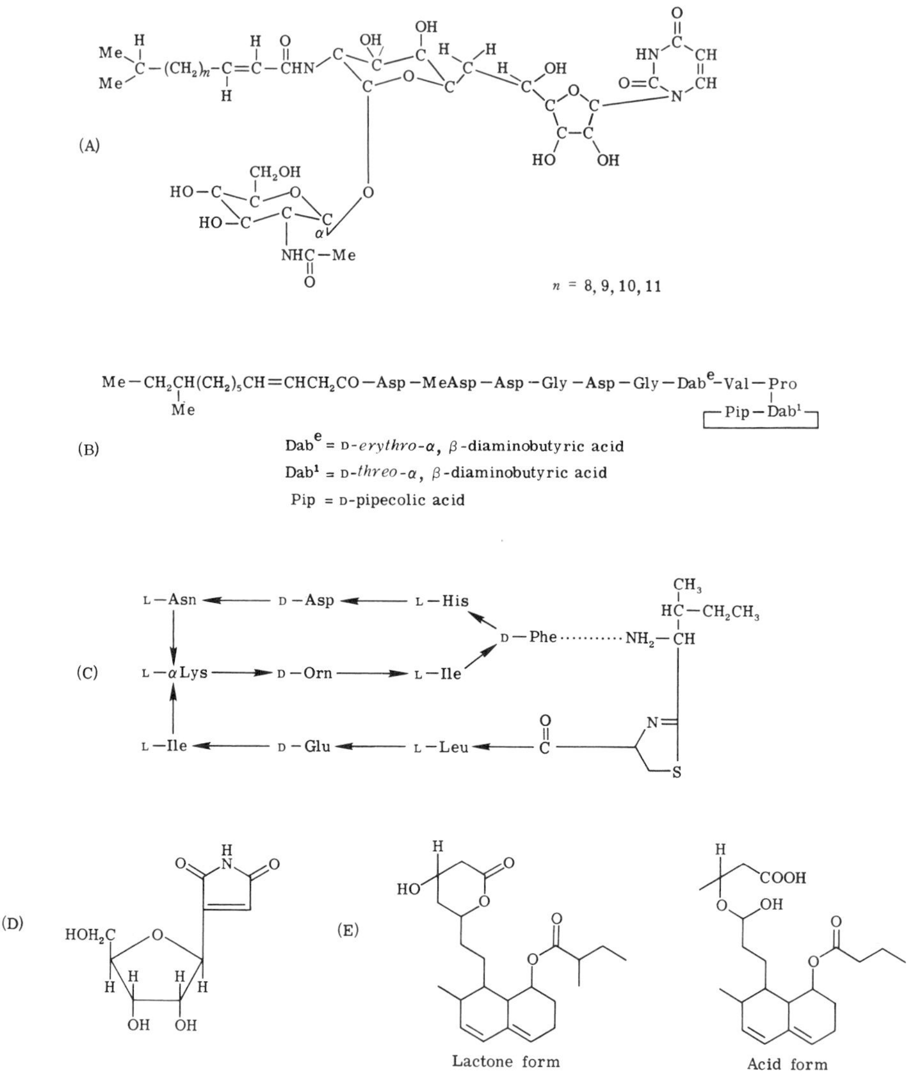

Figure 3 Formulas of inhibitors of protein glycosylation: (A) tunicamycin; (B) amphomycin; (C) bacitracin; (D) showdomycin; (E) acid and lactone form of compactin.

not 2-deoxy-2-amino-D-glucose (glucosamine), have to be metabolized to exert their inhibitory effects (Schwarz *et al.*, 1979). Direct proof that nucleotide esters of deoxyglucose (GDP-dGlc and UDP-dGlc) are inhibitory agents came from studies in cell-free systems (Schwarz *et al.*, 1978; Datema and Schwarz, 1978; Datema *et al.*, 1981). The results of these studies are summarized in Table I. In chick cells the inhibition caused by GDP-dGlc contributes

more to the inhibition of protein glycosylation than does the inhibition of formation of Glc-P-Dol by UDP-dGlc. The reason for this is unknown, but if the
inhibition of glycosylation *in vivo* by dGlc is reversed by the addition of mannose, glucosylated lipid-linked oligosaccharides are synthesized (Datema and
Schwarz, 1979), although UDP-dGlc is still present (Schmidt *et al.*, 1976a).
Probably, relatively large pools of Glc-P-Dol are present in these cells. Also, the
inhibition of formation of Glc-P-Dol by fluoroglucose (Datema *et al.*, 1980a)
does not block (Datema *et al.*, 1980b) the glucosylation of lipid-linked oligosaccharides [of $Man_5(GlcNAc)_2$-PP-Dol in this case]. A diminution of the pool of
UDP-Glc is often seen in chick embryo cells treated with sugar analogs (Table I),
but a decrease by a factor of 3 does not lead to the inhibition of protein glycosylation (Koch *et al.*, 1979).

If the concentration of Dol-P is rate limiting in protein glycosylation (for
discussions, see Hemming, 1977; Hubbard and Robbins, 1980), the trapping of
Dol-P as dGlc-P-Dol (see Table I) may be the major inhibitory step in the
inhibition of protein glycosylation by deoxyglucose. The complete inhibition of
formation of GlcNAc-PP-Dol by GDP-dGlc is difficult to achieve, however, *in
vitro* (Datema *et al.*, 1981). If this compound is still formed, or if existing
GlcNAc-PP-Dol is elongated to $(GlcNAc)_2$-PP-Dol, the assembly of the lipid-
linked oligosaccharide will still be prevented because of the formation of dGlc-
$(GlcNAc)_2$-PP-Dol from $(GlcNAc)_2$-PP-Dol and GDP-dGlc (Datema and
Schwarz, 1978). This trisaccharide–lipid, in effect, is a dead end in the assembly
pathway because it cannot be elongated by GDP-Man or Man-P-Dol (Datema
and Schwarz, 1978). A similar situation applies to dGlc-Man$(GlcNAc)_2$-PP-Dol.
Small, lipid-linked oligosaccharides containing deoxyglucose may indeed be
formed in intact cells treated with deoxyglucose, and it is likely that inhibition by
deoxyglucose of the alternate pathway of protein glycosylation (which does not
require Man-P-Dol) is caused by the formation of these abberrant, nonfunctional,
lipid-linked oligosaccharides (R. Datema and R. T. Schwarz, unpublished, 1981).

Deoxyglucose also interferes with the later steps in the assembly of lipid-
linked oligosaccharides, because the sugar analog is incorporated instead of
mannose (Datema and Schwarz, 1978) or glucose (Datema *et al.*, 1981) into
large lipid-linked oligosaccharides. Studies in cell-free systems indicated that, if
deoxyglucose were incorporated instead of mannose into these larger lipid-linked
oligosaccharides, the oligosaccharides were not transferred to protein (Datema
and Schwarz, 1978). Yet deoxyglucose can, under noninhibitory conditions, be
incorporated into glycoproteins (Kaluza *et al.*, 1973; Steiner *et al.*, 1973). This
phenomenon is not understood. Similarly, deoxyglucose can be incorporated into
yeast mannan or glucan (Biely *et al.*, 1974; Krátký, *et al.*, 1975). However, this
incorporation is not the cause of inhibition of protein glycosylation in yeast; also
in this organism, deoxyglucose inhibits glycosylation of nascent polypeptides
(Ruiz-Herrera and Sentandreu, 1975).

In yeast and chick embryo cells, both fluoroglucose and fluoromannose are metabolized to their UDP and GDP derivatives (Schmidt *et al.*, 1978). Apparently, these cells do not differentiate between these two fluoro sugars. Fluoroglucose, indeed, interferes with mannosylation reactions: It inhibits the formation of Man-P-Dol and therefore blocks the assembly of $Glc_3Man_9(GlcNAc)_2$-PP-Dol (Datema *et al.*, 1980a). The inhibition of this assembly can be reversed (Datema *et al.*, 1980a) or blocked (Datema and Schwarz, 1979) by mannose. Also, glycosylation of proteins, and therefore replication of viruses, are unaffected if the culture medium contains fluoroglucose and mannose (Schmidt *et al.*, 1976b). It is likely, but as yet unproved, that GDP-FGlc is the inhibitor of formation of Man-P-Dol, because no evidence has been obtained for incorporation of fluoroglucose into lipid-linked intermediates (Datema *et al.*, 1980a).

The formation of GDP-FGlc does not deplete the cells of GDP-Man (Datema *et al.*, 1980a); and, because $(GlcNAc)_2$-PP-Dol can still be formed, the formation of $Man_5(GlcNAc)_2$-PP-Dol still occurs in fluoroglucose-treated cells (Datema *et al.*, 1980b). The heptasaccharide can be transferred to protein *in vivo*, probably as a glucosylated derivative. The origin of the putative glucosyl residues is unknown, but we have suggested (Datema *et al.*, 1980) that Glc-P-Dol is a possible candidate because transfer of glucose from Glc-P-Dol to lipid-linked oligosaccharides is not impaired by FGlc. Since transfer of a mannosyl residue from Man-P-Dol to oligosaccharide–lipids is also not inhibited, membranes from fluoroglucose-treated cells have a decreased content of both Man-P-Dol and Glc-P-Dol (Datema *et al.*, 1980a).

A notable difference between the metabolic fates of fluoroglucose and deoxyglucose is that the fluoro sugar is not incorporated into lipid- and protein-linked oligosaccharides. Incorporation of fluoroglucose and fluoromannose occurs to a small extent in glycans of the yeast cell wall (Schmidt *et al.*, 1978). Although this incorporation may not be mediated by lipid-linked intermediates, this finding indicates that the nucleotide esters of fluoro sugars need not be the metabolic terminal stations (see Bernacki and Korytnyck, Chapter 4, Section 1, Volume IV).

Studies on the effects of the uncoupler of oxidative phosphorylation, *m*-chlorocarbonylcyanide phenylhydrazone (CCCP), on protein glycosylation have shown that the glycosylation of proteins via the alternate pathway [i.e., via $Glc_3Man_5(GlcNAc)_2$-PP-Dol; see Section II] is not a unique feature of fluoroglucose. If influenza-virus-infected cells are treated with CCCP to decrease the energy charge (ATP + 0.5ADP/ATP + ADP + AMP) from 0.9 to 0.3, the viral proteins are glycosylated in part via the endoglucosaminidase H-resistant high-mannose intermediates of the alternate pathway (Datema and Schwarz, 1981). These oligosaccharides are not processed to, for example, complex oligosaccharides because in these energy-depleted cells the rate of migration of the glycoproteins to the Golgi system is decreased (Datema and Schwarz, 1981; see also Tartakoff, 1980). In CCCP-treated cells, the formation of

Man-P-Dol is inhibited, but the formation of $Man_5(GlcNAc)_2$-PP-Dol from GDP-Man and $(GlcNAc)_2$-PP-Dol still occurs. Indeed, $(GlcNAc)_2$-PP-Dol - GDP-Man, and also Glc-P-Dol are still present in these energy-depleted cells. We assume that impaired protein glycosylation caused by omission of sugars or a carbon source from the medium of cultured cells ("starvation"), as described by Kaluza (1975), occurs via a similar mechanism (for a discussion, see Schwarz and Datema, 1981).

An investigation of the metabolism of glucosamine under conditions that inhibit protein glycosylation did not reveal any unusual metabolites, but the intracellular concentration of glucosamine was strongly increased (Koch et al., 1979). Washing these cells with glucosamine-free medium rapidly reversed the inhibition of glycosylation. During this reversion, only the intracellular concentration of glucosamine, and not of its metabolites, decreased. Also, the increased pool of GDP-Man and the decreased pool of UDP-Glc did not change during the short time (15 minutes) needed to reverse inhibition (Koch et al., 1979). Therefore, it was deduced that glucosamine itself is the inhibitor of protein glycosylation. Subsequent studies have shown that glucosamine inhibits an early step in the assembly of the lipid-linked oligosaccharide (Datema and Schwarz, 1979). Yet glucosamine does not inhibit the formation of lipid-linked oligosaccharides in a cell-free system. It has therefore been suggested that glucosamine indirectly inhibits dolichol-dependent glycosylation (Datema and Schwarz, 1979) because it affects the properties and structure (Molnar and Bekesi, 1972; Friedman and Skehan, 1980) of the endomembrane system (for a discussion, see Schwarz and Datema, 1982b). The potentiating of the membrane effects of local anesthetics by glucosamine (Friedman and Skehan, 1980) supports this idea.

Thus, it is not surprising that the extent and nature of inhibition by glucosamine differ with cell type (Nakamura and Compans, 1978; Marnell and Wertz, 1979). Furthermore, glucosamine may cause depletion of UTP and ATP pools when pyruvate, rather than glucose, serves as a carbon or energy source (Scholtissek, 1975). Uridylate trapping may lead to inhibition of RNA synthesis, and energy depletion leads to inhibition of protein glycosylation (see above). Also, metabolization of other sugar analogs may lead to metabolic alterations that indirectly affect protein glycosylation (Decker and Keppler, 1974). However, as we have seen, this need not be the case (see also Bernacki et al., 1978). We emphasize that metabolic alterations may also occur when antibiotics or inhibitors of HMG-CoA reductase are employed as inhibitors of protein glycosylation.

B. Antibiotics

The antibiotics tunicamycin, amphomycin, diumycin, and bacitracin inhibit not only peptidoglycan formation (Fig. 4a) but also reactions in the dolichol cycle of protein glycosylation (Fig. 4b). Both tunicamycin (Ward *et al.*, 1980) and amphomycin (Tanaka *et al.*, 1979) inhibit wall-polymer synthesis by blocking the phospho-*N*-acetylmuramoylpentapeptidyltransferase. Tunicamycin, in fact, is a rather specific inhibitor of enzymes that reversibly transfer GlcNAc-1-P (or its derivatives, such as phospho-*N*-acetylmuramoylpentapeptide) to polyprenol phosphates, such as dolichol or undecaprenol phosphate (for a discussion, see Suhadolnik, 1979). The antibiotics belonging to the tunicamycin family (called tunicamycin, streptovirudin, mycospocidin, and antibiotic 24010) all inhibit the formation of GlcNAc-PP-Dol in animal and plant systems (Struck and Lennarz, 1980; Elbein, 1979; Schwarz and Datema, 1982b). These antibiotics appear to be a series of homologs that differ from one another in the fatty acid (see Fig. 3A) and that can be separated (Mahoney and Duksin, 1979; Ito *et al.*, 1980; Eckhardt *et al.*, 1980) and obtained in pure form, free of activities that inhibit protein synthesis (Mahoney and Duksin, 1979).

A consideration of the structure of tunicamycin (Takatsuki *et al.*, 1977; Ito *et al.*, 1980) and the fact that tunicamycin is a tightly binding and most likely competitive inhibitor of a soluble UDP-GlcNAc : dolichol phosphate GlcNAc-1-P-transferase (Keller *et al.*, 1979b) has led to an appealing hypothesis regarding its mechanism of inhibition: It is a bisubstrate analog mimicking the transition state between substrates and product that is formed during catalysis (Keller *et al.*, 1979b). This hypothesis, which awaits testing, explains at least the relative specificity of inhibition by tunicamycin. Yet the tunicamycin-like antibiotics are also inhibitors of glucosyl transfer from UDP-Glc to Dol-P, although this reaction is 100-fold less sensitive (Elbein *et al.*, 1979).

Of the antibiotics, only tunicamycin has been shown to inhibit protein glycosylation in intact animal cells. It is generally assumed, but not proved, that the effects of tunicamycin on intact cells are caused only by its inhibition of formation of GlcNAc-PP-Dol. Indeed, the fact that O-glycosylation still proceeds in tunicamycin-treated cells (Gahmberg *et al.*, 1980; Butters and Hughes, 1980; Speak and White, 1979) indirectly indicates that the pool sizes of nucleotide sugars may not be significantly affected. However, if relatively high concentrations of tunicamycin are needed to achieve inhibition of glycosylation, for example, in human fibroblasts (Chatterjee *et al.*, 1979), other effects of tunicamycin (inhibition of Glc-P-Dol formation, inhibition of protein synthesis) may become important.

Amphomycin appears to inactivate many glycosyltransferases that transfer sugars to lipid acceptors, such as dolichol phosphate (Kang *et al.*, 1978). It thus resembles bacitracin, another polypeptide antibiotic, which inhibits the formation of lipid-linked sugars in many cell-free systems (Elbein, 1979; Schwarz and Datema, 1982b). However, in contrast to inhibition by bacitracin (Spencer *et al.*, 1978), inhibition by amphomycin is not reversed by Dol-P, manganese ions, or

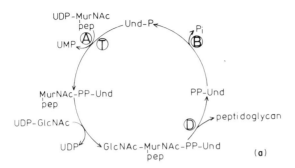

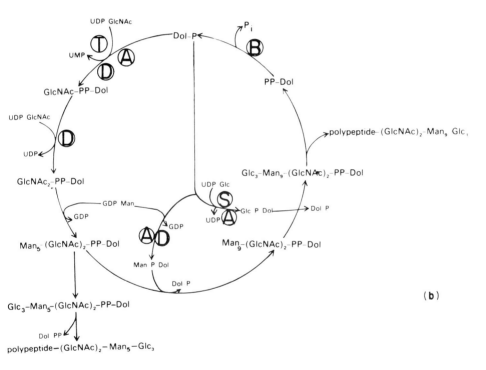

Figure 4 Lipid cycles of peptidoglycan synthesis (a) and of protein N-glycosylation (b), indicating the sites of the antibiotics amphomycin (A), bacitracin (B), diumycin (D), showdomycin (S), and tunicamycin (T).

both (Kang *et al.*, 1978). Probably, amphomycin does not form complexes with polyprenols (as does bacitracin), thus inhibiting the formation of lipid-linked sugars.

Amphomycin inhibits the formation *in vitro* of Man-P-Dol, Glc-P-Dol, and GlcNAc-PP-Dol but not the transfer of a mannosyl residue from GDP-Man or Man-P-Dol to lipid-linked oligosaccharides. Therefore, at antibiotic concentrations that completely inhibit the formation of Man-P-Dol, mannosylation by GDP-Man of endogenous acceptors to give the heptasaccharide–lipid $Man_5(GlcNAc)_2$-PP-Dol is still possible (Kang *et al.*, 1978). Also, when the formation of [^{3}H]Glc-P-Dol from UDP-[^{3}H]Glc is completely inhibited, incorporation of [^{3}H]Glc into lipid-linked oligosaccharides still proceeds. A similar result was obtained when showdomycin was used to inhibit the formation of [^{3}H]Glc-P-Dol (Kang *et al.*, 1979); only a slight inhibitory effect on incorporation into lipid-linked oligosaccharides was observed. Unfortunately, no structural data are available on the glucosylated oligosaccharide formed in the presence of these inhibitors. Nevertheless, these results suggest that glucosyl donors other than Glc-P-Dol may exist.

Showdomycin is a rather specific inhibitor of the formation of Glc-P-Dol in a solubilized enzyme preparation from pig aorta (Kang *et al.*, 1979). Yet at higher concentrations it also inhibits the formation of GlcNAc-PP-Dol and Man-P-Dol or the transfer of a mannosyl residue from Man-P-Dol to lipid-linked oligosaccharides. Showdomycin appears to be a noncompetitive inhibitor of the enzyme UDP-glucose : dolichol-phosphate glucosyltransferase, but how it affects the enzyme is unknown. Thus, the antibacterial effects of showdomycin are caused, in part, by the reaction of its maleimide moiety (see Fig. 3D) with sulfhydryl groups of proteins (Visser and Roy-Burman, 1979). In other words, the inhibitory effects of showdomycin are not necessarily related to its nucleoside structure.

Diumycin is not a very specific inhibitor because it inhibits the formation of GlcNAc-PP-Dol, Man-P-Dol, and $(GlcNAc)_2$-PP-Dol (Villemez and Carlo, 1980; Babczinski, 1980). Glucosyltransferase activity, however, is hardly affected by diumycin (Villemez and Carlo, 1980), and this drug may therefore be used in the search for roles of lipid-linked glucose in glucan formation. In a soluble enzyme preparation from *Acantamoeba castellani*, the formation of $(GlcNAc)_2$-PP-Dol from GlcNAc-PP-Dol and UDP-GlcNAc was most strongly inhibited (Villemez and Carlo, 1980). The elongation of $(GlcNAc)_2$-PP-Dol with a mannosyl residue from GDP-Man was unaffected in an enzyme preparation from yeast (Babczinski, 1980).

Bacitracin, by forming a complex with polyprenol diphosphates and a divalent cation, inhibits the dephosphorylation of dolichol diphosphate (Wedgwood and Strominger, 1980) and undecaprenol diphosphate (Siewert and Strominger,

1967; Stone and Strominger, 1972). The biosynthesis of polyprenol diphosphates is inhibited by bacitracin, because this polypeptide forms complexes with the isoprenyl diphosphate intermediates (Stone and Strominger, 1972; Schechter *et al.*, 1972, 1973).

The measurement of this inhibition can serve as a test for the uptake of bacitracin by cells (Mescher and Strominger, 1978). In the majority of cell systems tested, bacitracin does not enter the cytoplasm. Thus, inhibition of cell wall synthesis by bacitracin depends on (a) the generation of polyprenol diphosphate on the cell surface and (b) the regeneration of polyprenol phosphate from the diphosphate as the rate-limiting reaction in the synthesis of peptidoglycan (Strominger *et al.*, 1972). In eukaryotic systems, the dolichol cycle is located mainly in the endoplasmic reticulum (Hubbard and Ivatt, 1981; Struck and Lennarz, 1980; Schwarz and Datema, 1982b) and hence is inaccessible to bacitracin. Therefore, only dolichol-dependent glycosylation on plasma membranes (Hoflack *et al.*, 1980) may be susceptible to bacitracin. Thus, the incorporation of mannose into lipid-linked sugars by yeast protoplast was inhibited by bacitracin (Spencer *et al.*, 1978); and in this mannan-synthesizing organism, lipid-dependent glycosylation reactions may in part occur at the cell surface (Welten-Versteegen *et al.*, 1980). Recycling of Dol-P from Dol-PP apparently can occur in microsomes from rat liver, and this Dol-P can be used to form [^{14}C]Man-P-Dol from GDP-[^{14}C]Man. As expected, the formation of this [^{14}C]Man-P-Dol, and not that formed from endogenous Dol-P, was inhibited by bacitracin (Kato *et al.*, 1980).

There are three pathways leading to Dol-P, as shown in Figure 2. Bacitracin can inhibit one of these reactions: the formation of Dol-P from Dol-PP (see above). In addition the formation of Dol-P from dolichol diphosphate-linked oligosaccharides can be inhibited by bacitracin (Cacan *et al.*, 1980). It is not known which reaction is rate limiting in the synthesis of Dol-P under normal conditions. Thus, if inhibitors of HMG-CoA reductase are to inhibit the formation of Dol-P, they will be successful only in those systems in which HMG-CoA reductase is rate limiting for Dol-P synthesis. This appears to be the exception rather than the rule (Keller *et al.*, 1979a; James and Kandutsch, 1979, 1980). Yet in two systems, aortic smooth muscle cells (Mills and Adamany, 1978) and sea urchin embryos (Carson and Lennarz, 1979), a decrease in the activity of HMG-CoA reductase has led to the inhibition of formation of lipid-linked oligosaccharides (see Table III). As expected, the action of these inhibitors could be reversed by either mevalonate (Mills and Adamany, 1978) or dolichol (Carson and Lennarz, 1979). The latter finding accords with the result that extracellular, and possibly also intracellular, dolichol can be "activated" to dolichol phosphate (Chojnacki *et al.*, 1980). The contribution of this CTP-dependent phosphorylation of dolichol (Allen *et al.*, 1978; Burton *et al.*, 1979) to the pool size of Dol-P is not known, nor are any inhibitors of this reaction known. However, the finding that deazauridine and deoxyglucose act synergistically in inhibiting the replication of an encephalitis virus (Woodman and Williams, 1977) could be explained as follows. The trapping of Dol-P as dGlc-P-Dol is only a sufficient inhibitory condition in this system if the formation of CTP, and thus phosphorylation of dolichol is inhibited by deazauridine.

IV. BIOLOGICAL CONSEQUENCES OF INHIBITING GLYCOSYLATION

A. Contribution of Oligosaccharide Chains to the Conformation of Proteins

During or after their synthesis, proteins acquire a specific conformation according to the information given by the sequence of amino acids. Cotranslational or posttranslational modifications, such as phosphorylation or the attachment of fatty acids (Schmidt *et al.*, 1979) or carbohydrates, not only alter the molecular weight but may also change the way in which a protein molecule folds and may thus modulate its final conformation. For example, the solubility in and affinity to cellular membranes of these proteins may thus be different from that of their unmodified counterparts, and this may play a role in intracellular transport of the protein.

In the past the amino acid sequences of a few glycoproteins have been elucidated and their conformations determined. Independent of whether the glycosyl chain is attached via an O- or N-glycosydic bond to the protein backbone, it appears that the majority of the oligosaccharides are located in or near β turns in the protein. This assumption is based on results obtained by the method of Chou and Fasman (1974) for predicting the conformations of proteins. For instance, of 31 N-glycosydically linked oligosaccharides examined in one study, 30 occurred in sequences favoring turn or loop structures. Twenty-two of the glycosylated asparaginyl residues were detected in tetrapeptides predicted to have the β-turn conformation (Beeley, 1976, 1977). Similar conclusions were drawn in another investigation of 9 O-glycosidic linkages and 28 N-glycosidic linkages; in the case of O-glycosidic linkages, the seryl or threonyl residues that were involved in the linkages belonged to a β turn. Nineteen out of 28 asparaginyl residues substituted by glycosyl chains were part of a β turn. The N-glycosylated asparaginyl residues that were not located in β turns were situated in a random region near a β turn, in regions that are probably facing the solvent, or near the carboxy terminal end (Aubert *et al.*, 1976). However, not every potential glycosylation site that is part of a β turn must necessarily be glycosylated. Threonyl residue 55 of human casein, although situated in a β turn, is not glycosylated (Fiat *et al.*, 1980). It is interesting that the β turns are regions that also may be phosphorylated (Loucheux-Lefebvre *et al.*, 1978).

Similar characteristics hold true for viral envelope glycoproteins, which are integral constituents of the viral membrane (envelope of the viral particle). (For a review of the structure of influenza virus envelope glycoproteins, see Ward, 1981, and Chapter 6, Volume IV.) The hemagglutinin of the influenza virus envelope is composed of two subunits, designated HA_1 and HA_2, which are derived by proteolytic cleavage of the primary translation product HA. Both HA_1 and HA_2 are equipped with asparagine-linked oligosaccharides (Keil *et al.*, 1979; Ward, 1981) occurring with a high frequency in β turns (Ward, 1981). It is noteworthy that HA_2 subunits of three different strains are glycosylated near their carboxy terminus, which is inserted into the membrane bilayer. This means that carbohydrates do not necessarily occur exclusively at the end of the glyco-protein that faces the solvent but are also found close to membranes. A previous study showed that the high-mannose oligosaccharide chains of Sindbis virus glycoproteins are not accessible to endoglycosidases and may thus be buried in hydrophobic domains of the proteins (McCarthy and Harrison, 1977).

It appears reasonable to assume that carbohydrates may interact with domains of the protein or the surrounding water phase and may thus modulate the con-formation of a glycoprotein.

The roles of the carbohydrate residues of proteins are not yet clear but may best be recognized by generating glycoproteins that are devoid of carbohydrate and subsequently testing their remaining functions. One approach would be to cleave the carbohydrate using glycosidases, a procedure that usually does not deplete a protein completely of sugars. When this approach was used, yeast external invertase devoid of carbohydrate did not exhibit different catalytic prop-erties but was less stable toward multiple freeze-thaw treatment, incubation at 50°, acidic conditions, and trypsin digestion (Chou *et al.*, 1978). Other studies have demonstrated that there is little or no effect on the tertiary structure of a protein if the carbohydrate is removed from the native glycoprotein (Puett, 1973; Tarentino *et al.*, 1974; Trimble and Maley, 1977). Another method of producing carbohydrate-free proteins is to prevent the attachment of oligosaccharide chains during the biosynthesis of the protein molecule that has still to fold. This can be achieved using the inhibitors described in Section III. The results thus obtained have been diverse, ranging from no effect regarding, for example, stability against intracellular proteases to total proteolytic degradation of the polypeptide in the inhibitor-treated cell. The effects of depleting an envelope glycoprotein of carbohydrate has been extensively studied with the G protein of vesicular stomatitis virus (VSV). The nonglycosylated G protein of VSV (San Juan) aggregates when synthesized at 38° and at 30°. Nonglycosylated G protein of VSV (Orsay) aggregates when synthesized at 38° but functions normally when synthesized at 30°. From these studies it is concluded that carbohydrate or low temperature is required to stabilize the conformation of the G protein (Orsay) (Gibson et al., 1980).

In a similar investigation, further evidence was obtained that the amino acid sequence determines the extent of the influence of carbohydrates on the properties of G protein. Mutant ts 0110 was dependent at either temperature on the glycosylation of the G protein to form virus particles, whereas a pseudorevertant of ts 044 (ts 044R) could multiply at high or low temperature in the presence of tunicamycin (Chatis and Morrison, 1981). Also, a difference in the number of carbohydrate residues present on a folding protein can be crucial in attaining a proper conformation. G proteins containing either $Man_8(GlcNAc)_2$ or Man_5-$(GlcNAc)_2$ oligosaccharide chains were treated with guanidine hydrochloride, and their ability to refold during dialysis (to remove guanidine hydrochloride) was measured. San Juan G protein with $Man_5(GlcNAc)_2$ oligosaccharides aggregated during dialysis at 40° but remained in solution following dialysis at 30°. The Orsay G protein with $Man_5(GlcNAc)_2$ oligosaccharides and both proteins containing $Man_8(GlcNAc)_2$ oligosaccharides did not aggregate at either temperature (Gibson et al., 1981).

Thus, differences in the size of the oligosaccharides that exist during the trimming and maturation of glycosyl chains of glycoproteins may induce alterations in their conformation. Such an event is suggested by studies of Kaluza et al. (1980) in which changes in the antigenicity of maturating envelope glycoproteins of Semliki Forest virus were detected. It has been concluded that antigenic sites are exposed on immature glycoproteins which become hidden, whereas those found on mature forms become exposed due to conformational rearrangements (Kaluza et al., 1980) during processing of the oligosaccharide chains. Such a carbohydrate-dependent folding pattern is not necessarily be displayed by all glycoproteins. When a similar approach was used, no such rearrangement could be found with maturating envelope glycoproteins of influenza virus (Rott, 1980).

The establishment of a conformation of a nonglycosylated protein different from that of the corresponding glycosylated molecule may also interfere with the formation of the quaternary structures of proteins. Such an assumption appears to be justified by studies on the formation of thyroid-stimulating hormone (TSH). This hormone is composed of two glycosylated, noncovalently linked subunits, α and β. Within a species, the TSH α subunit is virtually identical with that of pituary and placental gonadotropins. The β subunit is unique and confers hormonal specificity. The inhibition of subunit glycosylation by tunicamycin apparently inhibited the combination of the subunits (Weintraub et al., 1980).

B. Role of Oligosaccharides in Protecting against or Specifying Proteolysis of Glycoproteins

A large number of proteins and glycoproteins arise by proteolytic cleavage of a high molecular weight precursor (Koch and Richter, 1980). In the simplest case two molecules are generated, but frequently one large precursor is cleaved in subsequent steps to give a higher number of proteins. The processing of the corticotropin (ACTH)/β-lipotropin (βLPH) precursor provides a good example. Its proteolytic cleavage yields ACTH, βLPH, and melanotropin. Among different species, the extent of glycosylation of the precursor and ACTH appears to differ. In man apparently only the amino terminal part of the molecule is glycosylated. Thus, nonglycoproteins can well be derived by cleavage from a glycoprotein (Miller *et al.*, 1980).

Among the best model systems for the study of precursor–product relationships of glycoproteins are cells infected with enveloped (glycoprotein-containing) viruses such as influenza, Semliki Forest, Sindbis, and avian sarcoma viruses. The common feature of the membrane glycoproteins of these viruses is that they are formed in close association with the intracellular membrane system. The amino acids that are translated first are part of a so-called leader sequence, which mediates the sequestration of the nascent peptide from the cytoplasma into the lumen of the endoplasmic reticulum. The glycoproteins are anchored with their carboxy terminus in the lipid bilayer while facing the lumen of the endoplasmic reticulum. These glycoproteins can be cotranslationally glycosylated and then transported from the endoplasmic reticulum to the Golgi apparatus and finally to the plasma membrane, the site of virus budding. It is during this transport that conversion of high-mannose oligosaccharides to complex ones and limited proteolysis of the glycosylated precursor proteins occur (Klenk and Rott, 1980).

The envelope glycoproteins of Semliki Forest virus E1, E2, and E3 are synthesized in equimolar amounts from a precursor that has a molecular weight of 130,000. The first cleavage generates the core protein C and a molecule with a molecular weight of 97,000, which upon further cleavage yields E1 and PE2 (MW 62,000); in a third cleavage, the latter is converted to E2 and E3 (Kääriäinen and Söderlund, 1978). A similar sequence has been described for Sindbis virus (a closely related virus); E3 is released into the medium and thus does not become an integral part of the sindbis viral envelope (Welch and Sefton, 1979). The influenza virus glycoproteins HA_1 and HA_2 are formed by the proteolytic cleavage of the precursor HA, the primary translation product (Klenk and Rott, 1980).

In a similar manner gp85 and gp37, which form the spike of Rous sarcoma virus, are derived from a glycoprotein precursor with a molecular weight of 92,000. The cleavages are specific, and it is assumed that oligosaccharides may contribute to specificity. They could do so by mere steric effects, e.g., blocking the access of the proteases to the potential cleavage site, or by stabilizing a conformation of the protein in which the cleavage sites are exposed. With these considerations in mind, some effects of reducing the carbohydrate content of glycoproteins by inhibiting glycosylation may be explained. For instance, β turns that have lost the oligosaccharide chains may be readily cleaved by protease (Geisow, 1979). It was therefore interesting to investigate how inhibitors of glycosylation, such as tunicamycin, 2-deoxy-D-glucose, 2-deoxy-2-fluoro-D-glucose, 2-deoxy-2-fluoro-D-mannose, and D-glucosamine, would interfere with the synthesis of viral glycoproteins. Would these proteins behave differently when devoid of oligosaccharide? In the course of such studies characteristic differences were found.

Thus, the processing of carbohydrate-free forms of Semliki Forest glycoproteins produced in the presence of tunicamycin apparently is not altered when compared with the glycosylated counterparts. Also, the nonglycosylated envelope proteins are properly inserted into the membranes (Garoff and Schwarz, 1978; Schwarz *et al.*, 1976). However, hemagglutinin synthesized in chick cells that had been infected with fowl plague virus (an influenza virus) and treated with tunicamycin was completely degraded unless a protease inhibitor was added (Schwarz *et al.*, 1976). In the presence of deoxyglucose instead of tunicamycin, processing of the nonglycosylated hemagglutinin (HA_0) yielded HA_{01} and HA_{02}, which were found in reduced amounts and were heterogeneous in size (Schwarz and Klenk, 1974). This difference is not understood. It should be pointed out, however, that such instability against proteases does not hold true for every hemagglutinin and host cell. Nakamura and Compans (1978) reported that HA_0 was detectable without the addition of a protease inhibitor in MDBK cells that had been infected with another strain of influenza virus (A/WSN) and maintained in tunicamycin-containing medium. Similarly, the differences found in studies on the processing of the ACTH/βLPH precursor may be due to different systems. Tunicamycin efficiently blocked glycosylation of the ACTH/βLPH precursor, and consequently it was processed to atypical peptides (Loh and Gainer, 1978, 1979). This also suggests that the carbohydrate protects the molecule from random intracellular degradation and directs the specificity of limited proteolysis. Herbert *et al.* (1980) also described the occurrence of a nonglycosylated ACTH/βLPH precursor that had a lower molecular weight owing to the absence of carbohydrate, but proteolytic cleavage of the precursor and ACTH intermediate was not grossly altered in the absence of carbohydrate side chains.

Enhanced intracellular proteolytic degradation of a variety of glycoproteins of cellular specificity has been described. For instance, the major surface glycoprotein of chick embryo fibroblast reached the plasma membrane of the cell and was exposed in correct form, but it showed a half-life reduced two- to threefold, most likely because of enhanced proteolytic degradation by cellular proteases (Olden et al., 1978).

Similarly, receptors and other proteins devoid of carbohydrate complement show decreased half-lives (Prives and Olden, 1980; Schwarz and Datema, 1982b).

C. Effects of Inhibiting Glycosylation on Routing and Secretion of Glycoproteins

As mentioned before, virus-coded membrane proteins are formed in close association with cellular membranes. Glycoproteins that will become part of a membrane and those that are secreted are synthesized on the intracellular membrane system. The interrelationship between membranes and glycoproteins is apparent from the fact that the lipid pathway of protein glycosylation is associated with the endoplasmic reticulum. Membrane glycoproteins and secretory proteins have common biosynthetic traits: Both types are synthesized by ribosomes attached to the endoplasmic reticulum. The first amino acids of the nascent glycoproteins are part of a hydrophobic sequence also frequently designated as a "signal" sequence. The signal theory assumes a specific interaction of this sequence with components of the membrane, which leads to the formation of a transient pore and active tunneling of the nascent protein into the lumen of the endoplasmic reticulum (Blobel and Dobberstein, 1975a,b; Blobel, 1980). The membrane-triggered folding hypothesis does not necessarily apply to a nascent peptide since it assumes specific interactions with hydrophobic regions of the protein to be sequestered and the membrane, resembling the events of self-assembly (Wickner, 1979, 1980). Active participation of carbohydrates in these processes is not postulated. However, with regard to these theories it has been argued that the carbohydrates of membrane proteins may keep the protein inserted into the lipid bilayer by acting like a lock after insertion (Bretscher and Raff, 1975). In many but not all cases reported, the signal or leader sequence is clipped off by a specific peptidase during or after transfer of the proteins.

It may be asked whether the nonglycosylated proteins are correctly positioned into the membrane lipid bilayer. An attempt to answer this question was made by testing the proteolytic stability of viral envelope glycoproteins in membrane vesicles prepared from cells that had been treated with tunicamycin or not treated. Upon treatment of such vesicles with tryspin, the viral glycoproteins were protected regardless of whether they originated from tunicamycin-treated or nontreated infected cells, indicating that the glycoproteins were correctly facing the luminal site of the vesicles (Garoff and Schwarz, 1978; Rothman et al., 1978; Wirth et al., 1979). Conclusions about the correct insertion of nonglycosylated glycoproteins into cellular membranes were drawn earlier from experiments using deoxyglucose or glucosamine because transport from rough to smooth endoplasmic reticulum had been demonstrated (Klenk *et al.*, 1974).

Although the insertion of proteins lacking the oligosaccharide moiety appears not to be disturbed, the intracellular transport and the secretion may be inhibited. Examples of this are summarized in Table IV and Schwarz and Datema (1982b). In the case of G protein of VSV a correlation between a suitable conformation and intracellular transport exists (Gibson *et al.*, 1981; Chatis and Morrison, 1981).

Taken together, these results suggest that it is the amino acid sequence of a protein which determines the requirement for carbohydrate and that simple mutational changes in a glycoprotein can drastically affect the requirement for oligosaccharide chains. The importance of the protein sequence of a protein to be transported is also apparent from experiments in which the migration of temperature-sensitive variants of viral envelope glycoproteins is blocked at the nonpermissive temperature and the glycoproteins do not leave the rough endoplasmic reticulum (Lohmeyer and Klenk, 1979; Pesonen *et al.*, 1981).

D. Effects of Inhibiting Glycosylation on Morphogenesis and Cell Differentiation

A large number of articles describing the effects of inhibitors on morphogenesis, for example, of virus particles and on cellular differentiation have appeared in the last couple of years, and it is to be expected that the literature will steadily grow. Many, but not all, of the observations in more complex biological systems may be explained by the factors that have been discussed earlier in this section, namely, alterations in the conformation of the nonglycosylated glycoproteins, their specific cleavage, or intracellular routing.

For instance, the formation of virus particles of VSV from infected cells that have been treated with inhibitors of glycosylation depends on whether the G protein devoid of oligosaccharides is transported to the site of virus budding, namely, the plasma membrane. The virus particles formed from tunicamycin-treated cells that had been infected with strain Orsay and maintained at 30° were fully infectious despite the lack of oligosaccharides on the G protein (Gibson *et al.*, 1978). Thus, they apparently are not essential for infection. The failure of infected cells treated with tunicamycin to form Sindbis or Semliki Forest virus particles is most likely also the result of the defective transport of the nonglycosylated envelope proteins. The cores of Semliki Forest virus still being formed can be visualized in microcrystalline arrays under the electron microscope (Ogura *et al.*, 1977). The proteolytic degradation of fowl plague nonglycosylated hemagglutinin in chick cells is possibly the reason for the inability of the cells to form virus particles. Although not necessary to become an integral part of the

TABLE IV

Effects of Inhibition of Glycosylation on Glycoproteins

System	Effects of inhibition of glycosylation	Reference
MIGRATION		
G protein of VSV		
San Juan	Migration of nonglycosylated protein to cell surface is prevented at temperatures above 38°	Gibson et al. (1980)
Orsay	Migration of nonglycosylated protein to cell surface is possible at 30°	Gibson et al. (1980)
ts 0110	Migration of nonglycosylated protein to cell surface is prevented at 39.5°	Chatis and Morrison (1981)
ts 044R	Migration of nonglycosylated protein to cell surface is independent of temperatures	Chatis and Morrison (1981)
SECRETION		
Immunoglobulins	81% inhibition of IgM secretion; 64% inhibition of IgA secretion; 28% inhibition of IgG secretion	Hickman et al. (1977), Hickman and Kornfeld (1978)
	No effect on secretion of nonglycosylated IgA	Williamson et al. (1980)
	Inhibition of transfer into endoplasmic reticulum; less inhibition of secretion once transferred	Melchers (1973)
	Extent of inhibition of secretion equals that of inhibition of protein synthesis	Eagon and Heath (1977)
Human immune interferon	No alteration of secretion or activity; hydrophobicity is changed	Mizrahi et al. (1978)
Procollagen from chick cells	Secretion not inhibited (proteolytic cleavage of carboxy-terminus prevented)	Duksin and Bornstein (1977)
Procollagen from human fibroblasts	Inhibition of secretion	Housley et al. (1980)
Transferrin, VLDL[a] apoprotein B α-acid glycoprotein	No effect on secretion	Struck et al. (1978); Edwards et al. (1979)
Fibronectin of chick tendon, fibroblasts, and 3T3 cells	Inhibition of secretion	Duksin and Bornstein (1977)

(Continued)

TABLE IV (*Continued*)

System	Effects of inhibition of glycosylation	Reference
Invertase and acid phosphatase of yeast	Inhibition of secretion	Kuo and Lampen (1974)
Alkaline phosphatase from mouse L cells	Nonglycosylated product 10-fold more sensitive against protease	Firestone and Heath (1981)
SPECIFIC PROTEOLYSIS		
Hemagglutinin of influenza virus	Increased proteolytic degradation; nonglycosylated hemagglutinin does not occur in virus particles	Schwarz and Klenk (1974); Schwarz et al. (1976); Nakamura and Compans (1978)
Envelope glycoproteins of Sindbis virus and Semliki forest virus	Nonglycosylated proteins are not transported to the plasma membrane	Schwarz et al. (1976); Leavitt et al. (1977a,b)
ACTH/βLPH common precursor from toad pituitary	Nonglycosylated precursor shows increased sensitivity against proteases, resulting in atypically processed peptides	Loh and Gainer (1978),
ACTH/βLPH common precursor from mouse pituitary cell line	Proteolytic cleavages not grossly altered	Herbert et al. (1980)
Fibronectin	Synthesis and exposure on outer cell membrane are not affected; however, half-life is reduced	Olden et al. (1978)
Acetylcholine receptor of chicken muscle cells	Enhanced susceptibility against proteases; half-life is reduced	Prives and Olden (1980)
Thyroid-stimulating hormone	Glycosylation not necessary for secretion but for combination between α- and β- subunit	Weintraub et al. (1980)
BIOSYNTHESIS		
Thyroglobulin	Proposal for a decreased synthesis due to a regulatory link between glycosylation and protein synthesis	Seagar et al. (1980)
Carboxypeptidase Y of yeast	Proposal for a decreased synthesis due to a regulatory link between glycosylation and protein synthesis	Hasilik and Tanner (1976, 1978)

[a] Very low density lipoprotein.

virus particle, nonglycosylated hemagglutinin when not degraded may support the budding of virus (Nakamura and Compans, 1978).

In the presence of tunicamycin, virus particles of Rous sarcoma virus are formed from chick cells, although with much reduced infectivity and devoid of their envelope glycoproteins gp85 and gp37, which are normally present in the spike (Schwarz *et al.*, 1976). This result is in line with results obtained in work with mutants of avian sarcoma viruses which showed that virus particles without envelope glycoproteins can be detected (Scheele and Hanafusa, 1971; Ogura and Friis, 1975; Halpern *et al.*, 1976; Schnitzer and Lodish, 1979).

Similarly, herpes virus particles with an altered morphology have been produced in the presence of inhibitors of glycosylation (Katz *et al.*, 1980). Additional examples may be found elsewhere (Schwarz and Datema, 1982b; Schwarz and Schmidt, 1982). It must be concluded from these reports that the requirements for the participation of envelope glycoproteins in the formation of virus particles are variable and that their roles are difficult to generalize or predict.

Inhibitors of glycosylation have been used to evaluate the possible involvement of glycoproteins in differentiation. Gastrulation of sea urchin embryos was prevented by such inhibitors, suggesting a requirement for active glycoprotein synthesis at this stage of development (Lennarz, 1980). Romanovski and Nosek (1980) drew similar conclusions in their work with *Xenopus laevis;* they attributed the failure of deoxyglucose to block this step in *X. laevis* to a decreased permeability of the cells to this compound.

Early cleavage divisions, adhesion, and compaction of mouse embryos are inhibited by tunicamycin, suggesting the involvement of glycoproteins in these processes (Surani, 1979; Atienza-Samols, *et al.*, 1980). A cell-surface glycoprotein possibly involved in the compaction of embryonal carcinoma cells and cleavage-stage embryos has been purified (Hyafil *et al.*, 1980).

The inhibition of glycosylation interferes with the formation of secondary palate formation, kidney tubule determination, and pairing between mating types of *Tetrahymena pyriformis* (Schwarz and Datema, 1982b). The lack of an effect of tunicamycin on cilia regeneration in *T. pyriformis* may rule out the possibility that reciliation requires *de novo* synthesis of glycoprotein (Keenan and Rice, 1980). Tunicamycin inhibits mouse tooth morphogenesis and odontoblast differentiation (Thesleff and Pratt, 1980a). It has been suggested that reduced amounts of fibronectin in the basement membrane and at the cell surfaces cause a disturbance of the necessary cell matrix interaction between the mesenchymal cells and the basement membrane (Thesleff and Pratt, 1980b). This supposition is in agreement with the work of Butters *et al.* (1980), which showed an inhibition of fibronectin-mediated adhesion of hamster fibroblasts to substratum. It must be pointed out that underglycosylated fibronectin is biologically active, making it unlikely that the carbohydrate portion of fibronectin plays a role in the interactions with the cell surface (Olden *et al.*, 1978, 1979; Pena and Hughes, 1979).

The differentiation of muscle cells is prevented by inhibiting glycosylation. Treatment of myoblasts with tunicamycin does not affect their growth and viability. Such treatment, however, prevent them from fusing, a step in the formation of muscle cells, suggesting the involvement of fusion glycoprotein that is not correctly synthesized (Gilfix and Sanwal, 1980). Olden *et al.* (1981) partially restored the ability of myoblasts to fuse by the additional administration of protease inhibitors. This result strongly favors the assumption of increased proteolytic sensitivity of the putative nonglycosylated fusion protein. Thus, degradation of the nonglycosylated fusion glycoprotein may be the basis for a halt in myoblast differentiation. Other examples are described by Schwarz and Datema (1982b).

From the work done so far it has become evident that inhibitors of glycosylation have attained a firm place among the standard tools in biochemistry and will increasingly be used in investigations on the synthesis and function of glycoproteins.

ACKNOWLEDGMENT

The work of the authors cited in this section was supported by Sonderforschungsbereich 47.

REFERENCES

Adair, W. L., and Trepanier, D. (1980). *Fed. Proc., Fed. Am. Soc. Exp. Biol.* **39,** Abstr. 1551.
Allen, C. M., Kalin, J. R., Sack, J., and Verizzo, D. (1978). *Biochemistry* **17,** 5020–5026.
Atienza-Samols, S. B., Pine, P. R., and Sherman, M. I. (1980). *Dev. Biol.* **79,** 19–32.
Aubert, J. P., Biserte, G., and Loucheux-Lefebvre, M. H. (1976). *Arch. Biochem. Biophys.* **175,** 410–418.
Babczinski, P. (1980). *Eur. J. Biochem.* **112,** 53–58.
Beeley, J. G. (1976). *Biochem. J.* **159,** 335–345.
Beeley, J. G. (1977) *Biochem. Biophys. Res. Commun.* **76,** 1051–1055.
Behrens, N. H., and Tábora, E. (1978). *In* "Methods in Enzymology" (V. Ginsburg, ed.), Vol. 50, Part C, pp. 402–439. Academic Press, New York.
Bergman, L. W., and Kuehl, W. M. (1978). *Biochemistry* **17,** 5174–5180.
Bernacki, R., Porter, C., Korytnik, W., and Mihich, E. (1978). *Adv. Enzyme Regul.* **16,** 217–237.
Biely, P., Krátký, Z., and Bauer, Š. (1974). *Biochim. Biophys. Acta* **352,** 268–274.
Blobel, G. (1980). *Proc. Natl. Acad. Sci. U.S.A.* **77,** 1496–1500.
Blobel, G., and Dobberstein, B. (1975a). *J. Cell Biol.* **67,** 835–851.
Blobel, G., and Dobberstein, B. (1975b). *J. Cell Biol.* **67,** 852–862.
Bretscher, M., and Raff, M. C. (1975). *Nature (London)* **258,** 43–49.
Burton, W. A., Scher, M. G., and Waechter, C. J. (1979). *J. Biol. Chem.* **254,** 7129–7136.
Butters, T. D., and Hughes, R. C. (1980). *Biochem. Soc. Trans.* **8,** 170–171.

Butters, T. D., Devalia, V., Aplin, J. D., and Hughes, R. C. (1980). *J. Cell Sci.* **44**, 33–58.

Cacan, R., Hoflack, B., and Verbert, A. (1980). *Eur. J. Biochem.* **106**, 473–479.

Carson, D. D., and Lennarz, W. J. (1979). *Proc. Natl. Acad. Sci. U.S.A.* **76**, 5709–5713.

Chambers, J., Forsee, W. T., and Elbein, A. D. (1977). *J. Biol. Chem.* **252**, 2498–2506.

Chapman, A., Li, E., and Kornfeld, S. (1979a). *J. Biol. Chem.* **254**, 10243–10249.

Chapman, A., Trowbridge, I. S., Hyman, R., and Kornfeld, S: (1979b). *Cell* **17**, 509–515.

Chapman, A., Fujimoto, K., and Kornfeld, S. (1980). *J. Biol. Chem.* **255**, 4441–4446.

Chatis, P. A., and Morrison, T. G. (1981). *J. Virol.* **37**, 307–316.

Chatterjee, S., Kwiterovich, P. O., and Sekerke, C. S. (1979). *J. Biol. Chem.* **254**, 3704–3707.

Chojnacki, T., Eckström, T., and Dallner, G. (1980). *FEBS Lett.* **113**, 218–220.

Chou, P. Y., and Fasman, G. D. (1974). *Biochemistry* **13**, 222–245.

Chu, F. K., Trimble, R. B., and Maley, F. (1978). *J. Biol. Chem.* **253**, 8691–8693.

Daleo, G. R., and Pont Lezica, R. (1977). *FEBS Lett.* **74**, 247–250.

Daleo, G. R., Hopp, H. E., Romero, P. A., and Pont Lezica, R. (1977). *FEBS Lett.* **81**, 411–414.

Das, R. C., and Heath, E. C. (1980). *Proc. Natl. Acad. Sci. U.S.A.* **70**, 3811–1815.

Datema, R., and Schwarz, R. T. (1978). *Eur. J. Biochem.* **90**, 505–516.

Datema, R., and Schwarz, R. T. (1979). *Biochem. J.* **184**, 113–123.

Datema, R., Schwarz, R. T., and Jankowski, A. W. (1980a). *Eur. J. Biochem.* **109**, 331–341.

Datema, R., Schwarz, R. T., and Winkler, J. (1980b). *Eur. J. Biochem.* **110**, 355–361.

Datema, R., Pont Lezica, R., Robbins, P. W., and Schwarz, R. T. (1981). *Arch. Biochem. Biophys.* **206**, 65–71.

Datema, R., and Schwarz, R. T. (1981). *J. Biol. Chem.* **256**, 11191–11198.

Decker, K., and Keppler, D. (1974). *Rev. Physiol. Biochem. Pharmacol.* **71**, 77–106.

Duksin, D., and Bornstein, P. (1977a). *J. Biol. Chem.* **252**, 955–962.

Duksin, D., and Bornstein, P. (1977b). *Proc. Natl. Acad. Sci. U.S.A.* **74**, 3433–3437.

Eagon, P. K., and Heath, E. C. (1977). *J. Biol. Chem.* **252**, 2372–2383.

Eckhardt, K., Wetzstein, H., Thrum, H., and Ihn, W. (1980). *J. Antibiot.* **33**, 908–910.

Edwards, K., Nagashima, M., Dryburgh, H., Wykes, A., and Schreiber, G. (1979). *FEBS Lett.* **100**, 269–272.

Elbein, A. D. (1979). *Annu. Rev. Plant Physiol.* **30**, 239–272.

Elbein, A. D., Gafford, J., and Kang, M. S. (1979). *Arch. Biochem. Biophys.* **196**, 311–318.

Endo, A. (1981). *Trends Biochem. Sci.* **6**, 10–13.

Fiat, A. -M., Jollès, J., Aubert, J. -P., Loucheux-Lefebvre, M. -H., and Jollès, P. (1980). *Eur. J. Biochem.* **111**, 333–339.

Firestone, G. L., and Heath, E. C. (1981). *J. Biol. Chem.* **256** 1404–1411.

Friedman, S. J., and Skehan, P. (1980). *Proc. Natl. Acad. Sci. U.S.A.* **77**, 1172–1176.

Gahmberg, C. G., Johinen, M., Karhi, K. K., and Andersson, L. C. (1980). *J. Biol. Chem.* **255**, 2169–2175.

Garoff, H., and Schwarz, R. T. (1978). *Nature (London)* **274**, 487–490.

Gateau, O., Morelis, R., and Louisot, P. (1978). *Eur. J. Biochem.* **80**, 613–622.

Geisow, M. (1979). *Nature (London)* **281**, 15–16.

Gibson, R., Leavitt, R., Kornfeld, S., and Schlesinger, S. (1978). *Cell* **13**, 671–679.

Gibson, R., Kornfeld, S., and Schlesinger, S. (1980). *Trends Biochem. Sci.* **5**, 290–293.

Gibson, R., Kornfeld, S., and Schlesinger, S. (1981). *J. Biol. Chem.* **256**, 456–462.

Gilfix, B. M., and Sanwal, B. D. (1980). *Biochem. Biophys. Res. Commun.* **96**, 1184–1191.

Grange, D. K., and Adair, W. L. (1977). *Biochem. Biophys. Res. Commun.* **79**, 734–740.

Halpern, M. S., Bolognesi, D. P., and Friis, R. R. (1976). *J. Virol.* **18**, 504–510.

Hanover, J. A., Lennarz, W. J., and Young, J. D. (1980). *J. Biol. Chem.* **255**, 6713–6716.

Hasilik, A., and Tanner, W. (1976). *Antimicrob. Agents Chemother.* **10**, 402–410.

Hasilik, A., and Tanner, W. (1978). *Eur. J. Biochem.* **91**, 567–575.

Hemming, F. W. (1977). *Biochem. Soc. Trans.* **5**, 1223-1231.

Herbert, E., Budarf, M., Phillips, M., Rosa, P., Policastro, P., Oates, E., Roberts, J. L., Seidah, N. G., and Chrétien, M. (1980). *Ann. N.Y. Acad. Sci.* **343**, 79-93.

Hickman, S., and Kornfeld, S. (1978). *J. Immunol.* **121**, 990-996.

Hickman, S., Kulczycki, A., Lynch, R. G., and Kornfeld, S. (1977). *J. Biol. Chem.* **252**, 4402-4408.

Hoflack, B., Cacan, R., and Verbert, A. (1980). *Eur. J. Biochem.* **112**, 81-86.

Hopp, H. E., Romero, P. A., Daleo, G., and Pont Lezica, R. (1978a). *Eur. J. Biochem.* **84**, 561-571.

Hopp, H. E., Romero, P. A., and Pont Lezica, R. (1978b). *FEBS Lett.* **86**, 259-262.

Hortin, G., and Boime, I. (1980). *J. Biol. Chem.* **255**, 8007-8010.

Housley, T. J., Rowland, F. N., Ledger, P. W., Kaplan, J., and Tanzer, M. L. (1980). *J. Biol. Chem.* **255**, 121-128.

Hubbard, S. C., and Ivatt, R. (1981). *Annu. Rev. Biochem.* **50**, 555-583.

Hubbard, S. C., and Robbins, P. W. (1980). *J. Biol. Chem.* **255**, 11782-11793.

Hyafil, F., Morello, P., Babinet, C., and Jacob, F. (1980). *Cell* **21**, 927-934.

Ito, T., Takatsuki, A., Kawamura, K., Sato, K., and Tamura, G. (1980). *Agric. Biol. Chem.* **44**, 695-698.

James, M. J., and Kandutsch, A. A. (1979). *J. Biol. Chem.* **254**, 8442-8446.

James, M. J., and Kandutsch, A. A. (1980). *J. Biol. Chem.* **255**, 8618-8622.

Jensen, J. W., Springfield, J. D., and Schutzbach, J. S. (1980). *J. Biol. Chem.* **255**, 11268-11272.

Kääriäinen, L., and Söderlund, H. (1978). *Curr. Top. Microbiol. Immunol.* **82**, 15-69.

Kaluza, G. (1975). *J. Virol.* **16**, 602-612.

Kaluza, G., Schmidt, M. F. G., and Scholtissek, C. (1973). *Virology* **54**, 179-189.

Kaluza, G., Rott, R., and Schwarz, R. T. (1980). *Virology* **102**, 286-299.

Kandutsch, A. A., and Chen, H. W. (1974). *J. Biol. Chem.* **249**, 6057-6061.

Kang, M. S., Spencer, J. P., and Elbein, A. D. (1978). *J. Biol. Chem.* **253**, 8860-8866.

Kang, M. S., Spencer, J. P., and Elbein, A. D. (1979). *J. Biol. Chem.* **254**, 10037-10043.

Kato, S., Tsuji, M., Nakanishi, Y., and Suzuki, S. (1980). *Biochem. Biophys. Res. Commun.* **95**, 770-776.

Katz, E., Margalith, E., and Duksin, D. (1980). *Antimicrob. Agents Chemother.* **17**, 1014-1022.

Keegstra, K., Sefton, B., and Burke, D. (1975). *J. Virol.* **16**, 613-620.

Keenan, R. W., and Rice, N. (1980). *Biochem. Biophys. Res. Commun.* **94**, 955-959.

Keil, W., Klenk, H. -D., and Schwarz, R. T. (1979). *J. Virol.* **31**, 253-256.

Keller, R. K., Adair, W. L., and Ness, G. C. (1979a). *J. Biol. Chem.* **254**, 9966-9969.

Keller, R. K. Boon, D. Y., and Crum, C. F. (1979b). *Biochemistry* **18**, 3946-3952.

Klenk, H. -D., and Rott, R. (1980). *Curr. Top. Microbiol. Immunol.* **90**, 19-48.

Klenk, H. -D., Scholtissek, C., and Rott, R. (1972). *Virology* **49**, 723-734.

Klenk, H. -D., Wöllert, W., Rott, R., and Scholtissek, C. (1974). *Virology* **57**, 28-41.

Koch, G., and Richter, D. (1980). "Biosynthesis, Modification, and Processing of Cellular and Viral Polyproteins." Academic Press, New York.

Koch, H. U., Schwarz, R. T., and Scholtissek, C. (1979). *Eur. J. Biochem.* **94**, 515-522.

Kornfeld, R. and Kornfeld, S. (1980). In "The Biochemistry of Glycoproteins and Proteoglycans" (W. J. Lennarz, ed.), pp. 1-34. Plenum, New York.

Kornfeld, S., Gregory, W., and Chapman, A. (1979). *J. Biol. Chem.* **254**, 11649-11654.

Krátký, Z., Biely, P., and Bauer, Š. (1975). *Eur. J. Biochem.* **54**, 459-467.

Kuo, S. -C., and Lampen, J. O. (1974). *Biochem. Biophys. Res. Commun.* **58**, 287-295.

Leavitt, R., Schlesinger, S., and Kornfeld, S. (1977a). *J. Virol.* **21**, 375-385.

Leavitt, R., Schlesinger, S., and Kornfeld, S. (1977b). *J. Biol. Chem.* **252**, 9018-9023.

Lennarz, W. J., ed. (1980). "The Biochemistry of Glycoproteins and Proteoglycans." Plenum, New York.

Li, E., and Kornfeld, S. (1979). *J. Biol. Chem.* **254**, 2754–2758.

Loh, Y. P., and Gainer, H. (1978). *FEBS Lett.* **96**, 269–272.

Loh, Y. P., and Gainer, H. (1979). *Endocrinology* **105**, 474–487.

Lohmeyer, J., and Klenk, H. -D. (1979). *Virology* **93**, 134–145.

Loucheux-Lefebvre, M. H., Aubert, J. P., and Jollès, P. (1978). *Biophys. J.* **23**, 323–336.

McCarthy, M., and Harrison, S. C. (1977). *J. Virol.* **23**, 61–73.

Maheshwari, R. K., Banerjee, D. K., Waechter, C. J., Olden, K., and Friedman, R. M. (1980). *Nature (London)* **287**, 454–456.

Mahoney, W. C., and Duksin, D. (1979). *J. Biol. Chem.* **254**, 6572–6576.

Marnell, L. L., and Wertz, G. W. (1979). *Virology* **98**, 88–98.

Martin, H. G., and Thorne, K. J. I. (1974). *Biochem. J.* **138**, 281–289.

Meeks, R. G., and Couri, D. (1980). *Biochim. Biophys. Acta* **630**, 238–245.

Melchers, F. (1973). *Biochemistry* **12**, 1471–1476.

Mescher, M. F., and Strominger, J. L. (1978). *FEBS Lett.* **89**, 37–41.

Miller, W. L., Johnson, L. K., Baxter, J. D., and Roberts, J. L. (1980). *Proc. Natl. Acad. Sci. U.S.A.* **77**, 5211–5215.

Mills, J. T., and Adamany, A. M. (1978). *J. Biol. Chem.* **253**, 5270–5273.

Mizrahi, A., O'Malley, J. A., Carter, W. A., Takatsuki, A., Tamura, G., and Sulkowski, E. (1978). *J. Biol. Chem.* **253**, 7612–7615.

Molnar, Z., and Bekesi, J. G. (1972). *Cancer Res.* **32**, 380–389.

Montreuil, J. (1980). *Adv. Carbohydr. Chem. Biochem.* **37**, 157–223.

Nakamura, K., and Compans, R. W. (1978). *Virology* **84**, 303–319.

Ogura, H., and Friis, R. R. (1975). *J. Virol.* **16**, 443–446.

Ogura, H., Schmidt, M. F. G., and Schwarz, R. T. (1977). *Arch. Virol.* **55**, 155–159.

Olden, K., Pratt, R. M., and Yamada, K. M. (1978). *Cell* **13**, 461–473.

Olden, K., Pratt, R. M., and Yamada, K. M. (1979). *Proc. Natl. Acad. Sci. U.S.A.* **76**, 3343–3347.

Olden, K., Law, J., Hunter, V. A., Romain, R., and Parent, J. B. (1981). *J. Cell Biol.* **88**, 199–204.

Parodi, A. J., and Leloir, L. F. (1979). *Biochim. Biophys. Acta* **559**, 1–37.

Pena, S. O. J., and Hughes, R. C. (1979). *Nature (London)* **276**, 80–83.

Pesonen, M., Saraste, J., Hashimoto, K., and Kääriäinen, L. (1981). *Virology* **109**, 165–173.

Pont Lezica, R., Romero, P. A., and Hopp, H. E. (1978). *Planta* **140**, 177–183.

Prives, J. M., and Olden, K. (1980). *Proc. Natl. Acad. Sci. U.S.A.* **77**, 5263–5267.

Puett, D. (1973). *J. Biol. Chem.* **248**, 3566–3572.

Reuvers, F., Habets-Willems, C., Reinking, A., and Boer, P. (1977). *Biochim. Biophys. Acta* **486**, 541–552.

Romanovsky, A., and Nosek, J. (1980). *Wilhelm Roux's Arch. Dev. Biol.* **189**, 81–82.

Rothman, J. E., Katz, F. N., and Lodish, H. F. (1978). *Cell* **15**, 1447–1454.

Rott, R. (1980). *In* "Structure and Variation in Influenza Virus" (G. Laver and G. Air, eds.), pp. 213–222. Elsevier/North-Holland Biomedical Press, Amsterdam.

Ruiz-Herrera, J., and Sentandreu, R. (1975). *J. Bacteriol.* **124**, 127–133.

Schachter, H., and Roseman, S. (1980). *In* "The Biochemistry of Glycoproteins and Proteoglycans" (W. . J. Lennarz, ed.), pp. 85–160. Plenum, New York.

Schechter, N., Momose, K., and Rudney, H. (1972). *Biochem. Biophys. Res. Commun.* **48**, 833–839.

Schechter, N., Nishino, T., and Rudney, H. (1973). *Arch. Biochem. Biophys.* **158**, 282–287.

Scheele, C. M., and Hanafusa, H. (1971). *Virology* **45**, 401–410.

Schmidt, M. F. G., Schwarz, R. T., and Scholtissek, C. (1976). *Eur. J. Biochem.* **70**, 55–62.

Schmidt, M. F. G., Schwarz, R. T., and Ludwig, H. (1976b). *J. Virol.* **18**, 819-823.

Schmidt, M. F. G., Biely, P., Krátký, Z., and Schwarz, R. T. (1978). *Eur. J. Biochem.* **87**, 55-68.

Schmidt, M. F. G., Bracha, M., and Schlesinger, M. J. (1979). *Proc. Natl. Acad. Sci. U.S.A.* **76**, 1687-1691.

Schmitt, J. W., and Elbein, A. D. (1979). *J. Biol. Chem.* **254**, 12291-12294.

Schnitzer, T. J., and Lodish, H. F. (1979). *J. Virol.* **29**, 443-447.

Scholtissek, C. (1975). *Curr. Top. Microbiol. Immunol.* **70**, 101-119.

Schutzbach, J. S., Springfield, J. D., and Jensen, J. W. (1980). *J. Biol. Chem.* **255**, 4170-4175.

Schwarz, R. T., and Datema, R. (1982a). *In* "Methods in Enzymology" (V. Ginsburg, ed.), Vol. 83: Complex Carbohydrates, Part D., pp. 432-443. Academic Press, New York.

Schwarz, R. T., and Datema, R. (1982b). *Adv. Carbohydr. Chem. Biochem.*

Schwarz, R. T., and Klenk, H. -D. (1974). *J. Virol.* **14**, 1023-1034.

Schwarz, R. T., and Schmidt, M. F. G. (1982). *In* "Tunicamycin" (G. Tamura, ed.) (in press).

Schwarz, R. T., Rohrschneider, J. M., and Schmidt, M. F. G. (1976). *J. Virol.* **19**, 782-791.

Schwarz, R. T., Schmidt, M. F. G., and Lehle, L. (1978). *Eur. J. Biochem.* **85**, 163-172.

Schwarz, R. T., Schmidt, M. F. G., and Datema, R. (1979). *Biochem. Soc. Trans.* **7**, 322-326.

Seagar, M. J., Miquelis, R. D., and Simon, C. (1980). *Eur. J. Biochem.* **113**, 91-96.

Sharon, N., and Lis, H. (1982). *In* "The Proteins" (H. Neurath and R. L. Hill, eds.), Vol. 5. Academic Press, New York.

Siewert, G., and Strominger, J. L. (1967). *Proc. Natl. Acad. Sci. U.S.A.* **57**, 767-773.

Speake, B. K., and White, D. A. (1979). *Biochem. J.* **180**, 481-489.

Spencer, J. C., Kang, M. S., and Elbein, A. D. (1978). *Arch. Biochem. Biophys.* **190**, 829-837.

Spencer, J. P., and Elbein, A. D. (1980). *Proc. Natl. Acad. Sci. U.S.A.* **77**, 2524-2527.

Spiro, M. J., Spiro, R. G., and Bhoyroo, V. P. (1979). *J. Biol. Chem.* **254**, 7668-7674.

Staneloni, R. J. Ugalde, R., and Leloir, L. F. (1980). *Eur. J. Biochem.* **105**, 275-278.

Stanley, P. (1980). *In* "The Biochemistry of Glycoproteins and Proteolglycans" (W. J. Lennarz, ed.), pp. 161-239. Plenum, New York.

Steiner, S., Courtney, R. J., and Melnick, J. L. (1973). *Cancer Res.* **33**, 2402-2407.

Stone, K. J., and Strominger, J. L. (1972). *Proc. Natl. Acad. Sci. U.S.A.* **69**, 1287-1289.

Strominger, J. L., Higashi, Y., Sanderman, H., Stone, K. J., and Willoughby, E. (1972). *In* "Biochemistry of the Glycosidic Linkage" (R. Piras and H. G. Pontis, eds.), pp. 135-154. Academic Press, New York.

Struck, D. K., and Lennarz, W. J. (1980). *In* "The Biochemistry of Glycoproteins and Proteoglycans" (W. J. Lennarz, ed.), pp. 35-83. Plenum, New York.

Struck, D. K., Siuta, P. B., Lane, M. D., and Lennarz, W. J. (1978). *J. Biol. Chem.* **253**, 5332-5337.

Suhadolnik, R. J. (1979). "Nucleosides as Biological Probes." Wiley, New York.

Surani, M. A. (1979). *Cell* **18**, 217-227.

Takatsuki, A., Kawamura, K., Okina, M., Kodama, Y., Ito, T., and Tamura, G. (1977). *Agric. Biol. Chem.* **41**, 2307-2309.

Tanaka, H., Oiwa, R., Matsukura, S., and Omura, S. (1979) *Biochem. Biophys. Res. Commun.* **86**, 902-908.

Tanzawa, K., and Endo, A. (1979). *Eur. J. Biochem.* **98**, 195-201.

Tarentino, A. L., Plummer, T. H., and Maley, F. (1974). *J. Biol. Chem.* **249**, 818-824.

Tartakoff, A. M. (1980). *Int. Rev. Exp. Pathol.* **22**, 227-251.

Thesleff, I., and Pratt, R. M. (1980a). *J. Embryol. Exp. Morphol.* **58**, 195-208.

Thesleff, I., and Pratt. R. M. (1980b). *Dev. Biol.* **80**, 175-185.

Trimble, R. B., and Maley, F. (1977). *Biochem. Biophys. Res. Commun.* **78**, 935-944.

Turco, S. J., Stetson, B., and Robbins, P. W. (1977). *Proc. Natl. Acad. Sci. U.S.A.* **74**, 4411-4414.

Vijay, I. K., and Perdew, G. H. (1980). *J. Biol. Chem.* **255**, 11221-11226.

Villemez, C. L., and Carlo, P. L. (1980). *J. Biol. Chem.* **255,** 8174–8178.

Visser, D. W., and Roy-Burman, S. (1979). *Antibiotics (N.Y.)* **5** (pt. 2), 363.

Ward, C. W. (1981). *Curr. Top. Microbiol. Immunol.* **94/95,** 1–74.

Ward, J. B., Wyke, A. W., and Curtis, C. A. M. (1980). *Biochem. Soc. Trans.* **8,** 164–166.

Wedgwood, J. F., and Strominger, J. L. (1980). *J. Biol. Chem.* **255,** 1120–1123.

Weintraub, B. D., Stannard, B. S., Linnekin, D., and Marshall, M. (1980). *J. Biol. Chem.* **255,** 5715–5723.

Welch, W. J., and Sefton, B. M. (1979). *J. Virol.* **29,** 1186–1195.

Wellner, R. B., and Lucas, J. J. (1979). *FEBS Lett.* **104,** 379–383.

Wellner, R. B., and Lucas, J. J. (1980). *Fed. Proc., Fed. Am. Soc. Exp. Biol.* **39,** Abstr. 1553.

Welten-Versteegen, G. W., Boer, P., and Steyn-Parvé, E. P. (1980). *J. Bacteriol.* **141,** 342–349.

White, D. A., and Speake, B. K. (1980). *Biochem. J.* **192,** 297–301.

Wickner, W. (1979). *Annu. Rev. Biochem.* **48,** 23–45.

Wickner, W. (1980). *Science* **210,** 861–868.

Williamson, A. R., Singer, H. H., Singer, P. A., and Mosmann, T. R. (1980). *Biochem. Soc. Trans.* **8,** 168–170.

Wirth, D. F., Lodish, H. F., and Robbins, P. W. (1979). *J. Cell Biol.* **81,** 154–162.

Woodman, D. R., and Williams, J. C. (1977). *Antimicrob. Agents Chemother.* **11,** 475–481.

SECTION 4

The Variable Temporal Relationship between Translation and Glycosylation and Its Effect on the Efficiency of Glycosylation

LAWRENCE W. BERGMAN AND W. MICHAEL KUEHL

Our understanding of the biosynthesis of glycoproteins containing *N*-asparagine-linked oligosaccharides and *O*-threonine- or *O*-serine-linked

81

THE GLYCOCONJUGATES, VOL. III

oligosaccharides has advanced substantially in the past few years. In this section, we discuss both *in vivo* and *in vitro* systems that have been employed to investigate the temporal and spatial relationship between the synthesis of the polypeptide and the attachment of the precursor oligosaccharide chain. Although this involves mainly a discussion of N-linked glycosylation, wherever possible results concerning O-linked glycosylation are included.

I. OVERVIEW OF GLYCOPROTEIN BIOSYNTHESIS

A. N-Linked Glycosylation

1. Common Core Structure of High-Mannose and Complex Oligosaccharides

The *N*-asparagine-linked oligosaccharides of cellular and extracellular glycoproteins are heterogeneous in structure but in general appear to fall into two classes: simple (high-mannose) and complex (for review, see Kornfeld and Kornfeld, 1976). Both types have a core structure consisting of three mannose residues and two *N*-acetylglucosamine residues. High-mannose oligosaccharides contain additional α-linked mannose residues and rarely *N*-acetylglucosamine and galactose attached to the invariant core, and complex oligosaccharides generally contain sialic acid, galactose, fucose, and *N*-acetylglucosamine attached to the invariant core. The presence of the invariant core structure led to the proposal that the high-mannose and complex N-linked oligosaccharides are derived from a common precursor oligosaccharide, which is subsequently modified to the two types of structures during glycoprotein maturation within the cell (Robbins *et al.*, 1977; Tabas *et al.*, 1978).

2. Transfer of a Preassembled Oligosaccharide from a Dolichol Donor

Experiments from a number of laboratories have established that a precursor oligosaccharide with a high-mannose structure can be assembled while linked to a polyisoprenoid lipid carrier through a pyrophosphate bond (for review, see Waechter and Lennarz, 1975). This lipid-linked oligosaccharide, which contains three glucose, nine mannose, and two *N*-acetylglucosamine residues (Li *et al.*, 1978; Liu *et al.*, 1979), has a structure consistent with its representing the common precursor for the high-mannose and complex oligosaccharides. Numerous experiments have provided evidence that this preassembled oligosaccharide is transferred *en bloc* from the dolichol lipid to an asparagine residue on the polypeptide (Waechter and Lennarz, 1975) in the rough endoplasmic reticulum. The process of N-linked glycosylation can be inhibited by the antibiotic tunicamycin (Takatsuki *et al.*, 1975). *In vitro* studies from several laboratories have shown that tunicamycin specifically inhibits the first step in the assembly of the dolichol lipid–oligosaccharide intermediate (i.e., transfer of *N*-acetylglucosamine 1-phosphate from UDP-*N*-acetylglucosamine to dolichyl phosphate to form dolichylpyrophosphoryl-*N*-acetylglucosamine) (Takatsuki *et al.*, 1975; Struck and Lennarz, 1977). If this step in the assembly process is blocked, the lipid-linked oligosaccharide is not formed and N-glycosylation is inhibited (for a review of inhibition of lipid-dependent glycosylation, see Chapter 1, Section 3).

3. Generation of Complex and High-Mannose Oligosaccharides by Sequential Processing and Stepwise Addition of Sugars

For both complex and high-mannose oligosaccharides, all of the glucose and some of the mannose residues are sequentially removed (processed) from the preassembled oligosaccharide after its transfer to the peptide asparagine. In the case of complex oligosaccharides, the peripheral sugar residues (additional *N*-acetylglucosamine and galactose) are then added in a stepwise manner catalyzed by specific glycosyltransferases in the smooth endoplasmic reticulum or Golgi apparatus (Schachter, 1974), and the terminal sugar residues (additional galactose, fucose, and sialic acid) are added at or very close to the time of secretion (for secreted glycoproteins) or insertion into the plasma membrane (for membrane glycoproteins). The molecular basis on which a decision is made to synthesize a complex rather than a high-mannose oligosaccharide from the common oligosaccharide precursor is unknown. Since some proteins contain both simple and complex oligosaccharides (Sefton, 1977; Kehry *et al.*, 1979), the information is most likely present in the microenvironment determined by the protein.

B. O-Linked Glycosylation

In contrast to N-linked glycosylation, little is known about the mechanism of attachment of *N*-acetylgalactosamine to serine or threonine residues of glycoproteins (O-linked glycosylation). Because the oligosaccharide chains of these glycoproteins show much variation in length and composition, it is unlikely that their biosynthesis takes place in a manner analogous to N-linked glycoprotein biosynthesis. Although the existence of lipid-linked oligosaccharides that could be involved in the synthesis of O-linked oligosaccharides has been reported (Hopwood and Dorfman, 1977; Peterson *et al.*, 1976), their involvement in the assembly of these chains has not been documented. Biosynthetic studies in a number of laboratories have suggested that initial glycosylation of epithelial glycoproteins takes place in microsomal fractions (e.g., Hill *et al.*, 1977), but the precise subcellular site has not been determined.

II. *IN VIVO* AND *IN VITRO* METHODS FOR INVESTIGATING THE TEMPORAL RELATIONSHIP BETWEEN TRANSLATION AND GLYCOSYLATION

Three main approaches have been used to analyze the relationship between polypeptide synthesis and glycosylation: (a) *in vivo* translation and glycosylation followed by isolation of nascent chains, (b) coupled *in vitro* translation and glycosylation, and (c) *in vitro* glycosylation systems utilizing endogenous or exogenous polypeptide acceptors.

A. Isolation of Nascent Chains Synthesized *in Vivo*

The first approach involves the isolation of nascent polypeptides from the cell by copurification with the appropriate polysomal fraction (e.g., Uhr, 1970). This approach is based on the assumption that, after short-term labeling, all protein-associated radioactivity (e.g., radiolabeled amino acids or sugars) in the polysomal region of a sucrose gradient is in nascent chains. However, the presence of contaminating completed polypeptides is a significant problem. For example, using the immunoglobulin-synthesizing mouse plasmacytoma system, we have found that approximately 70% of the [^{3}H]glucosamine-labeled material that copurifies with the ribosomal fraction derived from cells labeled for 5 minutes with [^{3}H]glucosamine is due to contaminating completed polypeptides (Bergman and Kuehl, 1977). To eliminate contamination by completed chains, peptidyl-tRNA (i.e., nascent chains) can be separated from completed chains by anionic ion-exchange chromatography of the solubilized ribosomal fraction in the presence of denaturants (Ganoza and Nakamota, 1966; Slabaugh and Morris, 1970; Cioli and Lennox, 1973). To minimize deacylation of nascent chains during the isolation procedure, it is important to avoid even slightly alkaline pH and the presence of buffers containing uncharged amino groups, both of which promote deacylation. During cell fractionations, for example, we have used solutions buffered with phosphate at pH 6.7, which resulted in better yields than fractionations performed in solutions buffered with Tris at pH 7.2 (M. Kuehl and L. Bergman, unpublished).

The ion-exchange chromatographic isolation of the peptidyl-tRNA is based on the multiple negative charges contributed by the tRNA moiety of the peptidyl-tRNA complex. For example, a ribosomal fraction is solubilized in QAE buffer [0.1% Brij 35–6 M urea–0.1 M NaCl–0.1 M ammonium formate (pH 4.7)] containing 0.05% SDS and 2 mM EDTA and then chromatographed on QAE-Sephadex in QAE buffer (Bergman and Kuehl, 1977). Peptidyl-tRNA quantitatively binds to the QAE-Sephadex, whereas completed chains do not bind. The peptidyl-tRNA is eluted nearly quantitatively by the addition of 1 M NaCl to the QAE buffer. The eluted peptidyl-tRNA can be diluted tenfold in the QAE buffer and reapplied to QAE-Sephadex for a second cycle of purification. Although DEAE-cellulose can be used instead of QAE-Sephadex, we have found that the latter gives higher yields of nascent chains with less contamination by completed chains (L. Bergman and M. Kuehl, unpublished); it is likely that hydrophobic peptides interact less with the hydrophilic Sephadex backbone than with the somewhat hydrophobic cellulose backbone. A variety of criteria indicate that two cycles of QAE-Sephadex chromatography produce a nascent chain fraction that contains no significant completed chains. The best documentation of the purity of specific nascent chains is that unlabeled nascent immunoglobulin light chains contain no detectable completed light chains as assayed by the absence of the carboxy terminal light chain tripeptide in a nascent chain fraction isolated from the second cycle of ion-exchange chromatography (Bergman and Kuehl, 1979b). Specific nascent chains have been isolated by either (a) immunoprecipitation of the peptidyl-tRNA fraction from the ion-exchange column (Bergman and Kuehl, 1977; Cioli and Lennox, 1973), or (b) immunoprecipitation of polysomes synthesizing the desired protein, followed by ion-exchange chromatography to isolate the peptidyl-tRNA species (Palacios *et al.,* 1972; Kiely *et al.,* 1976; Glabe *et al.,* 1980). The latter method has the potential advantages of including short nascent chains that contain no antigenic determinants and of minimizing the loss of antigenic determinants due to the denaturants used for ion-exchange purification of nascent chains. Possible disadvantages of the latter method include nonspecific immunoprecipitation of crude ribosomal or polyribosomal fractions and the potential difficulties of disrupting immune complexes before ion-exchange chromatography. For both methods involving immunoprecipitation of specific nascent chains, it is probably important that the specific antibodies be directed against as many antigenic determinants as possible, i.e., including determinants present in both native and denatured protein. In principle, other assays for specific nascent chains, e.g., enzymatic activity or affinity chromatography to an immobilized ligand, could be used.

B. *In Vitro* Coupled Translation–Glycosylation

The second approach involves the translation of mRNA coding for the desired polypeptide in an *in vitro* cell-free protein-synthesizing system supplemented with rough microsomal membranes capable of translocation and glycosylation. *In vitro* studies in which (a) translation of mRNA is partially synchronized (Szczesna and Boime, 1976; Rothman and Lodish, 1977; Rothman *et al.,* 1978) or (b) size separation of the *in vitro* translation product is done (Palmiter *et al.*, 1978) provide data that permit analysis of the *in vitro* temporal relationship between a specific modification event, such as glycosylation, and nascent chain length. This *in vitro* approach has been used extensively to study glycosylation (Rothman and Lodish, 1977; Rothman *et al.*, 1978; Toneguzzo and Ghosh, 1977; Katz *et al.*, 1977; Bielinska and Boime, 1978). The advantages of this approach include the following. (a) *In vitro* systems are usually simpler to manipulate than the *in vivo* system described in the first approach; (b) the use of partially purified mRNA minimizes the problem of eliminating contaminating proteins; and (c) the slow rate of *in vitro* elongation potentially allows precise timing of the glycosylation event. The disadvantages of this approach include the following. (a) The rates of *in vitro* elongation are significantly slower (20–50 times) than *in vivo* elongation rates (Palmiter *et al.*, 1978), so that glycosylation observed on nascent chains *in vitro* may not occur on nascent chains *in vivo*, and (b) there is some question as to whether a reconstituted *in vitro* system is a valid model of the *in vivo* system. In fact, results obtained using the coupled translation–glycosylation *in vitro* approach indicate that both potential asparagine acceptor sites (residues 293 and 312) may be glycosylated on some ovalbumin molecules synthesized *in vitro* (Palmiter *et al.*, 1980), yet only the first of these asparagine acceptor sites (residue 293) is glycosylated on ovalbumin molecules synthesized *in vivo* (Glabe *et al.*, 1980). Also, a similar *in vitro* analysis of the α subunit of human chorionic gonadotropin suggests that glycosylation of an asparagine acceptor 14 amino acids from the carboxy terminal end of the molecule occurs before the completion of translation (Bielinska and Boime, 1978). This *in vitro* result is unlikely to reflect the *in vivo* situation since *in vivo* evidence discussed later (Section II,A,1) suggests that a minimum of approximately 30 amino acid residues are required to clear the ribosome and span the membrane so that glycosylation can occur on the luminal side of the rough endoplasmic reticulum.

C. *In Vitro* Glycosylation of Endogenous or Exogenous Proteins

The third approach utilizes a membrane fraction that catalyzes the synthesis of an oligosaccharide–lipid and transfer of this oligosaccharide moiety to either endogenous microsomal protein acceptors or exogenously added proteins or peptides. Although this system has been used primarily to study the synthesis of the oligosaccharide moiety (Chen and Lennarz, 1978), the approach has also provided valuable information concerning the primary structural requirements for the occurrence of glycosylation (Hart *et al.*, 1979), as discussed later (Section V,B and V,D).

III. RESULTS OBTAINED ON THE TEMPORAL RELATIONSHIP BETWEEN TRANSLATION AND GLYCOSYLATION

A. *In Vivo* Timing of Glycosylation as Determined by Analysis of Nascent Chains

1. *Occurrence of N-Glycosylation on Nascent Chains*

Our results with immunoglobulin heavy chain (Bergman and Kuehl, 1977) and results from the laboratories of Schimke (Kiely *et al.*, 1976) and Lennarz (Glabe *et al.*, 1980) with ovalbumin have rigorously demonstrated that the initial glycosylation event occurs on nascent polypeptides. In the case of MPC 11 γ_{2b} heavy chain, analysis of the amino acid sequence (Tucker *et al.*, 1979) indicates that it is asparagine residue 299 which is glycosylated. Therefore, approximately 33,000 daltons of heavy chain mRNA must be translated before the glycosylation site is potentially available for the attachment of sugar residues. Analysis of the SDS gel electrophoresis patterns of immunoprecipitated deacylated nascent chains labeled with radioactive sugars reveals an apparent size distribution of 38,000–55,000 daltons (full size), indicating that glucosamine and mannose residues can be attached to nascent heavy chains very soon after the acceptor site is synthesized. A similar analysis of ovalbumin by Lennarz and co-workers (Glabe *et al.*, 1980) indicates that the smallest detectable glycosylated nascent chain of ovalbumin is approximately 37,000 daltons, suggesting that the nascent chain is glycosylated when the polypeptide is approximately 310 residues long. Ovalbumin is a polypeptide of 386 amino acid residues with a single N-linked oligosaccharide attached to asparagine residue 293.

Comparison of the minimum size of glycosylated nascent chain and the site of the acceptor asparagine residue indicates that for both immunoglobulin heavy chain and ovalbumin approximately 30 additional residues beyond the asparagine acceptor site must be synthesized before the asparagine site is glycosylated. Lennarz and co-workers (Glabe *et al.*, 1980) approached this question more directly by isolating from purified ovalbumin nascent chains a tryptic peptide of 32 amino acids (residues 292–323) that extends for a length of 30 residues on the carboxy terminal side of the glycosylated asparagine residue 293; no glycosylated tryptic peptide of smaller size was detected. This suggests that a minimum of 30 residues must be added to the carboxy terminal end of the acceptor site before glycosylation occurs. Additional studies by these workers indicate that a minimum of 15 residues of nascent ovalbumin molecules are protected by the ribosome from trypsin digestion. These results suggest that the nascent polypeptide must be extended approximately 15 residues to clear the ribosome and 15 residues to span the membrane so that the acceptor site becomes accessible on the luminal surface of the rough endoplasmic reticulum. Previous studies (for review, see Davis and Tai, 1980) suggest that glycosylation occurs on the luminal side of the rough endoplasmic reticulum, although this conclusion has not been unequivocally established (Snider *et al.*, 1980). Recent work has demonstrated that the oligosaccharide chains of endogenous protein glycosylated *in vitro* are accessible to the enzyme endoglycosidase H only when the microsomal membranes are made permeable by detergents to macromolecules (Hanover and Lennarz, 1980).

2. Rapid Quantitative N-Glycosylation after the Acceptor Site on Nascent Chains Becomes Available for Oligosaccharide Transfer

The data for immunoglobulin heavy chain and ovalbumin demonstrate that some nascent chains are glycosylated *in vivo* when the nascent chain contains a minimum of about 30 amino acid residues beyond the asparagine acceptor site. We attempted to determine the fraction of chains that are glycosylated when the nascent heavy chains reach this minimum size. We obtained the ratios of [^{3}H] glucosamine to [^{35}S]methionine for completed and nascent heavy chains isolated from the microsomal fraction of cells labeled for 5 minutes with [^{3}H]glucosamine and [^{35}S]methionine (Bergman and Kuehl, 1978). The results indicate that the ratios of ^{3}H to ^{35}S are very similar for both completed chains and nearly completed nascent chains, implying that the glycosylation of the nascent heavy chains is quantitative (i.e., essentially all nascent heavy chains receive the oligosaccharide precursor before chain completion). Since the ^{3}H/^{35}S ratios obtained for a nascent chain sample electrophoretically fractionated on SDS gels increases significantly in a single fraction (approximately 38,000 daltons), it is apparent that quantitative glycosylation of the nascent chain occurs within a very limited time frame as soon as the asparagine acceptor becomes available for glycosylation. That is, the heavy chain is quantitatively glycosylated when the nascent heavy chain contains about 30 amino acid residues beyond the asparagine acceptor site. The significance of this observation is discussed later. (Section V,D).

3. Possibility of O-Glycosylation Not Occurring on Nascent Chains

Strous (1979) reported that O-glycosylation of epithelial glycoproteins occurs at the level of nascent chains, which implies that this occurs in the rough endoplasmic reticulum. However, it should be noted that Strous did not rigorously exclude the possibility that the labeled *N*-acetylgalactosamine detected in the putative nascent chain preparation was derived from contaminating completed O-linked glycoproteins. Furthermore, several groups have investigated the subcellular localization of the N-acetylgalactosaminyltransferase that catalyzes the addition of the N-acetylgalactosamine residue to serine or threonine residues and have found that this enzyme is localized in a fraction consisting of smooth endoplasmic reticulum and Golgi membranes, with relatively little activity in the rough endoplasmic reticulum membrane fraction (Ko and Raghupathy, 1972; Hanover *et al.*, 1980).

B. Temporal Relationship between Translation and Glycosylation *in Vitro*

Several groups have investigated the *in vitro* temporal relationship between translation and glycosylation in cell-free protein-synthesizing systems in the presence of exogenous rough microsomal membranes (Toneguzzo and Ghosh, 1977; Katz *et al.*, 1977; Bielinska and Boima, 1978). Rothman and Lodish (1977) and Rothman *et al.* (1978), using a wheat germ cell-free protein-synthesizing system in the presence of exogenous pancreatic rough endoplasmic reticulum membranes (stripped of endogenous ribosomes), examined the sequence of glycosylation of nascent chains of the vesicular stomatitis virus G membrane glycoprotein, which contains two N-linked oligosaccharides when synthesized *in vivo*. To synchronize the synthesis of the G protein, they added a specific inhibitor of initiation, 7-methylguanosine 5'-phosphate, to the cell-free system a short time after protein synthesis was initiated to prevent subsequent initiations. These authors demonstrated that the addition of a nonionic detergent (Triton X-100) to the cell-free incubation blocks glycosylation (presumably by solubilizing the microsomal membranes that contain the glycosylation apparatus) but does not effect synthesis of the polypeptide. Therefore, by varying the time at which Triton X-100 was added, they could determine the precise timing of the glycosylation event. Since the addition of Triton X-100 blocks glycosylation but does not prevent the completion of translation, they were able to analyze completed chains by electrophoresis on SDS-polyacrylamide gels. If translation elongation rates were assumed to be constant, they were able to calculate the efficiency of glycosylation on nascent chains of a given size. If glycosylation was blocked (by the addition of Triton X-100) before one-third of the G protein was translated, no glycosylation occurred. If glycosylation was blocked when the G protein was one-third to two-thirds translated, the completed G protein contained only one oligosaccharide chain. If glycosylation was blocked when the G protein was more than two-thirds translated, the completed G protein contained two oligosaccharide chains. In other words, the first step in glycosylation occurs quantitatively when about one-third of the G mRNA has been translated. The second and final step of glycosylation occurs quantitatively when approximately two-thirds of the G mRNA has been translated. The DNA sequence of G protein, which was determined recently (J. Rose, personal communication), indicates that

G protein contains 508 amino acid residues, with potential asparagine glycosylation acceptor sites at residues 178 and 335. Thus, the timing of *in vitro* glycosylation is in reasonable agreement with the position of the glycosylation sites. As noted above (Section II,B), since *in vitro* translation rates are 20–50 times slower than *in vivo* translation rates, these results are not a rigorous demonstration of the timing of glycosylation that occurs on G protein *in vivo*.

C. Occurrence of Initial Processing of the Precursor Oligosaccharide on Nascent Chains

The initial glycosylation event involves the transfer of a large molecular weight oligosaccharide containing two *N*-acetylglucosamine residues, nine mannose residues, and three glucose residues as a unit from a lipid intermediate to an asparagine acceptor residue on the polypeptide (Robbins *et al.*, 1977; Tabas *et al.*, 1978; Waechter and Lennarz, 1975). Recent evidence has suggested that soon after transfer there is processing of the oligosaccharide to remove the glucose residues and the majority of the mannose residues (Robbins *et al.*, 1977; Tabas *et al.*, 1978). At present, the exact role of the glucose residues on the oligosaccharide chain is not known. Turco *et al.* have postulated from *in vitro* experiments (1977) that the glucose residues influence the rate and extent of transfer of the precursor oligosaccharide chain to the protein. Although no evidence is currently available, future studies should determine whether the glucose residues are present on the nascent polypeptide or whether some of the glucose and mannose residues are removed concomitantly with or shortly after transfer of the oligosaccharide to the nascent protein. It is interesting that the oligosaccharide chains of several of the lysosomal enzymes have a glucose constituent (Tulsiani *et al.*, 1977). Since these proteins are also synthesized on membrane-bound polyribosomes, it is probable that the blocking of the glucose residue from processing must take place soon after transfer of the oligosaccharide chain to the polypeptide. Possibly this blocking occurs on the nascent polypeptide, since several amino acid modification events have been shown to occur on nascent chains (e.g., Palmiter *et al.*, 1978; Bergman and Kuehl, 1979c).

IV. SLOWING OF PEPTIDE ELONGATION BY TRANSFER OF THE PRECURSOR OLIGOSACCHARIDE TO NASCENT CHAINS

We demonstrated an apparent block in the *in vivo* translation of heavy chain mRNA, which causes the accumulation of nonglycosylated nascent heavy chains of approximately 35,000 daltons (i.e., slightly smaller than the smallest detectable glycosylated nascent heavy chain, approximately 38,000 daltons) (Bergman and Kuehl, 1978; Bergman *et al.*, 1981). A similar observation was made by Lennarz and co-workers for ovalbumin nascent chains (Glabe *et al.*, 1980). Although the exact nature of the relationship between the translational block and glycosylation is not known, we have shown that inhibition of glycosylation with tunicamycin removes the translational block that results in the accumulation of nascent chains having a size of approximately 35,000 daltons. These results suggest that the apparent translational block is not inherent in the mRNA translation but is rather a secondary effect of the glycosylation process. It is possible that the binding of the oligosaccharide–lipid intermediate and the enzyme(s) involved in glycosylation to the asparagine acceptor site results in a localized slowing of polypeptide translation, thereby causing the observed accumulation of nascent polypeptides. These results with heavy chain, plus the preliminary results reported for ovalbumin, suggest that glycosylation may be a potential rate-limiting step in the translation of all glycoproteins containing N-linked oligosaccharides.

It is of interest that biosynthetic studies with myeloma tumors and cell lines, as well as with normal spleen cells, have demonstrated that generally there is synthesis of a significant molar excess of light chains compared to heavy chains (Baumal and Scharff, 1973; Kuehl, 1977). This is physiologically significant since synthesis of excess heavy chains, but not excess light chains, appears to be detrimental to the cell (since heavy chains are relatively insoluble and cannot be secreted unless associated with light chains). If there is a block in heavy chain mRNA translation related to the glycosylation process, one would expect a relative increase in heavy chain synthesis when this block is abolished. Analysis of the relative rates of synthesis of heavy chains and light chains in control and tunicamycin-treated mouse myeloma cells indicates that there is a relative increase in heavy chain (relative to light chain) synthesis when glycosylation is inhibited with tunicamycin (Bergman *et al.*, 1981). This suggests that the apparent translational block of heavy chain synthesis caused by the glycosylation process contributes significantly to the imbalance of heavy chain and light chain biosynthesis observed in normal and malignant lymphoid cells.

V. RECEPTOR SITES FOR TRANSFER OF THE PRECURSOR OLIGOSACCHARIDE OTHER THAN NASCENT CHAINS

A. Glycosylation on Completed Chains

Several groups have reported that the initial glycosylation event can occur on completed polypeptides after their release from the ribosome. Eagon *et al.* (1975) reported the *in vitro* transfer of the oligosaccharide core from a dolichol lipid intermediate to completed MOPC 46B κ light chains. Tucker and Pestka (1977), using an *in vitro* wheat germ synthesized MOPC 46B precursor light chain, have also shown *in vitro* glycosylation (2–3% efficient) of the completed precursor light chain. Also, the *in vitro* glycosylation system of Lennarz and co-workers (Chen and Lennarz, 1978) transfers the oligosaccharide core to several endogenous (presumably completed) polypeptides, the significance of which is not known.

Our laboratory investigated the question of whether *in vivo* glycosylation normally occurs on completed polypeptides in several ways (Bergman and Kuehl, 1978). First, we investigated the *in vivo* glycosylation of the MPOC 46B light chain to determine the time of transfer of the precursor oligosaccharide to the protein. Our results indicate that the majority of the glycosylation (85–90%) is a cotranslational event occurring on the nascent protein, whereas a minority of the glycosylation (10–15%) occurs on completed chains. Furthermore, the rate of MOPC 46B nascent chain glycosylation is much slower than that for the MPC 11 heavy chain in that glycosylation of nascent chains may take place at any time interval during translation of the light chain mRNA (i.e., on nascent chains from 12,000 to 25,000 daltons). The reason that quantitative glycosylation of the MOPC 46B light chain does not occur at a precise time, as shown for the MPC 11 heavy chain, may be the presence of the amino terminal precursor or signal sequence. Several groups have provided evidence that the removal of the signal sequence is also a cotranslational event (Blobel and Dobbenstein, 1975; Szczesna and Boime, 1976). The binding of the amino terminal region to the membrane or to a proteolytic enzyme for processing of the amino terminal signal sequence may interfere with the availability of the asparagine acceptor site (residue 28) for glycosylation, thus inhibiting glycosylation so that individual nascent chains may be glycosylated at various times after the acceptor site is potentially available.

Second, we studied a mouse myeloma mutant clone (M311), which was derived from the MPC 11 myeloma cells used for our studies on the temporal relationship between translation and glycosylation of normal heavy chains (Sections III,A,1 and 2). M311 cells synthesize a mutant heavy chain containing a carboxy terminal deletion (Kenter and Birshtein, 1979). Approximately one-third of M311 heavy chain exists as a 42,000-dalton glycosylated species, and the remainder exists as a 38,000-dalton nonglycosylated species (Weitzman and Scharff, 1976). When M311 cells were pulsed with [35S]methionine for 1.5 minutes, the glycosylated species represented 24% of the total heavy chain. After a 25-minute chase in the presence of excess unlabeled methionine, the glycosylated heavy chain species represented 36% of the total heavy chain (Bergman and Kuehl, 1978). Although this result does not indicate whether all M311 heavy chains are glycosylated as completed chains, it does demonstrate that a significant proportion of the M311 heavy chains are glycosylated as completed chains (i.e., at least one-third of glycosylated heavy chains are glycosylated as completed chains).

B. Proteins Containing a Potential Glycosylation Site That Is Not Glycosylated

Analysis of amino acid sequences of a large number of glycoproteins plus *in vitro* glycosylation studies using exogenous proteins or peptides, including a tripeptide with blocked amino and carboxy terminal ends, indicates that the tripeptide sequence Asn-X-Thr/Ser represents the signal for N-glycosylation (Hart *et al.*, 1979). Although asparagine is glycosylated in most instances when it is present in this tripeptide sequence in proteins not present in the cytosol (e.g., membrane, lysosomal, secreted proteins), there are a significant number of instances in which this is not true. For example, this tripeptide sequence is glycosylated in only about 10% of pancreatic RNase molecules, although denatured (but not native) RNase A (nonglycosylated form) can be glycosylated *in vitro* (Pless and Lennarz, 1977). Similarly, ovalbumin contains two potential asparagine acceptor sites for glycosylation, i.e., residues 193 and 312, but ovalbumin synthesized *in vivo* is glycosylated only at residue 293; denatured ovalbumin or peptide fragments of ovalbumin can be glycosylated *in vitro* at residue 312 (Glabe *et al.*, 1980). Also, as indicated above (Section II,B), some molecules of ovalbumin synthesized in a coupled translation–glycosylation *in vitro* system appear to be glycosylated at two sites (which would have to be residues 293 and 312 since these are the only potential asparagine glycosylation sites present in ovalbumin). Thus, in some cases a potential asparagine glycosylation site is not glycosylated *in vivo* either as a nascent chain or as a completed chain on any molecules (e.g., ovalbumin), or is glycosylated *in vivo* on only a fraction of molecules (e.g., RNase).

C. Occurrence of Glycosylation on Completed Chains if Prevented from Occurring on Nascent Chains

We have investigated the question of whether glycosylation can occur on completed heavy chains by inhibiting the initial glycosylation event with high concentrations of glucosamine (Bergman and Kuehl, 1978). Glycosylation is inhibited reversibly, perhaps as a result of depletion of uridine nucleotide pools by trapping of the cellular uridine nucleotide in the form of UDP-N-acetylglucosamine, UDP-glucose, and UDP-galatose (Bekesi and Winzler, 1969) or perhaps as an inhibiting effect of one of the UDP-monosaccharides. Our results indicate that nonglycosylated completed heavy chain synthesized in the presence of the glucosamine cannot be glycosylated during a 60-minute chase period in the absence of the glucosamine, although the cells regain the ability to glycosylate newly synthesized heavy chains by 20 minutes into the chase. This effect appears to be true for γ, μ, and α heavy chains synthesized by mouse plasmacytoma cell lines (Bergman and Kuehl, 1978, 1979c). It is noteworthy that this is true for μ and α heavy chains in that these two classes of heavy chains contain multiple oligosaccharide moieties per chain.

However, similar experiments with cells synthesizing either the MOPC 46B light chain (Section V,A) or the M311 variant heavy chain (Section V,A) demonstrate that some light and heavy chains synthesized in the presence of excess glucosamine to inhibit glycosylation can be glycosylated as completed chains when the glucosamine is removed (Bergman and Kuehl, 1978). These experiments provide additional evidence that the lack of glycosylation of the wild-type MPC 11 completed heavy chain is a function of the protein molecule itself and is not due either to the heavy chain molecules being spatially separated from the glycosylation apparatus or to a persistent inhibitory effect of the glucosamine. The variant M311 heavy chain is particularly noteworthy in that it includes only 38,000 daltons of amino acids due to a deletion, starting approximately 47 residues after the olgiosaccharide acceptor site, of the carboxy terminal region of the molecule (Kenter and Birshtein, 1979). This carboxy terminal deletion may account for some of the results seen with this cell line. First, since it is likely that the variant heavy chain is completed at approximately the same time that the asparagine acceptor site becomes available for glycosylation (see Sections III,A,1 and 2), it is not surprising that glycosylation is incomplete when translation of the chain is completed (Section V,A). Second, this deletion may affect the conformation of the asparagine acceptor site, thereby allowing glycosylation to occur to some extent, after chain completion and in contrast to the normal MPC 11 heavy chain (see above).

D. Physiological Significance of Glycosylation on Nascent Chains

The significance of N-glycosylation occurring on nascent polypeptides and the inability of the cell to glycosylate some completed polypeptides may be due to (a) intramolecular folding (secondary and tertiary structure) and/or (b) intermolecular assembly (quaternary structure) of the completed protein. In support of the first possibility, Lennarz and co-workers (Pless and Lennarz, 1977; Kronquist and Lennarz, 1978; Struck *et al.*, 1978) reported that the precursor oligosaccharide can be transferred *in vitro* to denatured and completed ovalbumin and RNase A but not to the native forms of these secretory proteins. Moreover, this simple tripeptide sequence, Asn-X-Ser/Thr, serves as an effective substrate for *in vitro* glycosylation provided that both the amino and carboxy termini of the tirpeptide are blocked. Presumably, nascent chains usually present the asparagine site to the glycosylation apparatus with little secondary, tertiary, or quaternary structure, so that glycosylation can occur before the protein folds and can no longer be glycosylated. In support of the second possibility, our experiments with MOPC 46B cells (Section V,C) indicate that a completed light chain can be glycosylated if it is in a monomeric form but cannot be glycosylated if it has been assembled into a dimer (Bergman and Kuehl, 1978). It is of interest that crystallographic studies of immunoglobulin molecules (Silverton *et al.*, 1977) have suggested that the carbohydrate moiety may play a central role as the principal contact between the second constant region domains of the heavy chains in intact molecules. Therefore, the unavailability of the glycosylation site in completed proteins may reflect an intermolecular assembly of the nonglycosylated proteins. In the case of heavy chains, we cannot distinguish between these two possibilities because of the very rapid intramolecular folding and intermolecular assembly of newly synthesized heavy chains (Bergman and Kuehl, 1979a,b). However, apparently the cell has evolved a system for efficiently glycosylating nascent chains as they pass through the membrane, thus usually avoiding the formation of secondary, tertiary, or quaternary structures in which the asparagine acceptor site might be unavailable for glycosylation. Apparently, the core oligosaccharide is available for processing on completed chains (Robbins *et al.*, 1977; Tabas *et al.*, 1978) and the type of processing (i.e., high-mannose versus complex) may be determined by the environment of the core oligosaccharide on the completed chain.

If the simple tripeptide sequence (i.e., Asn-X-Thr/Ser) lacking secondary, tertiary, or quaternary structure is a sufficient recognition signal for glycosylation, how can we explain the finding that some proteins containing this sequence are not glycosylated? Results from our laboratory indicate that glycosylation may be prevented or occur more slowly if the glycosylation site is near the carboxy terminal end of the protein (studies on M311 variant heavy chain, Section V,A) or near the amino terminal end of the chain (studies on MOPC 46B light chain, Section V,A). Also, as noted above (Section V,B), ovalbumin has two glycosylation acceptor sites (residues 293 and 312), both of which can be glycosylated *in vitro* and only one of which (residue 293) is glycosylated *in vivo*. Similarly, asparagine residue 332 is glycosylated, whereas the potential asparagine acceptor site at residue 347 is not glycosylated in mouse μ heavy chain (but asparagine residue 364 is glycosylated) (Kehry *et al.*, 1979). Perhaps glycosylation at one site can interfere with glycosylation of a second site located just beyond the first site.

Although we do not fully understand all of the factors that determine whether glycosylation occurs at a particular potential N-glycosylation site, the available data suggest possible explanations. First, a variety of interactions (e.g., a second site of glycosylation, signal sequence processing, association with other molecules in the cell, protein folding) near the potential glycosylation site on the nascent chain might interfere with transfer of the precursor oligosaccharide to the nascent chain. Even if the interactions are transient (e.g., glycosylation of an adjacent site), glycosylation at a later time might be prevented by the formation of secondary, tertiary, or quaternary protein structure in the interim. In other instances, the protein can be glycosylated after the transient interaction is terminated. Our experiments on the biosynthesis of the MOPC 46 glycosylated light chain (Sections V,A, and C) indicate that both of these results can occur. Second, if there are two adjacent potential N-glycosylation sites and only one glycosylation apparatus at the site where the nascent protein is being translocated, the glycosylation apparatus may not "reload" fast enough to be able to glycosylate the second site as it passes by (Palmiter *et al.*, 1980). This possibility is consistent with the results obtained for *in vitro* coupled translation–glycosylation (both sites glycosylated) versus *in vivo* glycosylation (one site glycosylated) of ovalbumin, in view of the much slower rates of translation *in vitro*. Third, if the potential glycosylation site is very close to the carboxy terminal end of a protein (approximately 30 residues or less), the protein is completed and released from the ribosome before glycosylation can occur. The kinetics of glycosylation may be much slower for completed than for nascent chains as a result of the lack of physical proximity of the glycosylation apparatus and the completed protein. Glycosylation of the completed chain may or may not occur subsequently, depending on the formation of secondary, tertiary, or quaternary protein structure, interaction with other molecular species, etc. The evolution of a glycosylation apparatus closely associated with the site at which protein translocation occurs may optimize the efficiency of the N-glycosylation process. The process of O-glycosylation may not require this coupling to translation since the hydrophilic threonine or serine acceptor site is more likely, presumably, to be available on the exterior of a protein than a potentially hydrophobic asparagine acceptor site.

VI. SUMMARY

The *in vivo* temporal and spatial relationships between translation and cotranslational modification events can be determined by isolating nascent chains from cells. An alternative approach to determining the temporal and spatial relationships between translational and cotranslational modification events is based on a coupled translation–cotranslational modification *in vitro* system, but this approach may not accurately reflect the *in vivo* situation. Using both approaches, it has been shown that the initial N-glycosylation event occurs on nascent chains for some proteins. The initial N-glycosylation event appears to require a minimum addition of about 30 amino acids beyond an asparagine acceptor site, presumably so that the acceptor site can clear the ribosome and pass through the membrane to the luminal side of the rough endoplasmic reticulum. In some instances, the initial N-glycosylation event occurs quantitatively within a very short time after the acceptor site first becomes available for transfer of the precursor oligosaccharide from a dolichol donor. In other instances, the transfer to nascent chains occurs over a longer period of time (i.e., at different times on individual nascent chains) and may not occur quantitatively on nascent chains at a given acceptor site. Transfer of the precursor oligosaccharide from the dolichol donor to molecules with an unglycosylated acceptor site may or may not occur subsequently on completed chains. Some completed proteins contain potential asparagine acceptor sites which remain in a totally nonglycosylated form *in vivo*. Possible interactions that govern the rate and efficiency of *in vivo* N-glycosylation on both nascent chains and completed chains have been discussed. The apparent difference in the temporal and spatial relationships between translation and N- versus O-glycosylation may reflect the different properties of the acceptor amino acids.

ACKNOWLEDGMENT

Research funds for work described in this section were provided through USPHS AI 12525 to one of the authors (WMK), who is also supported by an RCDA, AI 00293 USPHS.

REFERENCES

Baumal, R., and Scharff, M. D. (1973). *J. Immunol.* **111,** 448–456.
Bekesi, J. G., and Winzler, R. J. (1969). *J. Biol. Chem.* **244,** 5663–5668.
Bergman, L. W., and Kuehl, W. M. (1977). *Biochemistry* **16,** 4490–4497.
Bergman, L. W., and Kuehl, W. M. (1978). *Biochemistry* **17,** 5174–5180.
Bergman, L. W., and Kuehl, W. M. (1979a). *J. Biol. Chem.* **254,** 5690–5694.
Bergman, L. W., and Kuehl, W. M. (1979b). *J. Biol. Chem.* **254,** 8869–8876.
Bergman, L. W., and Kuehl, W. M. (1979c). *J. Supramol. Struct.* **11,** 9–24.
Bergman, L. W., Harris, E., and Kuehl, W. M. (1981). *J. Biol. Chem.* **256,** 701–706.
Bielinska, M., and Boime, I. (1978). *Proc. Natl. Acad. Sci. U.S.A.* **75,** 1768–1772.
Blobel, G., and Dobberstein, B. (1975). *J. Cell Biol.* **67,** 835–851.
Chen, W. W., and Lennarz, W. J. (1978). *J. Biol. Chem.* **253,** 5774–5779.
Cioli, D., and Lennox, E. L. (1973). *Biochemistry* **12,** 3204–3211.
Davis, B. D., and Tai, P. C. (1980). *Nature (London)* **283,** 433–438.
Eagon, P. K., Hsu, A. F., and Heath, E. C. (1975). *Fed. Proc., Fed. Am. Soc. Exp. Biol.* **34,** 678.
Ganoza, M. C., and Nakamota, T. (1966). *Proc. Natl. Acad. Sci. U.S.A.* **55,** 162–166.
Glabe, C. G., Hanover, J. A., and Lennarz, W. J. (1980). *J. Biol. Chem.* **255,** 9236–9242.
Hanover, J. A., and Lennarz, W. J. (1980). *J. Biol. Chem.* **255,** 3600–3604.
Hanover, J. A., Lennarz, W. J., and Young, J. D. (1980). *J. Biol. Chem.* **255,** 6713–6716.
Hart, G. W., Brew, K., Grant, G. A., Bradshaw, R. A., and Lennarz, W. J. (1979). *J. Biol. Chem.* **254,** 9747–9753.
Hill, H. D., Schwyzer, M., Steinman, H. M., and Hill, R. L. (1977). *J. Biol. Chem.* **252,** 3799–3804.
Hopwood, J. J., and Dorfman, A. (1977). *Biochem. Biophys. Res. Commun.* **75,** 472–479.
Katz, F. N., Rothman, J. E., Lingappa, V. R., Blobel, G., and Lodish, H. F. (1977). *Proc. Natl. Acad. Sci. U.S.A.* **74,** 3278–3282.
Kehry, M., Sibley, C., Fuhrman, J., Schilling, J., and Hood, L. E. (1979). *Proc. Natl. Acad. Sci. U.S.A.* **76,** 2932–2936.
Kenter, A., and Birshtein, B. K. (1979). *Science* **206,** 1307–1309.

Kiely, M. L., McKnight, G. S., and Schimke, R. T. (1976). *J. Biol. Chem.* **251**, 5490–5496.

Ko, G. K. W., and Raghupathy, E. (1972). *Biochim. Biophys. Acta* **264**, 129–143.

Kornfeld, R., and Kornfeld, S. (1976). *Annu. Rev. Biochem.* **45**, 217–237.

Kronquist, K. K., and Lennarz, W. J. (1978). *J. Supramol. Struct.* **8**, 51–65.

Kuehl, W. M. (1977). *Curr. Top. Microbiol. Immunol.* **76**, 1–47.

Li, E., Tabas, I., and Kornfeld, S. (1978). *J. Biol. Chem.* **253**, 7771–7778.

Liu, T., Stetson, B., Turco, S. J., Hubbard, S. C., and Robbins, P. W. (1979). *J. Biol. Chem.* **254**, 4554–4559.

Palacios, R., Palmiter, R. D., and Schimke, R. T. (1972). *J. Biol. Chem.* **247**, 2316–2321.

Palmiter, R. D., Gagnon, J., and Walsh, K. A. (1978). *Proc. Natl. Acad. Sci. U.S.A.* **75**, 94–98.

Palmiter, R. D., Thibodeau, S. N., Rogers, G., and Boime, I. (1980). *Ann. N.Y. Acad. Sci.* **343**, 192–209.

Peterson, P. A., Rask, L., Helting, T., Ostberg, L., and Fernstedt, Y. (1976). *J. Biol. Chem.* **251**, 4986–4995.

Pless, D. D., and Lennarz, W. J. (1977). *Proc. Natl. Acad. Sci. U.S.A.* **74**, 134–138.

Robbins, P. W., Hubbard, S. C., Turco, S. J., and Wirth, D. F. (1977). *Cell* **12**, 893–900.

Rothman, J. E., and Lodish, H. F. (1977). *Nature (London)* **269**, 775–780.

Rothman, J. E., Katz, F. N., and Lodish, H. F. (1978). *Cell* **15**, 1447–1454.

Schachter, H. (1974). *Adv. Cytopharmacol.* **2**, 207–218.

Sefton, B. M. (1977). *Cell* **10**, 659–667.

Silverton, E. W., Navia, M. A., and Davies, D. R. (1977). *Proc. Natl. Acad. Sci. U.S.A.* **74**, 5140–5144.

Slabaugh, R. C., and Morris, A. J. (1970). *J. Biol. Chem.* **245**, 6182–6189.

Snider, M. D., Sultzman, L. A., and Robbins, P. W. (1980). *Cell* **21**, 385–392.

Strous, G. J. A. M. (1979). *Proc. Natl. Acad. Sci. U.S.A.* **76**, 2694–2698.

Struck, D. K., and Lennarz, W. J. (1977). *J. Biol. Chem.* **252**, 1007–1013.

Struck, D. K., Lennarz, W. J., and Brew, K. (1978). *J. Biol. Chem.* **253**, 5786–5794.

Szczesna, E., and Boime, I. (1976). *Proc. Natl. Acad. Sci. U.S.A.* **73**, 1179–1183.

Tabas, I., Schlesinger, S., and Kornfeld, S. (1978). *J. Biol. Chem.* **253**, 716–722.

Takatsuki, A., Kohno, K., and Tamura, G. (1975). *Agric. Biol. Chem.* **39**, 2089–2091.

Toneguzzo, F., and Ghosh, H. P. (1977). *Proc. Natl. Acad. Sci. U.S.A.* **74**, 1516–1520.

Tucker, P., and Petska, S. (1977). *J. Biol. Chem.* **252**, 4474–4484.

Tucker, P. W., Marcu, K. B., Newell, N., Richards, J., and Blattner, F. R. (1979). *Science* **206**, 1303–1306.

Tulsiani, D. R., Opheim, D. J., and Touster, O. (1977). *J. Biol. Chem.* **252**, 3227–3233.

Turco, S. J., Stetson, B., and Robbins, P. W. (1977). *Proc. Natl. Acad. Sci. U.S.A.* **74**, 4411–4414.

Uhr, J. W. (1970). *Cell. Immunol.* **1**, 228–255.

Waechter, C. J., and Lennarz, W. J. (1975). *Annu. Rev. Biochem.* **45**, 95–127.

Weitzman, S., and Scharff, M. D. (1976). *J. Mol. Biol.* **102**, 237–248.

SECTION 5

Intracellular Transport of Glycoproteins

GEIR O. GOGSTAD AND LIV HELGELAND

I. INTRODUCTION

Mammalian cells contain numerous glycoproteins, of which some are destined for secretion, some are components of the plasma membrane and various intracellular membranes, and some are intracellular nonmembranous glycoproteins. The majority of these glycoproteins are synthesized on polysomes attached to the rough endoplasmic reticulum and subsequently transported in various membranous compartments of the cell to their final destinations. During this migration, the oligosaccharide moieties are attached and processed, the polypeptide chain is cleaved, and some glycoproteins undergo other modifications as well before they become mature glycoproteins (for review, see Chapter 2, Volume II). In the course of maturation, the glycoproteins are apparently sorted out and follow various routes of transport. Mechanisms for these transport and sorting processes obviously exist, and apparently the necessary information resides in the structure of the glycoprotein. This section deals with the present knowledge of intracellular transport of various glycoproteins from their apparent common site of synthesis to their destinations in or outside the cell. Intracellular transport of lysosomal enzymes and viral membrane glycoproteins is also dealt with by W. S. Sly (Chapter 1, Section 1, Volume IV) and L. Kääriäinen and M. Pesonen (Chapter 3, Volume IV).

THE GLYCOCONJUGATES, VOL. III

II. INTRACELLULAR TRANSPORT OF
NONMEMBRANOUS GLYCOPROTEINS

Nonmembranous glycoproteins appear as secretory products or as material enclosed in vesicular structures such as the lysosomes or specialized granules. The transport of secretory glycoproteins is by far the most studied system and may serve as a basic model for the first steps of intracellular transport of all nonmembranous glycoproteins.

A. Secretory Glycoproteins

A wide variety of tissues secrete glycoproteins to the extracellular space (e.g., parenchymal liver cells, plasma cells, thyroid gland cells, fibroblasts, pancreatic cells, and salivary gland cells). The techniques most frequently employed to investigate the intracellular transport of secretory glycoproteins are kinetic studies following the administration of radioactive carbohydrates and amino acids in whole animals, isolated organs, tissue slices, and cell cultures. The experimental evidence obtained from such studies is reviewed by H. Schachter in Chapter 2, Volume II. The transport route of secretory glycoproteins seems to follow a pattern that is common to all secretory proteins and will therefore be briefly described in light of the work performed on protein secretion in general (for review, see Palade, 1975).

Most of the process takes place in the membranous compartments represented by the endoplasmic reticulum and the Golgi apparatus. Amino terminal extensions of the polypeptide chains, termed "signal sequences," are responsible for the binding of the ribosomes to receptors on the reticulum membrane. The initial step in intracellular transport, the transfer of the nascent polypeptide chain across the reticulum membrane to the luminal space, is possibly mediated by these polypeptide extensions (Blobel and Dobberstein, 1975; Austen, 1979). Subsequently, the nascent proteins migrate to the smooth endoplasmic reticulum and farther on to the Golgi apparatus. The secretory proteins are then transported in vesicles to the cell-surface membrane and released into the extracellular space.

Generally, all secretory glycoproteins seem to follow the transport pattern described above. During the transport, stepwise glycosylations and trimming of the growing oligosaccharide chain occur, the processing being completed in the Golgi apparatus. The glycosylation process is strictly connected with enzymes located in the membranes of the endoplasmic reticulum and the Golgi apparatus. It has therefore been proposed that the nascent glycoproteins are associated with the reticulum membranes during transport. The study of this potential membrane association has been based on short-time administration of [^{14}C]glucosamine *in vivo,* a technique that almost selectively labels the secretory glycoproteins (Kreibich *et al.,* 1973). The system often studied is the transport of secretory glycoproteins in rat liver (the nascent serum glycoproteins). After the isolation of liver microsomes, the intravesicular contents are released by limited solubilization of the membranes with detergents or by such mechanical treatment as ultrasonication and decompression (Kreibich and Sabatini, 1974; Gogstad and Helgeland, 1978). The intravesicular content of the microsomes is believed to correspond to the luminal content of the endoplasmic reticulum (Palade and Siekevitz, 1956). The radiolabeled glycoproteins were measured in the membrane and luminal content concomitant with immunological measurements of serum albumin, a nonglycosylated protein. The newly synthesized glycoproteins were found to be partially associated with the membrane fractions of both the rough and the smooth microsomes (about 50% in each fraction), whereas more than 90% of serum albumin was recovered in the luminal content. This indicates that the secretory glycoproteins being synthesized in rat liver interact with the reticulum membranes during transport (Gogstad and Helgeland, 1978). The finding that the microsomal membranes contain considerably less serum glycoprotein than the luminal content when examined by immunoprecipitation, led to the suggestion that the membrane association is selective (Kreibich and Sabatini, 1974). This was contradicted by later findings that SDS–polyacrylamide gel electrophoresis of microsomal membranes and luminal contents containing labeled secretory glycoproteins had close similarities in the distribution of radioactivity (Gogstad and Helgeland, 1978). Furthermore, the membrane and luminal content pools of labeled secretory glycoproteins appeared to be interchangeable (Gogstad and Helgeland, 1978). This indicates that the secretory glycoproteins of rat liver interact with the membranes during their transport through the endoplasmic reticulum, without being firmly bound, however. Accordingly, the transport of such proteins cannot be explained by a lateral transport alone as previously suggested for the transport of secretory glycoproteins in the rough endoplasmic reticulum (Redman and Cherian, 1972). The finding of albumin almost exclusively in the luminal content evidently implies that an association of protein with the reticulum membranes is not a requirement for transport.

The interplay of biosynthetic events and the transfer of glycoprotein precursors from the rough to the smooth endoplasmic reticulum is apparently of importance. Thus, it has been shown that if triple helix formation of procollagen (Prockop *et al.*, 1979) and the carboxylation of prothrombin (Helgeland, 1977) are prevented from occurring, the precursors accumulate in the rough endoplasmic reticulum. The prothrombin precursor was predominantly recovered in the intravesicular content of the rough microsomes (Helgeland, 1977). Certain molecular characteristics of at least some secretory glycoproteins therefore seem to be required for normal transport and secretion. The mechanism underlying this inhibition of intracellular transport is not known, however.

The mode of transfer of secretory glycoproteins from the rough endoplasmic reticulum to the Golgi apparatus is not fully understood. Jamieson and Palade (1967) have documented discontinuities in the membrane system of the pancreatic exocrine cell. Tubular connections between the rough and smooth endoplasmic reticulum, and between the smooth endoplasmic reticulum and the Golgi apparatus, have, however, been demonstrated in liver parenchymal cells (for review, see Morré *et al.*, 1979). The general appearance of discontinuities in the membrane system is therefore still a matter of controversy (for reviews, see Geuze *et al.*, 1977; Morré *et al.*, 1979). As yet, there is no evidence for transport of secretory glycoproteins by such tubular connections. The transport between discontinuous membrane systems seems to be mediated by vesicles covered with a fuzzy coating (Morré *et al.*, 1979). Energy-dependent steps in the secretory pathway of pancreatic zymogens (Jamieson and Palade, 1968) and procollagen (Prockop *et al.*, 1979) have been revealed. The nature of this energy requirement is not known but may be connected to vesicular transport of material between compartments of the endoplasmic reticulum and the Golgi apparatus. It has been suggested that contractile proteins (actomyosin filaments) may be involved in this process (Palade, 1975).

The transport of secretory glycoproteins in the cell is presumed to be unidirectional. All experiments performed to demonstrate a reversal migration have failed, even if the protein synthesis or the energy metabolism has been inhibited (Geuze *et al.*, 1977). Most investigators seem to favor a draining of the endoplasmic reticulum by the budding off of protein-containing vesicles as the driving force for a unidirectional transport. The vesicles later migrate to the Golgi apparatus. Recently, it has been suggested that the driving force of lateral transport in the endoplasmic reticulum of myeloma cells is a high lipid turnover (Cohen and Phillips, 1980) producing a unidirectional movement of membrane proteins (Cohen *et al.*, 1979). This concept may well correspond to the vesicle transfer hypothesis since the force produced by these mechanisms evidently may transmit to the luminal content of the endoplasmic reticulum, explaining how proteins not associated with the endoplasmic reticulum membranes also maintain a unidirectional transport. Other force-producing mechanisms have been suggested to depend on a continuous membrane system from the rough endoplasmic reticulum to the Golgi apparatus (for review, see Geuze *et al.*, 1977). The two prevailing hypotheses in this respect predict the driving force to be produced either by polymerization of proteins in the Golgi apparatus, or by a hydrostatic pressure gradient. As yet, it cannot be decided whether more than one of the mechanisms is operating in different cells or even in the same cell.

When the nascent secretory proteins reach the Golgi apparatus, they also reach the final site for posttranslational modifications such as glycosylations. The fully glycosylated proteins are secreted from the Golgi apparatus by different mechanisms related to the function of the actual cell. Cells that store their secretory products for some time form condensing vacuoles from the Golgi apparatus (Jamieson and Palade, 1967). These vacuoles concentrate the secretory material and finally reach the state termed "secretory granules" (Caro and Palade, 1964). These granules migrate to the plasma membrane, and their contents are emptied in the extracellular space in an energy- and calcium-dependent exocytotic process (for review, see Schramm, 1967). The secretory granules seem to be specific for fusion with the plasma membrane. The condensing step is omitted in such cells as fibroblasts, plasma cells, liver parenchymal cells, and some exocrine gland cells, which maintain a more continuous secretion. However, equivalents of the secretory granules seem to be present in such cells (Weinstock and Leblond, 1974; Olsen and Prockop, 1974). Coated vesicles may also be involved in the secretory pathway of many cells, in the final steps as well as in the transition of material between separate membranous compartments (Jamieson and Palade, 1971; Castle et al., 1972; Franke et al., 1976). Such vesicles are further described in Section III,A.

B. Nonsecretory Glycoproteins

Some nonmembranous glycoproteins synthesized by mammalian cells are not secreted but rather are retained within the cells. Among these, the lysosomal enzymes are most extensively studied with regard to intracellular transport. Since this topic is thoroughly reviewed by W. S. Sly in Chapter 1, Section 1, Volume IV, only a brief description is given here.

The lysosomal enzymes catalyze hydrolytic processes in the acid environment inside the lysosomes. They all appear to be glycoproteins and are synthesized by a mechanism similar to that of the secretory glycoproteins (Erickson and Blobel, 1979). As to the route of transport to the lysosomes, three hypotheses have been suggested: (a) intracellular transfer, (b) transfer by secretion to the extracellular medium followed by reuptake into the cell (secretion–recapture hypothesis), and (c) transport via the cell surface.

The intracellular transfer hypothesis is founded on cytological observations, obtained mostly from studies of actively phagocytotic cells such as the polymorphonuclear leukocytes (for review, see Novikoff, 1976). According to this hypothesis, the lysosomal enzymes migrate from the rough endoplasmic reticulum to the Golgi apparatus or a specialized membrane system termed GERL (Golgi–endoplasmic reticulum–lysosomes) (for review, see Morré et al., 1979). Primary lysosomes are then formed by the budding off of enzyme-containing vesicles from the GERL. Further transport of enzymes to the lysosomes may be mediated by coated vesicles (Goldstein et al., 1979) (see Sections III,A and B for further description of coated vesicles). It has been suggested that membrane receptors that recognize a common marker on the lysosomal enzymes are involved in the intracellular transfer from the endoplasmic reticulum to the lysosomes (Gonzales-Noriega et al., 1980).

The secretion–recapture hypothesis suggests a secretion of lysosomal enzymes similar to that of the secretory glycoproteins, followed by reuptake into the cell by binding to specific cell-surface receptors (for review, see Neufeld *et al.*, 1977). This hypothesis seeks to explain the finding that in normal fibroblast cultures 5–50% of the newly synthesized lysosomal enzymes may accumulate in the extracellular pool. Furthermore, it was found that the lysosomal enzymes were taken up by the cells provided that a recognition marker, mannose 6-phosphate, was present in the enzyme. In the disease mucolipidosis, nearly all the newly formed lysosomal enzymes are secreted and accumulated in the extracellular medium. No uptake of enzymes into the cells occurs, and it was suggested that this is due to the lack of the recognition marker, mannose 6-phosphate. In addition, the cells fail to form functional lysosomes. More recent studies have shown that a normally occurring proteolytic processing of the lysosomal enzymes is also incomplete in mucolipidosis, suggesting that this failure as well as the reduced amount of recognition marker are secondary consequences of a common, primary molecular defect (Bach *et al.*, 1979; Hasilik, 1980; Hasilik and Neufeld, 1980b). The secretion of lysosomal enzymes appears to be mainly a one-way process in which only a minor part of the secreted enzymes is recaptured (Hasilik and Neufeld, 1980a). Accordingly, the secretion-recapture route for transport of enzymes to the lysosomes represents a minor pathway, and the condition occurring in mucolipidosis may be a consequence of defects in the intracellular transport, directing the lysosomal enzymes to be secreted rather than stored in the lysosomes.

The third transport hypothesis suggests that the lysosomal enzymes occur transiently on the cell surface during their transport to the lysosomes, probably tightly complexed with receptors of intracellular origins (Hasilik and Neufeld, 1980a). According to this model, the lysosomal enzymes follow a pathway almost identical to that of secretory glycoproteins until the secretory vesicles reach the state of exocytosis. Subsequently, the lysosomal enzymes are retained, probably by binding to membrane receptors, and carried to the lysosomes. Immunochemical studies of lysosomal enzyme transport in cultured fibroblasts indicate that at least a part of the lysosomal pool at some stage is located on the cell surface (Rome *et al.*, 1977), supporting this model.

It is not yet clear which of the three listed mechanisms for lysosomal enzyme transport dominates in the formation of lysosomes. It may well be that all three mechanisms are of importance, depending on the type of cell under consideration.

III. INTRACELLULAR TRANSPORT OF MEMBRANE-BOUND GLYCOPROTEINS

The term "membrane-bound glycoprotein" is used here to indicate those glycoproteins which are structural constituents of membranes by being integrated into the phospholipid bilayer. Such glycoproteins occur in the plasma membrane as well as in various intracellular membranes of mammalian cells.

A. Cell-Surface Glycoproteins

The surface membrane of mammalian cells contains several integral glycoproteins disposed so that the carbohydrate residues are located on the exterior surface of the membrane. The initial stages of their biosynthesis and the route of intracellular transport from the rough endoplasmic reticulum to the Golgi apparatus via the smooth endoplasmic reticulum appear to be similar to those of the secretory glycoproteins. The mechanism of transport between these compartments is not fully understood, but vesicles are evidently involved in the transport of the membrane-embedded glycoproteins to the cell surface, where they fuse with the plasma membrane by a process similar to the exocytosis of secretory granules (for review, see Morré *et al.,* 1979). Protein asymmetry is preserved during the intracellular migration of the membrane-bound glycoproteins, which means that the luminal surface of the endoplasmic reticulum membrane is topologically equivalent to the exterior surface of the plasma membrane.

Because of the great variety of proteins in the plasma membrane and the small quantity of each protein species, it is difficult to study the intracellular transport of individual plasma membrane glycoproteins. Apart from a recent study of nucleotide pyrophosphatase and dipeptidyl peptidase in rat liver (Elovson, 1980), very little is known about the biosynthesis and intracellular transport of such glycoproteins in normal cells. It was found by Elovson (1980) that dipeptidyl peptidase precursor was translated and inserted into the membranes of rough endoplasmic reticulum, which shows that the transport process begins in this compartment. The two plasma membrane glycoproteins, nucleotide pyrophophatase and dipeptidyl peptidase, appeared to turn over independently, their half-lives being about 1 and 5 days, respectively. This favors a heterogeneous turnover of plasma membrane glycoproteins in rat liver (Elovson, 1980). On the other hand, Tweto and Doyle (1976) found that plasma membrane proteins and glycoproteins of a hepatoma tissue culture cell line turned over at very similar rates. It is doubted, however, that the hepatoma tissue culture system represents a good model for studying plasma membrane biogenesis in normal tissues (Elovson, 1980).

Certain viral glycoproteins, such as those in vesicular stomatitis virus (VSV) and Sindbis virus, have proved to be useful models for studying the biosynthesis and intracellular transport of plasma membrane glycoproteins. During the replication of VSV, the viral transmembrane glycoprotein designated the "G protein," is synthesized in the rough endoplasmic reticulum of the infected cell and transported via the Golgi apparatus to the cell surface, where it is embedded in the plasma membrane (for review, see L. Kääriäinen and M. Pesonen, Chapter 3, Volume IV). Since the virus uses the enzyme machinery of the host cell, the pathway and mechanism of intracellular transport are believed to resemble that of the cells' own plasma membrane glycoproteins. Furthermore, since the G protein is the only glycoprotein being synthesized in the virus-infected cells, the system is most useful. The mechanism of translocating portions of the G protein to the luminal side of the rough endoplasmic reticulum membrane is similar to that of secretory glycoproteins. To initiate transfer across the membrane, a signal sequence of 16 amino acids at the amino terminal end of the G protein is required (Lingappa et al., 1978). A tight coupling between polypeptide synthesis and membrane insertion has been found (Rothman and Lodish, 1977). Whereas the entire polypeptide chains of secretory glycoproteins are transported through the membrane, the G protein is integrated asymmetrically in the phospholipid bilayer. Thus, the carbohydrate-bearing portion of the G protein is oriented within the lumen of the endoplasmic reticulum, whereas about 30 amino acid residues at the carboxy terminal end remain exposed on the cytoplasmic surface of the membrane (Rothman and Lodish, 1977). About 30 minutes after the protein part of the G protein is synthesized on the ribosomes in the rough endoplasmic reticulum, the G protein is embedded in the plasma membrane.

The transport of this integral plasma membrane glycoprotein to the cell surface takes place in two successive stages. Rothman and Fine (1980) have presented evidence for transport of the G protein in clathrin-coated vesicles, which are vesicles surrounded by a basketlike polyhedral network of the protein clathrin (Goldstein et al., 1979; Pearse, 1980). Two classes of clathrin-coated vesicles are involved in the transport of the G protein. In the first stage, clathrin-coated vesicles that carry the pre-Golgi form of the G protein apparently bud off from the rough endoplasmic reticulum membrane and are transferred to the Golgi apparatus, where they fuse with the Golgi membrane. In the Golgi apparatus the core oligosaccharides of the G protein are modified by a replacement of mannose residues by terminal carbohydrates. In the second stage, the mature G protein is transported from the Golgi apparatus to the plasma membrane in clathrin-coated vesicles that appear to be structurally distinct from those involved in the first stage (Rothman et al., 1980). Rothman and collaborators have suggested that there may be several species of coated vesicles and that intracellular transport of plasma membrane glycoproteins in general takes place in such vesicles. The attachment of fatty acids to VSV and Sindbis virus membrane glycoproteins in the Golgi apparatus may possibly have a function by anchoring the glycoprotein to the membrane (Schmidt and Schlesinger, 1980).

B. Intracellular Membrane Glycoproteins

The intracellular membranes of mammalian cells contain numerous glycoproteins as structural constituents. In the hepatocyte, glycoproteins have been identified in the membranes of mitochondria (Martin and Bosmann, 1971), lysosomes (Goldstone and Koenig, 1974), nucleus (Kawasaki and Yamashina, 1972), Golgi apparatus (Bergman and Dallner, 1976), and endoplasmic reticulum (Miyajima et al., 1969; Helgeland et al., 1972; Autuori et al., 1975a). With regard to the mechanisms and pathways of intracellular transport of these glycoproteins, little experimental evidence is available. Membrane glycoproteins appear to contain carbohydrate moieties that are similar to those found in non-membranous glycoproteins (Kornfeld and Kornfeld, 1976), indicating that they follow a sequence of carbohydrate attachment and processing similar to that found for other glycoproteins. The sequence of enzymatic processes is apparently closely related to the route of transport of the nascent glycoproteins, and the transport of the glycoproteins to the Golgi apparatus seems to be a prerequisite for the completion of their carbohydrate moieties (see also the review on glycoprotein biosynthesis by H. Schachter, Chapter 2, Volume II). Therefore, the distribution to their appropriate destinations in the cell would be expected to occur via the Golgi apparatus. In histochemical and autoradiographic investigations, glycoproteins from the Golgi region were found to be incorporated into lysosomal membranes, dense body membranes, basement membranes, and plasma membranes in the rat duodenal columnar cells and hepatocytes (Bennett and Leblond, 1971; Bennett *et al.*, 1974).

The lysosomal membrane may originate from the Golgi apparatus, the GERL, or the plasma membrane, depending on the mechanism involved in the formation of the lysosomes (see Section II,B). A specific set of glycoproteins may follow the lysosomal membrane when it is budded off from its originating structure. The incorporation of glycoproteins peculiar to the endoplasmic reticulum membranes seems to require a different mechanism. These glycoproteins evidently pass their target membranes upon glycosylation but probably continue to the Golgi apparatus for completion of their carbohydrate moieties, as previously mentioned. Such membrane glycoproteins must therefore be transferred back to the membranes of the endoplasmic reticulum, but this transfer can hardly involve migration through the membrane system since this continuously maintains a unidirectional protein and lipid flow in the opposite direction. The mechanism for some membrane glycoproteins seems rather to be loosening from the Golgi membranes as protein–lipid complexes that migrate back to the reticulum membranes through the cytosol. This mechanism is proposed for certain glycoproteins in the endoplasmic reticulum of the rat hepatocyte (Autuori *et al.*, 1975a,b; Elhammer *et al.*, 1975). A similar mechanism may also work in the transport of membrane glycoproteins to destinations other than the reticulum membranes. Cytochrome b_5, which is an integral membrane glycoprotein (Elhammer *et al.*, 1978) of several intracellular membranes, is synthesized in the endoplasmic reticulum. Its transfer to the mitochondria seems to occur through the aqueous phase in the cytosol (Leto *et al.*, 1980). Other mitochondrial glycoproteins may also be transported through the cytosol from the endoplasmic reticulum.

There is growing evidence that the main transfer of proteins between intracellular membranes is mediated by coated vesicles similar to those described in Section III,A (for review, see Pearse, 1980). Certain membrane proteins concentrate in coated areas of the reticulum membrane, which in turn are budded off as coated vesicles. These proteins seem to include those being transferred to other compartments and certain membrane-bound receptors recycling between two or more membrane systems. Proteins directing the transfer of the coated vesicles may also be included. The occurrence of coated vesicles may explain the finding that membrane systems such as the endoplasmic reticulum and the Golgi apparatus contain two sets of proteins: one set of mature proteins peculiar to the membrane itself and one set of proteins in transit.

The biogenesis of the intracellular membranes has been proposed to be a complex, multistep process involving several different transport mechanisms (Dallner et al., 1966). The transport mechanisms mentioned may all participate in the biogenesis, but at present it is not known which mechanism dominates in the formation of different membranes.

IV. CONTROL OF INTRACELLULAR TRANSPORT OF GLYCOPROTEINS

The rough endoplasmic reticulum represents the common site from which a large number of different glycoproteins migrate to their final destinations. It seems obvious that each step in the transport must be under strict control to ensure that the glycoproteins are directed to the appropriate places. The control apparently resides in the structure of the glycoproteins, i.e., in the amino acid sequence, in the carbohydrate moieties, in other posttranslational modifications, and in the tertiary structure.

Intracellular transport of glycoproteins is initiated by the attachment of ribosomes to receptors on the membranes of the rough endoplasmic reticulum with subsequent transfer of the growing polypeptide chain through the membrane. The signal for ribosome attachment resides in an amino terminal, hydrophobic extension of the polypeptide chain. It has been proposed that this "signal sequence" is also involved in the transfer of the polypeptide chain across the membrane, directing the polypeptide to the luminal compartment of the endoplasmic reticulum. Nonmembranous glycoproteins are transferred into the lumen of the reticulum, whereas the translocation of membrane-bound glycoproteins is interrupted in such a way that the polypeptide chain is integrated in the membrane. The finding that the sequences have been highly conserved during evolution (Shields and Blobel, 1977) underlines the importance of their structures for normal transport in the cell. The signal sequences vary considerably (for reviews, see Austen, 1979; Davis and Tai, 1980), but homology has been demonstrated among protein precursors with the same destinations, e.g., secretory glycoproteins (Devillers-Thiery et al., 1975). The signal sequences might thus act in the further segregation of glycoproteins with different destinations, e.g., secretory and integral membrane glycoproteins, by directing them to individual receptors on the reticulum membrane. Since the signal sequence of secretory (Jackson and Blobel, 1977) as well as membrane protein precursors (Lingappa et al., 1978) is removed by endopeptidases immediately after the transfer of the polypeptide chain through the reticulum membrane, their potential regulatory properties must be expressed at an early stage. This could imply that these glycoproteins are synthesized on different regions of the rough endoplasmic reticulum. Nascent chain competition experiments performed by Lingappa *et al.* (1978) demonstrated, however, that the VSV glycoprotein and a secretory glycoprotein (bovine pituitary prolactin) were bound to the same receptors, the hypothetical transport sites on the reticulum membrane. This finding indicates that the segregation of different proteins occurs at a later stage. Another suggestion for a sorting mechanism was made by Blobel (1980), who predicted the existence of additional topogenic sequences in the polypeptide chains. These are "stop transfer sequences," which interrupt the polypeptide chain translocation across the membranes to yield integral proteins, and "sorting sequences," which act in directing intracellular traffic. No experimental evidence for stop transfer sequences has yet been presented, however. The study of intracellular transport of lysosomal enzymes makes the existence of sorting sequences in the polypeptide chains probable. As mentioned in Section II,B, the lysosomal enzymes are proteolytic-processed in the cell during their transport to the lysosomes (Hasilik and Neufeld, 1980a). This proteolytic maturation is clearly distinct from the cleavage of signal peptides and seems to be a stepwise process that is similar in different cell types (Hasilik *et al.*, 1980; Hasilik, 1980). Studies performed on the lysosomal enzyme α-glucosidase indicate that the proteolytic cleavage may follow alternative pathways that are connected to different routes of transport (Hasilik and Neufeld, 1980a). It was found that lysosomal enzymes that are secreted into the extracellular space are less proteolyzed than their counterparts in the lysosome. These observations thus indicate that transport-directing information may reside in the polypeptide chains and that different proteolytic cleavages may alter the route of intracellular transport.

The carbohydrate moieties of glycoproteins were for some time believed to act as intracellular transport markers necessary for secretion (Eylar, 1965). This hypothesis was eventually proved to be wrong since the lysosomal enzymes and membrane glycoproteins are all glycoproteins retained within the cell. Furthermore, secretory proteins such as serum albumin and a number of hydrolytic enzymes in the pancreatic exocrine cells are nonglycosylated proteins. It was possible to block glycosylation of immunoglobulin light chain in myeloma cells (Eagon and Heath, 1977) and of immunoglobulin M in plasmablasts (Tartakoff and Vassalli, 1979) by the administration of 2-deoxyglucose and tunicamycin. Nonglycosylated polypeptides were correctly assembled, transported, and secreted, although at a reduced rate. Similar results were obtained with transferrin (Struck *et al.*, 1978). These observations indicate, therefore, that the carbohydrate moieties as such have no function as transport-directing factors. Specific carbohydrate residues may be involved, however, in directing the transport of intracellular glycoproteins. Studies of lysosomal enzymes have revealed that mannose 6-phosphate in the high-mannose parts of the carbohydrate moieties may participate in directing the enzymes into the lysosomes (see Chapter 1, Section 1, Volume IV). It seems that both the high-mannose part and the specific phosphorylation of mannose residues are characteristics of the lysosomal enzymes that distinguish them from other glycoproteins. Furthermore, processing of the carbohydrate moieties with the removal of glucose residues and the subsequent addition of terminal carbohydrates has been demonstrated (see Chapter 1, Sections 1 and 4, of this volume). Variations in trimming of the carbohydrate chains may also prove to represent transport-directing elements.

The formation of the proper tertiary structure has proved to be of importance for normal transport, as shown for VSV glycoprotein (Gibson *et al.*, 1979) and procollagen (Jimenez and Yankowski, 1978; Prockop *et al.*, 1979). When the hydroxylation of proline residues in procollagen is prevented form occurring, the triple helical tertiary structure is not formed and the rate of intracellular transport is markedly reduced. A similar effect on the transport of VSV glycoprotein has been observed when glycosylation is inhibited. It has been proposed that the reduced rate of transport is the result of an aggregation of the nonglycosylated virus protein, which might be explained by an alteration in the tertiary structure (Gibson *et al.*, 1979). It therefore seems that the posttranslational modifications required for the formation of the proper tertiary structure may be necessary to ensure normal transport.

The transfer of membrane glycoproteins from one membrane compartment to another seems to be mediated by coated vesicles in at least some systems (see Sections III,A and B). The coated vesicles allow sets of proteins to be transferred between membranes. The study of viral membrane proteins has revealed the presence of at least two kinds of such vesicles. It has been suggested that several types of coated vesicles, each specialized for a certain transport operation, may exist in mammalian cells (Rothman *et al.*, 1980). The signals directing the transfer of coated vesicles to their respective destinations, however, are still unknown. The nature of the precise movement of vesicles between membrane compartments may involve microtubuli, connecting areas of one membrane to certain areas in another, and the driving force of vesicle movement may be related to contractile proteins (actomyosin). Present knowledge in this field, however, is meager.

NOTE ADDED IN PROOF

Since this review was completed, in the fall of 1980, Amar-Costesec (1981) has found that the endoplasmic reticulum-derived elements of the microsome fraction do not contain protein-bound sialic acid. This finding casts doubt upon the theory for a transfer of sialoglycoproteins from the

Golgi apparatus membranes back to the endoplasmic reticulum through the cytosol (see Section III B).

REFERENCES

Amar-Costesec, A. (1981). *J. Cell Biol.* **89**, 62–69.

Austen, B. M. (1979). *FEBS Lett.* **103**, 308–313.

Autuori, F., Svensson, H., and Dallner, G. (1975a). *J. Cell Biol.* **67**, 687–699.

Autuori, F., Svensson, H., and Dallner, G. (1975b). *J. Cell Biol.* **67**, 700–714.

Bach, G., Bargal, R., and Cantz, M. (1979). *Biochem. Biophys. Res. Commun.* **91**, 976–981.

Bennett, G., and Leblond, C. P. (1971). *J. Cell Biol.* **51**, 875–881.

Bennett, G., Leblond, C. P., and Haddad, A. (1974). *J. Cell Biol.* **60**, 258–281.

Bergman, A., and Dallner, G. (1976). *Biochim. Biophys. Acta* **433**, 496–508.

Blobel, G. (1980). *Proc. Natl. Acad. Sci. U.S.A.* **77**, 1496–1500.

Blobel, G., and Dobberstein, B. (1975). *J. Cell Biol.* **67**, 835–851.

Caro, L. G., and Palade, G. E. (1964). *J. Cell Biol.* **20**, 473–495.

Castle, J. D., Jamieson, J. D., and Palade, G. E. (1972). *J. Cell Biol.* **53**, 290–311.

Cohen, B. G., and Phillips, A. H. (1980). *J. Biol. Chem.* **255**, 3075–3079.

Cohen, B. G., Mosler, S., and Phillips, A. H. (1979). *J. Biol. Chem.* **254**, 4267–4275.

Dallner, G., Siekevitz, P., and Palade, G. E. (1966). *J. Cell Biol.* **30**, 97–115.

Davis, B. D., and Tai, P.-C. (1980). *Nature (London)* **283**, 433–438.

Devillers-Thiery, A., Kindt, T., Scheele, G., and Blobel, G. (1975). *Proc. Natl. Acad. Sci. U.S.A.* **72**, 5016–5020.

Eagon, P. K., and Heath, E. C. (1977). *J. Biol. Chem.* **252**, 2372–2383.

Elhammer, Å., Svensson, H., Autuori, F., and Dallner, G. (1975). *J. Cell Biol.* **67**, 715–724.

Elhammer, Å., Dallner, G., and Omura, T. (1978). *Biochem. Biophys. Res. Commun.* **84**, 572–580.

Elovson, J. (1980). *J. Biol. Chem.* **255**, 5807–5815.

Erickson, A. H., and Blobel, G. (1979). *J. Biol. Chem.* **254**, 11771–11774.

Eylar, E. H. (1965). *J. Theor. Biol.* **10**, 89–113.

Franke, W. W., Lüder, M. R., Kartenbeck, J., Zerban, H., and Keenan, T. W. (1976). *J. Cell Biol.* **69**, 173–196.

Geuze, J. J., Kramer, M. F., and deMan, J. H. C. (1977). *In* "Mammalian Cell Membranes" (G. A. Jamieson and D. M. Robinson, eds.), Vol. 2, pp. 55–107. Butterworth, London.

Gibson, R., Schlesinger, S., and Kornfeld, S. (1979). *J. Biol. Chem.* **254**, 3600–3607.

Gogstad, G. O., and Helgeland, L. (1978). *Biochim. Biophys. Acta* **508**, 551–564.

Goldstein, J. L., Anderson, R. G. W., and Brown, M. S. (1979). *Nature (London)* **279**, 679–685.

Goldstone, A., and Koenig, H. (1974). *FEBS Lett.* **39**, 176–181.

Gonzales-Noriega, A., Grubb, J. H., Talkad, V., and Sly, W. S. (1980). *J. Biol. Chem.* **255**, 5069–5074.

Hasilik, A. (1980). *Trends. Biochem. Sci.* **9**, 237–240.

Hasilik, A., and Neufeld, E. F. (1980a). *J. Biol. Chem.* **255**, 4937–4945.

Hasilik, A., and Neufeld, E. F. (1980b). *J. Biol. Chem.* **255**, 4946–4950.

Hasilik, A., Voss, B., and von Figura, K. (1980). *Hoppe-Seyler's Z. Physiol. Chem.* **361**, 262.

Helgeland, L. (1977). *Biochim. Biophys. Acta* **499**, 181–193.

Helgeland, L., Christensen, T. B., and Janson, T. L. (1972). *Biochim. Biophys. Acta* **286**, 62–71.

Jackson, R. C., and Blobel, G. (1977). *Proc. Natl. Acad. Sci. U.S.A.* **74**, 5598–5602.

Jamieson, J. D., and Palade, G. E. (1967). *J. Cell Biol.* **34**, 577–596.

Jamieson, J. D., and Palade, G. E. (1968). *J. Cell Biol.* **39**, 589–603.

Jamieson, J. D., and Palade, G. E. (1971). *J. Cell Biol.* **50**, 135–158.

Jimenez, S. A., and Yankowski, R. (1978). *J. Biol. Chem.* **253**, 1420–1426.

Kawasaki, T., and Yamashina, I. (1972). *J. Biochem. (Tokyo)* **72**, 1517–1525.

Kornfeld, R., and Kornfeld, S. (1976). *Annu. Rev. Biochem.* **45**, 217–237.

Kreibich, G., and Sabatini, D. D. (1974). *J. Cell Biol.* **61**, 789–807.

Kreibich, G., Debey, P., and Sabatini, D. D. (1973). *J. Cell Biol.* **58**, 436–463.

Leto, T. L., Roseman, M. A., and Holloway, P. W. (1980). *Biochemistry* **19**, 1911–1916.

Lingappa, V. R., Katz, F. N., Lodish, H. F., and Blobel, G. (1978). *J. Biol. Chem.* **253**, 8667–8670.

Martin, S. S., and Bosmann, H. B. (1971). *Exp. Cell Res.* **66**, 59–64.

Miyajima, N., Tomikawa, M., Kawasaki, T., and Yamashina, I. (1969). *J. Biochem. (Tokyo)* **66**, 711–732.

Morré, D. J., Kartenbeck, J., and Franke, W. W. (1979). *Biochim. Biophys. Acta* **559**, 71–152.

Neufeld, E. F., Sando, G. N., Garvin, A. J., and Rome, L. H. (1977). *J. Supramol. Struct.* **6**, 95–101.

Novikoff, A. B. (1976). *Proc. Natl. Acad. Sci. U.S.A.* **73**, 2781–2787.

Olsen, B. R., and Prockop, D. J. (1974). *Proc. Natl. Acad. Sci. U.S.A.* **71**, 2033–2037.

Palade, G. E. (1975). *Science* **189**, 347–358.

Palade, G. E., and Siekevitz, P. (1956). *J. Biophys. Biochem. Cytol.* **2**, 171–199.

Pearse, B. (1980). *Trends Biochem. Sci.* **5**, 131–134.

Prockop, D. J., Kivirikko, K. I., Tuderman, L., and Guzman, N. A. (1979). *N. Engl. J. Med.* **301**, 13–23.

Redman, C. M., and Cherian, M. G. (1972). *J. Cell Biol.* **52**, 231–245.

Rome, L., Garvin, A. J., and Neufeld, E. F. (1977). *Fed. Proc., Fed. Am. Soc. Exp. Biol.* **36**, 749.

Rothman, J. E., and Fine, R. E. (1980). *Proc. Natl. Acad. Sci. U.S.A.* **77**, 780–784.

Rothman, J. E., and Lodish, H. F. (1977). *Nature (London)* **269**, 775–780.

Rothman, J. E., Bursztyn-Pettegrew, H., and Fine, R. E. (1980). *J. Cell Biol.* **86**, 162–171.

Schmidt, M. F. G., and Schlesinger, M. J. (1980). *J. Biol. Chem.* **255**, 3334–3339.

Schramm, M. (1967). *Annu. Rev. Biochem.* **36**, 307–320.

Shields, D., and Blobel, G. (1977). *Proc. Natl. Acad. Sci. U.S.A.* **74**, 2059–2063.

Struck, D. K., Siuta, P. B., Lane, M. D., and Lennarz, W. J. (1978). *J. Biol. Chem.* **253**, 5332–5337.

Tartakoff, A., and Vassalli, P. (1979). *J. Cell Biol.* **83**, 284–299.

Tweto, J., and Doyle, D. (1976). *J. Biol. Chem.* **251**, 872–882.

Weinstock, M., and Leblond, C. P. (1974). *J. Cell Biol.* **60**, 92–127.

SECTION 6

Nonenzymatic Glycosylation of Proteins *in Vitro* and *in Vivo*

SUZANNE R. THORPE AND JOHN W. BAYNES

I. INTRODUCTION

A. Historical Perspective

In current research on the chemistry and sequelae of nonenzymatic glycosylation* of protein *in vivo,* contemporary scientists are reinvestigating and extending studies initiated many decades ago. The classical studies on the chemical interactions of reducing sugars and amino acids, and subsequent rearrangements leading to brown degradation products, were carried out by Maillard between

---

*As a point of clarification, "glycosylation" is used to refer to the general process of protein modification by a reducing sugar, whereas "glucosylation" is used to refer to specific modification of protein by glucose. Thus, "glycosylated" is used in reference to hemoglobins, in general, since some hemoglobins contain carbohydrate other than glucose, whereas "glucosylated" is used in reference to proteins, such as plasma proteins, for which only glucosylated derivatives have been described.

113

THE GLYCOCONJUGATES, VOL. III

1912 and 1920 (reviewed by Gottschalk, 1972a). This work provided the foundation for applied research begun in the 1940s, when workers in the dairy and food industries studied the loss in nutritional value and development of brown color associated with the storage of milk and other protein foods containing reducing sugar (reviewed by Reynolds, 1963, 1965). The nutritional deficit in milk protein was shown to arise from the loss of available lysine in casein as a result of the formation of lactose adducts with the ϵ-amino group of lysine residues. The subsequent browning of dried milk and other proteins *in vitro* is now known to result from a series of dehydration, condensation, and polymerization reactions which occur after the initial, nonenzymatic glycosylation step. The discovery, in the late 1960s, of nonenzymatically glycosylated derivatives of hemoglobin in human red blood cells and their occurrence in elevated amounts in patients with diabetes (reviewed by Bunn *et al.*, 1978; Bunn, 1981a) led to the current interest in the chemistry of the Maillard reaction of proteins in living systems.

This section summarizes present concepts of the chemistry of nonenzymatic glycosylation of protein and reviews recent research on this reaction in mammalian systems. Possible consequences of nonenzymatic glycosylation to the structure and function of body proteins and the potential significance of this process in the development of the chronic complications of diabetes are also discussed. Finally, methods for quantitating the extent of nonenzymatic glycosylation of proteins and their usefulness in the clinical management of diabetes are described.

B. Chemistry of Nonenzymatic Glycosylation

In contrast to the enzymatic attachment of carbohydrate to protein in glycosidic linkage, nonenzymatic glycosylation proceeds by a direct chemical reaction between reducing sugars and protein, forming a ketoamine derivative. The initial step in the reaction (Gottschalk, 1972a) is the formation of a Schiff base between free amino groups on protein and the open-chain, carbonyl form of a reducing sugar, as illustrated in Figure 1 for glucose. The aldimine exists in equilibrium with the N-substituted aldosylamine structure.

Bunn and Higgins (1981) studied the reaction of various sugars with hemoglobin and showed that the rate of glucosylation is directly dependent on the extent to which the sugar exists in the open-chain structure. They suggested that glucose may have emerged as the primary metabolic fuel because of its high stability in the ring structure which limits potentially deleterious nonenzymatic glycosylation of proteins. The amine participates in the reaction as the nucleophilic, free base, so that low-pK_a amino groups on protein should be most reactive in forming adducts with reducing sugars, barring steric or electronic complications. Thus, the amino terminal amino acid and intrachain lysine residues which titrate with low pK_a values are likely sites for Schiff base formation.

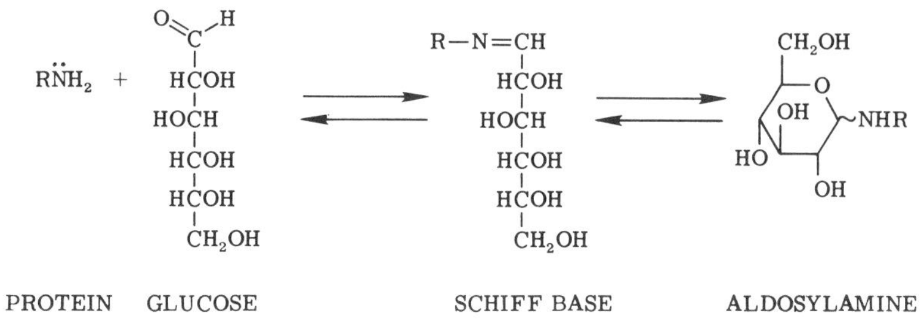

Figure 1 Initial stage in nonenzymatic glucosylation of protein.

The equilibrium constant for the Schiff base reaction is highly unfavorable for aliphatic amines in neutral, aqueous solution, e.g., physiological fluids or buffers. However, the aldimine can undergo a slow Amadori rearrangement to a ketoamine (Fig. 2), the stable, isolable form of nonenzymatically glycosylated protein. Formally, once the rearrangement has occurred, the ketoamine becomes the 1-deoxyhexulose derivative of the amino acid. Thus 1-(N^5-lysino)-1-deoxy-D-fructose (fructoselysine) arises from the reaction between glucose and lysine.

The Amadori rearrangement is acid-catalyzed, and Shapiro *et al.* (1980) reported that glucosylation of lysines in hemoglobin occurs at lysines in the vicinity of carboxyl groups on the protein surface. Unexpectedly, there was not a direct correspondence between the pK_a of the amino group and its extent of glucosylation, suggesting, therefore, that the kinetics of the Amadori rearrangement rather than the pK_a of the amino group may be the crucial factor in determining sites of glycosylation in protein. The ketoamine derivatives of protein have not been characterized structurally but probably exist largely as the cyclized ketosylamine (Fig. 2). Model studies with hexoses and amino acids indicate that ketosylamines crystallize in the pyranose form, preferentially as the β anomer (Gottshalk, 1972a). Brownlee *et al.* (1980), however, have suggested the existence of furanose structures in solution, based on the strength of interaction between fructoselysine and a phenyl-boronic acid resin.

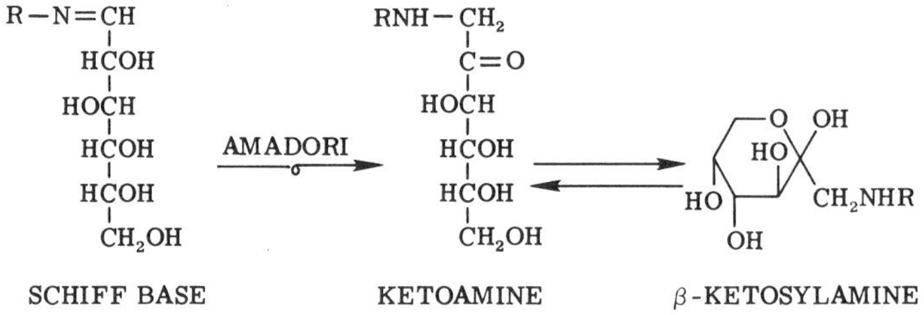

Figure 2 Amadori rearrangement of Schiff base to ketoamine derivative of protein.

The Amadori product is unstable to acid or alkaline hydrolysis but can be reduced to the stable hexitolamino acid by $NaBH_4$. The radioactive C-2 epimers, mannitol- and glucitollysine, have been recovered from acid and alkaline hydrolysates of a variety of proteins following reduction with $[^3H]NaBH_4$. The epimers are separable on the amino acid analyzer, along with a rearrangement or dehydration product whose concentration increases with extended acid hydrolysis (Robins and Bailey, 1972; Bailey *et al.*, 1976; Stevens *et al.*, 1978). Reduced glucitol and mannitol derivatives of hydroxylysine in collagen (Tanzer *et al.*, 1972) and the amino terminal valine residue in hemoglobin (Koenig *et al.*, 1977) have also been isolated and characterized. It should be noted that $[^3H]NaBH_4$ will reduce either aldimine or ketoamine adducts to hexitol derivatives, however, tritium would be incorporated at C-1 or C-2 of the aldimine or ketoamine, respectively.

C. Detection and Quantitation of Ketoamine-Linked Carbohydrate in Proteins

The chemical basis for quantitating ketoamine derivatives of protein was developed from model studies by Gottschalk and Partridge (1950), who showed that, on mild acid hydrolysis of fructoseglycine, the sugar was released as 5-hydroxymethylfurfural (HMF). Keeney and Bassette (1959) introduced the use of thiobarbituric acid (TBA) as a reagent for determining HMF released from sugars in ketoamine linkage to milk protein. This TBA assay was first applied to measure the extent of nonenzymatic glycosylation of a human protein, hemoglobin, by Flückiger and Winterhalter (1976). Briefly, the carbohydrate is released from protein by mild acid hydrolysis, yielding HMF, and the residual protein, which interferes with the subsequent colorimetric reaction, is then removed by precipitation with tricholoroacetic acid. The soluble HMF is reacted with TBA and the adduct, shown in Figure 3 (Dox and Plaisance, 1916), is measured spectrophotometrically at 443 nm.

Optimal yields of HMF are obtained by hydrolysis in organic acids, e.g., 0.3–1.0 N oxalic (Keeney and Bassette, 1959; Flückiger and Winterhalter, 1976) or 2 N acetic acid (Gottschalk, 1952; Dolhofer and Wieland, 1979, 1981).

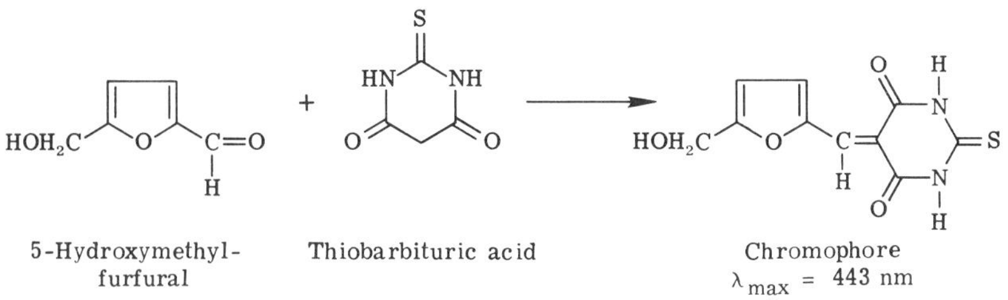

5-Hydroxymethyl-   Thiobarbituric acid   Chromophore
furfural λ_{max} = 443 nm

Figure 3 Formation of chromophore in the thiobarbituric acid assay.

Harsher conditions, such as those used for the release of glycosidically bound carbohydrate, cause HMF to decompose (Gottschalk, 1952). Strict temperature control during the hydrolysis step is essential since the color yield, not all of which is due to HMF, is very sensitive to temperature (Dolhofer and Wieland, 1980). Finally, since the TBA assay is widely used for the determination of sialic acid in glycoproteins (Warren, 1959), the HMF and sialic acid assays should be distinguished. In the sialic acid assay, TBA forms a $2:1$ adduct with β-formylpyruvic acid released following HIO_4 cleavage of sialic acid (Gottschalk, 1972b). This adduct absorbs maximally at 549 nm; it is not formed during the assay for ketoamines, and sialic acid does not interfere with quantitation of HMF.

Although glucose itself is unreactive in the TBA assay, Kennedy et al. (1980) have shown that free glucose in serum can produce artificially high results in the assay for ketoamine derivatives of serum proteins. The problem is most pronounced in hyperglycemic sera from diabetic patients and is probably caused by reaction between glucose and protein during the hydrolysis step. The interference is readily eliminated by dialysis of serum before hydrolysis. In our laboratory we separate serum protein from glucose by ethanol precipitation and then redissolve the protein pellet directly in oxalic acid for hydrolysis. The ethanol precipitation has the added benefit of removing lipid and eliminates interference from the turbidity of highly lipemic sera. Even after removal of glucose there are unidentified substances formed in serum and plasma during the hydrolysis step, which produce absorbance at 443 nm, so that a sample blank must be used when these fluids are tested. Two types of blanks can be used with comparable results. In one method, the serum sample is split and one portion is treated with $NaBH_4$ to reduce ketoamine adducts to nonreactive hexitolamines. The A_{443} is then determined as the difference between unreduced and reduced sample (McFarland et al., 1979; Kennedy et al., 1980; Dolhofer and Wieland, 1981). In the second procedure, the serum hydrolysate is split and the A_{443} is determined as the difference between sample with and without added TBA (authors' unpublished data). With some purified proteins, e.g., albumin (Dolhofer and Wieland, 1980), the blank is sufficiently low that it can be ignored, but with others, e.g., hemoglobin, a standardization (Gabbay et al., 1979a) or correction (Pecoraro et al., 1979) protocol has been employed. The need for the sample blank must be carefully evaluated before total A_{443} in any sample is attributed to the HMF product alone.

Reactions of a variety of proteins with aldoses, ketoses, reducing disaccharides, and sugar phosphates proceed slowly at physiological pH and temperature. The kinetics of glycosylation may be followed by the TBA assay, by quantitating radioactive glycitolamino acid following [^3H]$NaBH_4$ reduction, or, most conveniently, by using ^3H- or ^{14}C-labeled sugars. In the latter case, protein-bound hexose can be measured by acid precipitation or column chromatographic separation of labeled protein from free sugar. However, in a recent and important study, Trüeb *et al.* (1980) have clearly shown that commercial preparations of radioactive glucose contain various amounts of nonglucose, radioactive impurities that are highly reactive with a variety of proteins and form stable, covalent adducts. Since many studies of glucosylation *in vitro* have involved the incorporation of only small percentages of added tracer, the work of Trüeb *et al.* suggests that the radioactive contaminants may have contributed significantly to the observed apparent rates and extents of glucosylation. Future work based on the use of radioactive sugars will have to address carefully the purity of the tracer and to confirm the identity of the putative glycosyl derivatives of protein.

II. NONENZYMATIC GLYCOSYLATION OF HEMOGLOBIN AND OTHER PROTEINS

A. Glycosylated Hemoglobins

1. Isolation and Characterization

Nonenzymatic glycosylation under physiological conditions occurs at a relatively slow rate compared to enzymatic reactions. Therefore, ketoamine derivatives will be most readily detected in those proteins with long half-lives and continuous exposure to blood glucose, such as hemoglobin. Nonenzymatic glycosylation of hemoglobin has been discussed thoroughly in several recent and excellent reviews (Bunn *et al.*, 1978; Garel *et al.*, 1979; Koenig and Cerami, 1980; Bunn, 1981a,b), to which the reader is referred for more details. The following discussion summarizes current information on hemoglobin glycosylation and its relevance to the diagnosis and clinical management of diabetes.

Human hemoglobin A (HbA) can be separated by cation-exchange chromatography into two major fractions: a fast-moving component termed HbA_1, and the main fraction designated HbA_0. The HbA_1 fraction (4–8% of total hemoglobin) can be further separated into HbA_{1a1}, HbA_{1a2}, HbA_{1b}, and HbA_{1c}, with HbA_{1c} accounting for more than 75% of the HbA_1 fraction, or about 3–5% of total hemoglobin in normal blood (Allen and Balog, 1958; Huisman and Meyering, 1960; McDonald *et al.*, 1978). Rahbar (1968) detected an electrophoretic variant of hemoglobin in blood from patients with diabetes and later showed that this "diabetic component" was identical with HbA_{1c} but present in higher concentrations in diabetic blood (Rahbar *et al.*, 1969). Independently, Holmquist and Schroeder (1966) had established that HbA_{1c} contained an alkali-labile blocking group at the β-chain amino terminal valine residue. The blocking group could be stabilized by reduction with $NaBH_4$ and was tentatively identified as a Schiff base. Bookchin and Gallop (1968) later identified the blocking group as a hexose, and Dixon (1972) synthesized the blocked amino terminal tryptic peptide *de novo* from glucose and valylhistidine. Bunn *et al.* (1975) reduced HbA_{1c} with $[^3H]NaBH_4$ and then oxidized the hexitolvaline with HIO_4. The isolation of radioactive formate, rather than formaldehyde, established that the hexose was bound in ketoamine rather than Schiff base linkage. Koenig *et al.* (1977) identified both mannitol- and glucitolvalylhistidine from an enzymatic digest of $NaBH_4$-reduced HbA_{1c}. All of the hemoglobins, including HbA_0, also contain glucose bound in ketoamine linkage to lysine residues in the peptide chains, but this modification does not appear to affect the chromatographic properties of the hemoglobins (Bunn *et al.*, 1979; Shapiro *et al.*, 1980). The other minor hemoglobins, A_{1a1}, A_{1a2}, and A_{1b}, are also blocked at their β-chain amino termini and contain sugar in ketoamine linkage, based on their reactivity in the TBA assay (McDonald *et al.* 1978). The HbA_{1a1} and HbA_{1a2} contain 2 and 1 moles of phosphate per mole, respectively, and HbA_{1a2} chromatographs identically with the glucose 6-phosphate adduct of HbA_0 prepared *in vitro* (McDonald *et al.*, 1978). Haney and Bunn (1976) reported that glycosylation of hemoglobin by glucose 6-phosphate was about 20 times as rapid as the reaction with glucose *in vitro*. The reaction with hemoglobin was specific for the β-chain amino terminal valine residue, and the authors concluded that glucose 6-phosphate was actually serving as an affinity label for the allosteric, organic phosphate binding site on hemoglobin.

There is no evidence that hemoglobin glycosylation has any pathologic hematological consequences in diabetes (Jones and Peterson, 1981). In fact, whole-blood oxygen saturation curves for normal and diabetic populations are essentially identical (Ditzel and Standl, 1975; Ditzel *et al.*, 1975). However, it is clearly documented that glycosylated hemoglobins have altered oxygen saturation curves and/or decreased sensitivity to the allosteric effects of organic phosphates (McDonald *et al.*, 1979). This results from the fact that the amino terminal valine residue in hemoglobin, the site of glycosylation in HbA_1, is normally involved in the binding of 2,3-diphosphoglycerate, the physiologically important regulator of the oxygen affinity of hemoglobin.

2. Quantitation

A variety of techniques are available for the quantitation of glycohemoglobins. Besides the manual TBA (Gabbay *et al.*, 1979a; Pecoraro *et al.*, 1979; Fischer *et al.*, 1980) and the phenol sulfuric acid (Nayak and Pattabiraman, 1981) assays, automated TBA assays have been developed for the Technicon autoanalyzer (Rose and Gibson, 1979; Burrin *et al.*, 1980). HbA_{1c} itself can be determined by several methods, including isoelectric focusing (Jeppsson *et al.*, 1978; Spicer *et al.*, 1978; Mortensen, 1980a,b) and radioimmunoassay (Javid *et al.*, 1978), as well as ion-exchange chromatography. The original chromatographic procedure of Allen and Balog (1958) has been automated for an amino acid analyzer (Wacjman *et al.*, 1979) and also adapted for high-pressure liquid chromatography (Cole *et al.*, 1978; Davis *et al.*, 1978). In the most commonly used clinical procedures, the percent total HbA_1 is measured by various modifications of a "short-column" method developed by Trivelli *et al.* (1971). HbA_1 and HbA_0 are eluted sequentially from a carboxymethylcellulose column by two different buffers and quantitated by their absorbance at 415 nm.

With time, two problems with the short-column technique have become apparent. First, variations in the temperature at which the chromatography is conducted markedly affect the separation and thus the long-term reproducibility of the assay, so that rigorous control of column temperature is required (Dix *et al.*, 1978; Hankins and Holladay, 1980; Johnson *et al.*, 1980; Worth *et al.*; 1980). Second, several laboratories reported that HbA_1 values determined by the short-column method change rapidly in response to acute changes in blood glucose concentration or following short term incubation of erythrocytes in hyperglycemic media (Svendsen *et al.*, 1979a,b; Goldstein *et al.*, 1980; Mortensen, 1980a,b; Widness *et al.*, 1980). This observation is inconsistent with the slow rate of glucosylation of hemoglobin, which has been documented both *in vivo* and *in vitro* (Bunn *et al.*, 1976; Spicer *et al.*, 1979). The dilemma was resolved when it was recognized that the Schiff base adduct of hemoglobin, "pre-HbA_{1c}," has a significant half-life *in vitro* and has chromatographic and electrophoretic properties similar to those of HbA_{1c} itself (Mortensen, 1980a,b). The interference by pre-HbA_{1c} with short-column HbA_1 measurements can be corrected by preincubation of red cells or their lysates in saline for a few hours before assay (Goldstein *et al.*, 1980; Svendsen *et al.*, 1980; Widness *et al.*, 1980; Compagnucci *et al.*, 1981). During this time the Schiff base dissociates while the concentration of the ketoamine is essentially unchanged. Nathan *et al.* (1981) have shown that the Schiff base form of glycohemoglobin can also be discharged by a 30 minute incubation with semicarbazide and aniline. Menard *et al.* (1980) have recently described an agar gel electrophoresis procedure for separation and quantitation of HbA_1.

3. Correlations with Hyperglycemia in Diabetes

The formation of HbA_{1c} occurs slowly ($\sim 0.05\%/$day) and continually during the 120-day lifetime of the red blood cell (Bunn *et al.*, 1976; Higgins and Bunn, 1981) at a rate that is approximately first order with respect to blood glucose concentration. Thus, the concentration of glycohemoglobin is higher in older cells (Fitzgibbons *et al.*, 1976), and the cellular glycohemoglobin concentration reflects both the age of the cell and the mean blood glucose to which it has been exposed. Chronic elevations in blood glucose, as seen in diabetes, lead to an increase in the overall percent glycohemoglobins in blood, assuming a normal red blood cell lifetime. In fact, glycohemoglobins may account for as much as

20% of total hemoglobins in poorly controlled diabetic patients. The majority of the increase in glycohemoglobins in diabetes occurs in the HbA_{1c} fraction; a slight increase is observed in HbA_{1b}, whereas HbA_{1a} appears to be unaffected (McDonald *et al.*, 1978).

Measurements of either the percent total glycohemoglobins (HbA_{1a} + HbA_{1b} + HbA_{1c}), percent HbA_{1c}, or glycosylated hemoglobin by the TBA assay have proved useful to physicians as long-term integrators of blood glucose concentration. In practice, glycohemoglobin concentration appears to correlate linearly with mean blood glucose in diabetic patients during the previous 1–2 months (Koenig *et al.*, 1976; Ditzel and Kjaergaard, 1978; Kennedy and Merimee, 1981). Gabbay *et al.* (1977) also reported an optimal correlation between HbA_1 and urinary glucose excretion in patients during the previous two months. Several studies have compared the results of glycohemoglobin and glucose tolerance tests in normal and diabetic patients (reviewed by Bunn, 1981; Jovanovic and Peterson, 1981). In general, glycohemoglobin measurements are not considered to be as sensitive as the glucose tolerance test for the diagnosis of diabetes. However, with improvement and standardization of the procedure, measurement of total HbA_1 or HbA_{1c} may be accepted for the diagnosis as well as the long-term management of this disease. Glycohemoglobin levels should be indicative of blood glucose regulation under normal physiological conditions, while the glucose tolerance test should be more indicative of the capacity for glycemic control under stressful conditions, i.e., a heavy glucose load.

B. Glucosylated Plasma Proteins

1. Total Plasma Proteins

A large fraction of plasma proteins have circulating half-lives in man in excess of 1–2 weeks (Schultze and Heremans, 1966) and thus may be subject to detectable levels of nonenzymatic glucosylation *in vivo*. Several laboratories have recently studied the glucosylation of serum proteins by incubating serum with radioactive glucose *in vitro* (Day *et al.*, 1979a,b; Dolhofer and Wieland, 1979); Yue *et al.*, 1980). In retrospect, these experiments must be interpreted cautiously because of the possibility of reactive impurities in the radioactive glucose preparations, as discussed by Trüeb *et al.* (1980) (see Section I,C above). The various groups reporting glucosylation of serum proteins *in vitro* have used a variety of glucose concentrations, reaction temperatures, and times of incubation. In all cases, the incorporation of radioactive glucose showed no evidence of saturation, even after 200 hours at 25° or 50 hours at 37° and over a wide, 5–278 mM, range of glucose concentrations. When the radioactive serum proteins were fractionated by gel-permeation chromatography, allowing for differences in the molecular sieves used, the profiles were quite similar, showing all molecular weight classes of serum protein to have been labeled. Yue *et al.* (1980) have provided the most convincing evidence for serum protein glucosylation since they also measured TBA-reactive material in the column fractions obtained from chromatography of serum protein, both before and after incubation with added glucose. There were differences between the radioactivity and TBA-reactive patterns, but all classes of serum protein were found to contain ketoamine-linked carbohydrate. Dolhofer and Wieland (1979, 1980) also found HMF released from hydrolysates of both albumin and nonalbumin serum proteins. Using the TBA assay with dialyzed serum samples, Kennedy *et al.* (1980) found ~ 0.3 nmole HMF per milligram serum protein, equivalent to at least 3% glucosylation of serum protein with an average molecular weight of 100,000.

2. Albumin

Nonenzymatic glucosylation of albumin, the most abundant serum protein, has been investigated in some detail. Again, however, many of the *in vitro* kinetic data are compromised because they are based on amounts of acid-precipitable radioactivity in purified albumin after incubation with [^{3}H]- or [^{14}C]glucose. Thus, the presence of radioactive impurities probably accounts for the large discrepancies in the rate and extent of albumin glucosylation reported from our laboratory (Day *et al.*, 1979a,b) and by Dolhofer and Wieland (1979). We observed 1 mole glucose per mole albumin after incubation of albumin for 3 days in 20 mM glucose at pH 7.5 and 25°. In contrast, Dolhofer and Wieland observed 9 moles glucose per mole albumin after incubation of albumin for 1 day in 27 mM glucose at pH 7.4 and 37°. Even allowing for differences in reaction parameters, these data show a large variation in the apparent reactivity of albumin to chemical glucosylation. Both reports, however, significantly overestimate the actual rate of glucosylation of human serum albumin *in vitro*. In recent (unpublished) work we studied the reaction rate using [^{3}H]glucose purified by paper chromatography immediately before use. These experiments yielded a second-order rate constant, $k \approx 1$ liter/mole/day, for the reaction at pH 7.4 and 37°. This *in vitro* rate constant would account for conversion of about 0.5% of human albumin to glucosylated albumin per day in plasma at 5 mM glucose. Assuming approximately an 18-day half-life for albumin in man (Schultze and Heremans, 1966), it can be estimated that about 12% of human albumin should exist in the glucosylated form *in vivo*. This is in reasonable agreement with reports that indicate ~ 60–80 mmoles HMF released per mole albumin isolated from normal human serum (Day *et al.*, 1979a; Dolhofer and Wieland, 1980), i.e., 6–8% glucosylated albumin.

Bailey *et al.* (1976) reduced human albumin with [^{3}H]KBH$_4$, identified radioactive mannitol- and glucitollysine by amino acid analysis, and confirmed the identifications by mass spectrometry. We have also found hexitollysine to be the major radioactive product on amino acid analysis of both NaBH$_4$-reduced

[^{3}H]glucose-labeled albumin (Day *et al.*, 1979b) and [^{3}H]NaBH$_4$-reduced albumin isolated from normal and diabetic human and rat sera. In our initial reports on the detection of glucosylated albumin in human and rat sera, we described the separation of glucosylated and unglucosylated albumin by carboxymethyl cellulose chromatography (Day *et al.*, 1979a,b). We regret to report that this separation has not been reproducible and can no longer be offered as a means of preparing these two forms of albumin. Thus, glucosylation of lysine residues in albumin, as in hemoglobin (Shapiro *et al.*, 1980), does not appear to alter the chromatographic behavior of the protein significantly.

3. Clinical Correlations

Using the TBA assay, several laboratories have found a two- to threefold increase in total serum protein and albumin glucosylation in diabetic patients (Guthrow *et al.*, 1979; McFarland *et al.*, 1979; Dolhofer and Wieland, 1979, 1980; Kennedy *et al.*, 1980; Yue *et al.*, 1980). Rats with alloxan-induced diabetes also showed similar elevations in these glucosylated proteins (Day *et al.*, 1980). In the human studies, good correlations ($r = 0.7–0.9$) were found in various comparisons of levels of glucosylated serum proteins, glucosylated albumin, HbA$_1$, and fasting and mean blood glucose concentrations.

Because of the shorter circulating half-life of serum proteins and albumin compared to hemoglobin, changes in percent glucosylation of these proteins should occur more rapidly than changes in glycohemoglobins, in response to alterations in mean blood glucose concentration in diabetes. Thus, measurements of glucosylated serum proteins may prove useful as an index of recent or short-term glucose control and would supplement information provided by glycohemoglobin and blood sugar measurements. Kennedy and Merimee (1981) have shown that glucosylated serum protein values accurately reflect changes in mean blood glucose within 1–2 weeks after improvement in glycemic control in diabetic patients. In their studies glycosylated hemoglobin levels were not an effective indicator of improved control within two weeks.

C. Other Glucosylated Proteins

1. Collagens

Although collagen is the major extracellular protein in the body and has a relatively long metabolic half-life, limited information on collagen glucosylation has appeared since the early studies by Robins and Bailey (1972) and Tanzer *et al.* (1972). These workers isolated glucitollysine and glucitolhydroxylysine from reduced bovine skin collagen. They had also shown earlier that levels of the material, subsequently identified as glucitollysine, increased in collagen as a function of age (Bailey and Shimokamaki, 1971).

More recently HMF was measured after hydrolysis of normal and diabetic rat aortal (Rosenberg *et al.*, 1979) and glomerular basement membrane (Cohen *et al.*, 1980) collagens. The methods used for collagen preparation, i.e., autoclaving of the aorta for 18 hours or strong base solubilization of urea-extracted glomerular components, may have led to significant loss of ketoamine-linked carbohydrate and thus to an underestimate of the actual amount of glucosylation. On the other hand, the failure to employ a blank in these TBA assays may have led to an overestimate of HMF released. Despite these limitations and the differences in the collagens, both groups observed an approximate twofold increase in glucosylation of collagen within 4–6 weeks after induction of diabetes in rats. Cohen and Yu-Wu (1981) also detected increased amounts of glucitollysine and glucitolhydroxylysine in basement membrane collagen of diabetic rats by amino acid analysis.

Schnider and Kohn (1980, 1981) used the TBA assay to study glucosylation of human diaphragmatic tendon and skin collagens obtained at autopsy. Glucosylation was found to increase with age and was elevated up to twofold in people with juvenile-onset diabetes, compared to age-matched controls. The level of glucosylation of collagen from older individuals with maturity-onset diabetes was not distinguishable from that of controls, but the history of blood glucose control in these patients was not reported. Both Hamlin *et al.* (1975) and Schnider and Kohn (1981) reported that, with increasing age and diabetes, collagens are increasingly glucosylated and become more insoluble and resistant to proteolytic digestion. However, the possible role of nonenzymatic glucosylation in the altered properties of aged and diabetic collagens has not been determined.

2. Lens Crystallins

Because of the high frequency of cataracts in both diabetes and galactosemia, Stevens *et al.* (1978) investigated the effects of nonenzymatic glycosylation on the properties of bovine lens crystallins and rat lenses. Their data suggest that glucosylation may induce the aggregation and precipitation of crystallins in lens, leading to lens opacification and the development of cataracts. These authors observed that the purified crystallins aggregated more rapidly in the presence of glucose and glucose 6-phosphate than in sugar-free media and that this aggregation could be inhibited by reducing agents such as glutathione and dithiothreitol. Aggregation in an oxidizing environment was also about 80% reversible by these reducing agents. Stevens *et al.* (1978) concluded that glycosylation increased the susceptibility of sulfhydryl groups in crystallins to oxidation, leading to the formation of disulfide cross-linked aggregates. Support for this hypothesis was drawn from the observation that glycosylation by glucose 6-phosphate proceeded at equal rates under O_2 or N_2, whereas aggregation was two or three times more rapid in the O_2 atmosphere. About 20% of the aggregated protein was resistant to solubilization by reducing agents, suggesting, as the authors indicate, either incomplete reduction or the presence of other types of intermolecular cross-links. Unfortunately, harsher reducing conditions, e.g., in the presence of urea or sodium dodecyl sulfate, were not tested, so that it is difficult to differentiate between these possibilities. Stevens *et al.* (1978) also reported increased incorporation of tritium from [^{3}H]NaBH$_4$ into cataractous lenses and crystallins from diabetic rats. Monnier *et al.* (1979) reported tenfold increases in glucitollysine in lens protein from diabetic rats. Pande *et al.* (1979), however, were unable to detect increases in hexitollysine in human diabetic lens proteins, and have questioned the role of glucosylation in cataract formation.

Osmotic stress, induced by the accumulation of sorbitol in the lens in diabetes (or galactitol in the lens in galactosemia), has also been proposed as an alternative mechanism for inducing sugar cataracts. Aldose reductase inhibitors have, in fact, been shown to be effective in preventing the development of cataracts in both diabetic and galactosemic animals (Dvornik *et al.*, 1973; Kinoshita *et al.*, 1979; Peterson *et al.*, 1979b), even in the presence of hyperglycemia. Chiou *et al.* (1980) showed that the amount of hexitollysine (in this case, galactitollysine) was increased several-fold in galactosemic rat lens protein, and did not decrease during treatment with aldose reductase inhibitors. They concluded that nonenzymatic glycosylation was not responsible for sugar-induced cataracts. Cerami *et al.* (1979) proposed that a combination of mechanisms may be involved. Thus, increased activity of the sorbitol pathway may lead to reduced levels of lens glutathione, creating a more oxidizing environment, which favors the aggregation of glycosylated crystallins.

3. Peripheral Nerve Protein

Peripheral neuropathy in diabetes is characterized functionally by decreased sensory and motor nerve conduction velocity, and morphologically by patchy or segmental demyelination. Graf *et al.* (1979) observed a correlation between the decrease in motor nerve conduction velocity and fasting plasma glucose in a group of maturity-onset diabetics, and proposed that the degree of hyperglycemia may be a contributing factor in diabetic neuropathy. Vlassara *et al.* (1981) observed two- to fourfold increases in the amount of glucitollysine in sciatic nerve protein of diabetic rats and dogs, suggesting a possible role for nonenzymatic glucosylation in the development of the neuropathy. Some support for this hypothesis was drawn from earlier work on cyanate-induced neuropathy (Tellez-Nagel *et al.*, 1977) which has similar morphological findings. Thus carbamylation and glucosylation of protein may have similar effects, and the neuropathy might be attributed, in both cases, to chronic chemical modifications of peripheral nerve protein. As in the lens, however, sorbitol also accumulates in nerve tissue in diabetes, and it has been difficult to assess the relative contributions of nonenzymatic glucosylation and sorbitol accumulation to the development of diabetic neuropathy. Aldose reductase inhibitors prevent sorbitol accumulation in sciatic nerves of diabetic rats (Peterson *et al.*, 1979) and decrease motor nerve pathology in galactosemic rats (Gabbay, 1973) but clinical trials with these inhibitors in man have been inconclusive (Gabbay *et al.*, 1979b).

4. Erythrocyte Membrane Proteins

Erythrocyte membrane proteins are probably among the longest-lived plasma membrane proteins in the body. Bailey *et al.* (1976) originally reported the presence of glucitollysine in these proteins, and Miller *et al.* (1980) studied their glucosylation in several patients with diabetes. Erythrocyte ghosts were labeled by reduction with [^{3}H]NaBH$_4$, and the majority of the radioactivity in protein was recovered as glucitollysine. The membrane proteins from the diabetic patients incorporated two or three times as much radioactivity as those from controls. After separation of the proteins by polyacrylamide gel electrophoresis in sodium dodecyl sulfate, comparison of dye staining and autoradiographic patterns revealed that all proteins were labeled roughly in proportion to their concentration in the membrane. Radioactivity in all protein bands also appeared to be proportionately elevated in diabetes, and there was no evidence for selective glucosylation of specific membrane proteins. There have been no reports of nonenzymatic glucosylation of membrane proteins from other cell types. Considering the rate of turnover of plasma membrane proteins of nucleated cells, however, the extent of their glucosylation may be as much as an order of magnitude lower than that of erythrocyte membrane proteins.

Enhanced glucosylation of membrane proteins may affect the structural properties of the erythrocyte membrane. Evidence is accumulating which indicates that erythrocytes are less elastic and deformable in diabetes and that deformability is most significantly reduced in poorly controlled patients. This decreased deformability has been corroborated independently by studies on the resistance of red blood cells to filtration through small-pore membranes (Schmid-Schönbein and Volger, 1976; Barnes *et al.*, 1977), by measurements of flow resistance through capillary pipets (McMillan *et al.*, 1978), and by slow-motion photography studies of the rate at which erythrocytes regain their original shape after deformation (McMillan *et al.*, 1978). Recently, Baba *et al.* (1979) used fluorescent depolarization to verify that the intrinsic viscosity of the red cell membrane itself is increased in diabetes; these authors further observed a correlation between membrane microviscosity and the degree of metabolic control in diabetic patients. Peterson *et al.* (1977) reported that the red cell life span in diabetes is at the low range of normal and that institution of strict control of blood glucose in a group of hospitalized patients led to normalization of red cell survival. There is no information, however, on the relationship between nonenzymatic glucosylation and alterations in the physical and functional properties of the red cell or its survival in the circulation.

III. RELEVANCE OF NONENZYMATIC GLUCOSYLATION TO PATHOPHYSIOLOGY OF DIABETES

There is an increasing body of evidence, based on both clinical experience with patients (Gabbay, 1975; West, 1978) and animal model studies (Fox *et al.*, 1977), which indicates that chronic, subclinical hyperglycemia, rather than the insulin deficiency itself, may be the major factor contributing to the long-term complications of diabetes. Increased nonenzymatic glucosylation of protein is now known to be one of the chemical consequences of hyperglycemia. Thus, a major challenge for scientists today is to ascertain how nonenzymatic glucosylation may be involved in the sequelae of diabetes.

A. Catabolism of Nonenzymatically Glucosylated Protein

Early investigations on the loss of nutritional value of milk and food proteins as a result of the Maillard reaction suggested that mammalian systems have only a limited ability to hydrolyze the ketoamine linkage. In model studies, Folk (1956) showed that glucosylation inhibited trypsin and pancreatin hydrolysis of the peptide bond on the carboxyl side of glucosylated lysine residues. There have been no comparable studies, however, on the degradation of glucosylated proteins by lysosomes or lysosomal enzymes. Brownlee *et al.* (1980) measured fructoselysine in normal human urine, ~ 3 μmoles/kg body weight per day, or ~ 60 mg per 70 kg person per day. This quantity may be sufficient to account for the total body turnover of ketoamine derivatives of protein per day, if one assumes, as in the case of albumin (Section I,C,2 above), that $\sim 0.5\%$ of an average body protein of 100,000 molecular weight is converted to a glucosylated derivative per day. Two- to threefold increases in urinary fructoselysine were also observed in a group of diabetic patients, but there is no indication at present that the demand for increased catabolism of glucosylated amino acids or peptides in diabetes has any pathological consequences in man.

B. Direct Effects of Glucosylation on Protein Structure and Function

Nonenzymatic glucosylation is a continuous, cumulative chemical insult to body proteins and perhaps also to phospholipids and nucleic acids. However, with the exception of hemoglobin, there is little information available concerning the effects of glucosylation on the function of specific proteins or enzymes. It seems reasonable, however, that glucosylation may interfere with the Schiff base binding of pyridoxal phosphate to some enzymes or that glucosylation of active-site lysine residues may inhibit enzyme activity and affect the function of numerous metabolic pathways. These possibilities, along with potential effects of glucosylation on the endogenous rates of turnover of proteins, remain to be evaluated.

C. Glucose-Dependent Cross-Linking of Protein

In addition to direct effects of glucose on protein structure and function, it has also been proposed that hyperglycemia in diabetes may lead to an increase in protein cross-linking and polymerization by Maillard-type browning reactions *in vivo* (Cerami *et al.*, 1979). Ketoamines are known to be essential intermediates in the Maillard reaction between protein and sugar *in vitro*. Anet proposed (1959, 1964) that the ketoamine condenses with a second molecule of sugar, forming an unstable diketoamine. The diketoamine then decomposes rapidly to yield the original ketoamine and a dicarbonyl sugar, 3-deoxyhexosulose, in equilibrium with its enolic α,β-unsaturated isomer (Fig. 4). It is these compounds, and further dehydration products, which are thought to be the actual initiators of the browning reactions. These dehydration products have been detected in browning reactions of protein and are more reactive than free glucose or ketoamine in catalyzing the browning of amino acid solutions (Anet, 1964; Reynolds, 1963, 1965). The details of the further polymerization reactions in this complex system of amino, carbonyl, and enolic compounds are not well understood. The end products of the reactions, brown "melanoidin" pigments, are also poorly described. They are resistant to solubilization or reduction by $NaBH_4$ and form insoluble humins on acid or alkaline hydrolysis (Reynolds, 1963, 1965).

Since the Maillard products formed *in vitro* are so poorly characterized, it has been difficult to identify them unequivocally *in vivo*. Thus, there is no clear evidence at present that nonenzymatically glucosylated proteins undergo further cross-linking and polymerization reactions *in vivo*. On the other hand, Bjorksten (1977) has proposed, in his "cross-linkage theory of aging," that continuous

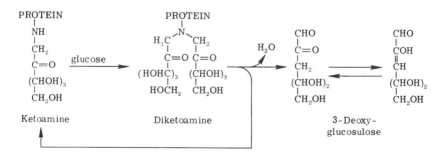

Figure 4 Formation and decomposition of diketoamines.

low-level cross-linking, or "cumulative tanning," of biopolymers has a major role in the aging of living systems. Diabetes is, in fact, a disease in which the appearance of age-related changes and debilities seems to be accelerated at both the whole-body and microscopic level. The accelerated aging in diabetes is characterized by premature atherosclerosis and microangiopathy, stiffening of skin and joints, decreased lung elasticity, and, in general, age-like insolubilization and thickening of connective tissue (Hamlin *et al.*, 1975; Kohn and Hamlin, 1978). The possibility that Maillard-type reactions are involved in the cross-linking and accelerated aging of tissues in diabetes deserves serious study.

ACKNOWLEDGMENTS

This work was supported by U.S. Public Health Services research grants AM19971 and AM25373 and a research grant from the American Diabetes Association.

REFERENCES

Allen, D. W., and Balog, J. (1958). *J. Am. Chem. Soc.* **80**, 1628–1634.

Anet, E. F. L. J. (1959). *Aust. J. Chem.* **12**, 491–496.

Anet, E. F. L. J. (1964). *Adv. Carbohydr. Chem.* **19**, 181–218.

Baba, Y., Kai, M., Kamada, T., Setoyama, S., and Otsuji, S. (1979). *Diabetes* **28**, 1138–1140.

Bailey, A. J., and Shimokamaki, M. (1971). *FEBS Lett.* **16**, 86–88.

Bailey, A. J., Robins, S. P., and Tanner, M. J. A. (1976). *Biochim. Biophys. Acta* **434**, 51–57.

Barnes, A. J., Locke, P., Scudder, P. R., Dormandy, T. L., Dormandy, J. A., and Slack, J. (1977). *Lancet* **2**, 789–791.

Bjorksten, J. (1977). *Adv. Exp. Med. Biol.* **86B**, 579–602.

Bookchin, R. M., and Gallop, P. M. (1968). *Biochem. Biophys. Res. Commun.* **32**, 86–93.

Brownlee, M., Vlassara, H., and Cerami, A. (1980). *Diabetes* **29**, 1044–1047.

Bunn, H. F. (1981a). *Am. J. Med.* **70**, 325–330.

Bunn, H. F. (1981b). *Diabetes* **30**, 613–617.

Bunn, H. F., and Higgins, P. J. (1981). *Science* **213**, 222–224.

Bunn, H. F., Haney, D. N., Gabbay, K. H., and Gallop, P. M. (1975). *Biochem. Biophys. Res. Commun.* **67**, 103–109.

Bunn, H. F., Haney, D. N., Kamin, S., Gabbay, K. H., and Gallop, P. M. (1976). *J. Clin. Invest.* **57**, 1652–1659.

Bunn, H. F., Gabbay, K. H., and Gallop, P. M. (1978). *Science* **200**, 21–27.

Bunn, H. F., Shapiro, R., McManus, M., Garrick, L., McDonald, M. J., Gallop, P. M., and Gabbay, K. (1979). *J. Biol. Chem.* **254**, 3892–3898.

Burrin, J. M., Worth, R., Ashworth, L. A., Currie, S., and Alberti, K. G. M. M. (1980). *Clin. Chim. Acta* **106**, 45–50.

Cerami, A., Stevens, V. J., and Monnier, V. M. (1979). *Metab., Clin. Exp.* **28**, Suppl. 1, 431–437.

Chiou, S. H., Chylack, L. T., Bunn, H. F., and Kinoshita, J. H. (1980). *Biochem. Biophys. Res. Commun.* **95**, 894–901.

Cohen, M., Urdanivia, E., Surma, M., and Wu, V.-Y. (1980). *Biochem. Biophys. Res. Commun.* **95**, 765–769.

Cohen, M. P., and Yu-Wu, V. (1981). *Biochem. Biophys. Res. Commun.* **100**, 1549-1554.

Cole, R. A., Soeldner, J. S., Dunn, P. J., and Bunn, H. F. (1978). *Metab., Clin. Exp.* **27**, 289-292.

Compagnucci, P., Cartechini, M. G., Bolli, G., DeFeo, P., Santeusanio, F., and Brunetti, P. (1981). *Diabetes* **30**, 607-612.

Davis, J. E., McDonald, J. M., and Jarett, L. H. (1978). *Diabetes* **27**, 102-106.

Day, J. F., Thorpe, S. R., and Baynes, J. W. (1979a). *J. Biol. Chem.* **254**, 595-597.

Day, J. F., Thornburg, R. W., Thorpe, S. R., and Baynes, J. W. (1979b). *J. Biol. Chem.* **254**, 9394-9400.

Day, J. F., Ingebretsen, C. G., Ingebretsen, W. R., Jr., Baynes, J. W., and Thorpe, S. R. (1980). *Diabetes* **29**, 524-527.

Ditzel, J., and Kjaergaard, J.-J. (1978). *Br. Med. J.* **1**, 741-742.

Ditzel, J., and Standl, E. (1975). *Acta Med. Scand., Suppl.* **578**, 59-68.

Ditzel, J., Andersen, H., and Peters, N. D. (1975). *Acta Paediatr. Scand.* **64**, 355-361.

Dix, D., Cohen, P., Kingsley, S., Senkbeil, J., and Sexton, K. (1978). *Clin. Chem. (Winston-Salem, N.C.)* **24**, 2073-2078.

Dixon, H. B. F. (1972). *Biochem. J.* **129**, 203-208.

Dolhofer, R., and Wieland, O. H. (1979). *FEBS Lett.* **103**, 282-286.

Dolhofer, R., and Wieland, O. H. (1980). *Diabetes* **29**, 417-421.

Dolhofer, R., and Wieland, O. H. (1981). *Clin. Chim. Acta* **112**, 197-204.

Dox, A. W., and Plaisance, G. P. (1916). *J. Am. Chem. Soc.* **38**, 2156-2166.

Dvornik, D., Simard-Duquesne, N., Krami, M., Sestanj, K., Gabbay, K. H., Kinoshita, J. H., Varma, S. D., and Merola, L. O. (1973). *Science* **182**, 1146-1148.

Fischer, R. W., DeJong, C., Voigt, E., Berger, W., and Winterhalter, K. H. (1980). *Clin. Lab. Haematol.* **2**, 129-138.

Fitzgibbons, J. F., Koler, R. P., and Jones, R. T. (1976). *J. Clin. Invest.* **58**, 820-824.

Folk, J. E. (1956). *Arch. Biochem. Biophys.* **64**, 6-18.

Flückiger, R., and Winterhalter, K. H. (1976). *FEBS Lett.* **71**, 356-360.

Fox, C. J., Darby, S. C., Ireland, J. T., and Sonksen, P. H. (1977). *Br. Med. J.* **2**, 605-607.

Gabbay, K. H. (1973). *In* "Vascular and Neurological Changes in Early Diabetes. Advances in Metabolic Disorders" (R. A. Camerini-Davalos and H. S. Cole, eds.) Suppl. 2, pp. 417-429. Academic Press, New York.

Gabbay, K. H. (1975). *Annu. Rev. Med.* **26**, 521-550.

Gabbay, K. H., Hasty, K., Breslow, J. L., Ellison, R. C., Bunn, H. F., and Gallop, P. M. (1977). *J. Clin. Endocrinol. Metab.* **44**, 859-864.

Gabbay, K. H., Sosenko, J. M., Banuchi, G. A., Minnisohn, M. J., and Flückiger, R. (1979a). *Diabetes* **28**, 337-340.

Gabbay, K. H., Spack, N., Loo, S., Hirsch, H. J., and Ackil, A. A. (1979b). *Metab., Clin. Exp.* **28**, Suppl. 1, 471-476.

Garel, M. C., Blauquist, Y., Molko, F., and Rasa, J. (1979). *Biomedicine* **30**, 234-240.

Goldstein, D. E., Peth, S. B., England, J. D., Hess, R. L., and DaCosta, J. (1980). *Diabetes* **29**, 623-628.

Gottschalk, A. (1952). *Biochem. J.* **52**, 455-460.

Gottschalk, A. (1972a). *In* "The Glycoproteins" (A. Gottschalk, ed.), Part A, pp. 141-157. Am. Elsevier, New York.

Gottschalk, A. (1972b) *In* "The Glycoproteins" (A. Gottschalk, ed.), Part A, pp. 434-437. Am. Elsevier, New York.

Gottschalk, A., and Patridge, S. M. (1950). *Nature (London)* **165**, 684-685.

Graf, R. J., Halter, J. B., Halar, E., and Porte, D. (1979). *Ann. Int. Med.* **90**, 298-303.

Guthrow, C. E., Morris, M. A., Day, J. F., Thorpe, S. R., and Baynes, J. W. (1979). *Proc. Natl. Acad. Sci. U.S.A.* **76**, 4258-4261.

Hamlin, C. R., Kohn, R. R., and Luschin, J. H. (1975). *Diabetes* **24**, 902–904.
Haney, D. N. and Bunn, H. F. (1976). *Proc. Natl. Acad. Sci. U.S.A.* **73**, 3534–3538.
Hankins, W. D., and Holladay, L. (1980). *Clin. Chim. Acta* **104**, 251–257.
Higgins, P. J., and Bunn, H. F. (1981). *J. Biol. Chem.* **256**, 5204–5208.
Holmquist, W. R., and Schroeder, W. A. (1966). *Biochemistry* **5**, 2489–2503.
Huisman, T. H. J., and Meyering, C. A. (1960). *Clin. Chim. Acta* **5**, 103–123.
Javid, J., Pettis, P. K., Koenig, R. J., and Cerami, A. (1978). *Br. J. Haematol.* **38**, 329–337.
Jeppsson, J.-O., Franzén, B., and Nilsson, K. O. (1978). *Sci. Tools* **25**, 69–72.
Johnson, M. W., Dobrea, G. M., Bendezu, R., and Wieland, R. G. (1980). *Clin. Chim. Acta* **104**, 319–328.
Jones, R. L., and Peterson, C. M. (1981). *Am. J. Med.* **70**, 339–352.
Jovanovic, L., and Peterson, C. M. (1981). *Am. J. Med.* **70**, 331–338.
Keeney, M., and Bassette, R. (1959). *J. Dairy Sci.* **42**, 945–960.
Kennedy, A. L., Mehl, T. D., and Merimee, T. J. (1980). *Diabetes* **29**, 413–415.
Kennedy, A. L., and Merimee, T. J. (1981). *Ann. Int. Med.* **95**, 56–58.
Kinoshita, J. H., Fukushi, S., Kador, P., and Merola, L. O. (1979). *Metab., Clin. Exp.* **28**, Suppl. 1, 462–469.
Koenig, R. J., and Cerami, A. (1980). *Ann. Rev. Med.* **31**, 29–34.
Koenig, R. J., Peterson, C. M., Jones, R. L., Saudek, C., Lehrman, M., and Cerami, A. (1976). *N. Engl. J. Med.* **295**, 417–420.
Koenig, R. J., Blobstein, S. H., and Cerami, A. (1977). *J. Biol. Chem.* **252**, 2992–2997.
Kohn, R. R., and Hamlin, C. R. (1978). *Birth Defects, Orig. Artic. Ser.* **14**, 387–401.
McDonald, M. J., Shapiro, R., Bleichman, M., Solway, J., and Bunn, H. F. (1978). *J. Biol. Chem.* **253**, 2327–2332.
McDonald, M. J., Bleichman, M., Bunn, H. F., and Noble, R. (1979). *J. Biol. Chem.* **254**, 702–707.
McFarland, K. F., Catalano, E. W., Day, J. F., Thorpe, S. R., and Baynes, J. W. (1979). *Diabetes* **28**, 1011–1013.
McMillan, D. E., Utterback, N. G., and LaPuma, J. (1978). *Diabetes* **27**, 895–901.
Menard, L., Dempsey, M. E., Blankstein, L. A., Aleyassine, H., Wacks, M., and Soeldner, J. S. (1980). *Clin. Chem. (Winston-Salem, N.C.)* **26**, 1598–1602.
Miller, J. A., Gravallese, E., and Bunn, H. F. (1980). *J. Clin. Invest.* **65**, 896–901.
Monnier, V. M., Stevens, V. J., and Cerami, A. (1979). *J. Exp. Med.* **150**, 1098–1107.
Mortensen, H. B. (1980a). *J. Chromatogr.* **182**, 325–330.
Mortensen, H. B. (1980b). *Sci. Tools* **27**, 21–22.
Nathan, D. M., Avezzano, E. S., and Palmer, J. L. (1981). *Diabetes* **30**, 700–701.
Nayak, S. S., and Pattabiraman, T. N. (1981). *Clin. Chim. Acta* **109**, 267–274.
Pande, A., Garner, W. H., and Spector, A. (1979). *Biochem. Biophys. Res. Commun.* **89**, 1260–1266.
Pecoraro, R. A., Graf, R. J., Halter, J. B., Beiter, H., and Porte, D., Jr. (1979). *Diabetes* **28**, 1120–1125.
Peterson, C. M., Jones, R. L., Koenig, R. J., Melvin, E. T., and Lehrman, M. L. (1977). *Ann. Intern. Med.* **86**, 425–429.
Peterson, M. J., Sarges, R., Aldinger, C. E., and MacDonald, D. P. (1979). *Metab., Clin. Exp.* **28**, Suppl. 1, 456–461.
Rahbar, S. (1968). *Clin. Chim. Acta* **22**, 296–301.
Rahbar, S., Blumenfeld, O., and Ranney, H. M. (1969). *Biochem. Biophys. Res. Commun.* **36**, 838–843.
Reynolds, T. M. (1963). *Adv. Food Res.* **12**, 1–52.
Reynolds, T. M. (1965). *Adv. Food Res.* **14**, 167–283.

Robins, S. P., and Bailey, A. J. (1972). *Biochem. Biophys. Res. Commun.* **48**, 76–84.

Rosenberg, H., Modark, J. B., Hassing, J. M., Al-Turk, W. A., and Stohs, S. J. (1979). *Biochem. Biophys. Res. Commun.* **91**, 498–501.

Ross, M. S., and Gibson, P. F. (1979). *Clin. Chim. Acta* **98**, 53–59.

Schmid-Schönbein, H., and Volger, E. (1976). *Diabetes* **25**, Suppl. 2, 897–902.

Schnider, S. L., and Kohn, R. R. (1980). *J. Clin. Invest.* **66**, 1179–1181.

Schnider, S. L., and Kohn, R. R. (1981). *J. Clin. Invest.* **67**, 1630–1635.

Schultze, H. E., and Heremans, J. F. (1966). *In* "Molecular Biology of Human Proteins," pp. 473–477. Am. Elsevier, New York.

Shapiro, R., McManus, M. J., Zalut, C., and Bunn, H. F. (1980). *J. Biol. Chem.* **255**, 3120–3127.

Spicer, K. M., Allen, R. C., and Buse, M. G. (1978). *Diabetes* **27**, 384–388.

Spicer, K., Allen, R. C., Hallett, D., and Buse, M. G. (1979). *J. Clin. Invest.* **64**, 40–48.

Stevens, V. J., Rouzer, C. A., Monnier, V. M., and Cerami, A. (1978). *Proc. Natl. Acad. Sci. U.S.A.* **75**, 2918–2922.

Svendsen, P. A., Christiansen, J. S., Welinder, B., and Nerup, J. (1979a). *Lancet* **1**, 603.

Svendsen, P. A., Christiansen, J., Andersen, A. R., Welinder, B., and Nerup, J. (1979b). *Lancet* **1**, 1142–1143.

Svendsen, P. A., Christiansen, J. S., Søegaard, U., Welinder, B. S., and Nerup, J. (1980). *Diabetologia* **19**, 130–136.

Tanzer, M. L., Fairweather, R., and Gallop, P. M. (1972). *Arch. Biochem. Biophys.* **151**, 137–141.

Tellez-Nagel, I., Korthals, J. K., Vlassara, H., and Cerami, A. (1977). *J. Neuropathol. Exp. Neurol.* **43**, 351–363.

Trivelli, L. A., Ranney, H., and Lai, H. T. (1971). *N. Engl. J. Med.* **284**, 353–357.

Trüeb, B., Holenstein, C. G., Fischer, R. W., and Winterhalter, K. H. (1980). *J. Biol. Chem.* **255**, 6717–6720.

Vlassara, H., Brownlee, M., and Cerami, A. (1981). *Proc. Natl. Acad. Sci. U.S.A.* **78**, 5190–5192.

Wacjman, H., Dastugue, B., and Labie, D. (1979). *Clin. Chim. Acta* **92**, 33–39.

Warren, L. (1959). *In* "Methods in Enzymology" (S. P. Colowick and N. O. Kaplan, eds.), Vol. 6, pp. 463–464. Academic Press, New York.

West, K. M. (1978). "Epidemiology of Diabetes and Its Vascular Lesions," Chapter 6. Am. Elsevier, New York.

Widness, J. A., Rogler-Brown, T. L., McCormick, K. L., Petzold, K. S., Susa, J. B., Schwartz, H. C., and Schwartz, R. (1980). *J. Lab. Clin. Med.* **95**, 386–394.

Worth, R. C., Ashworth, L. A., Burrin, J. M., Johnston, D. G., Skillen, A. W., Anderson, J., and Alberti, K. G. M. M. (1980). *Clin. Chim. Acta* **104**, 401–404.

Yue, D. K., McLennan, M. S., and Turtle, J. R. (1980). *Diabetes* **29**, 296–300.

2

Glycosylation and Development

SECTION 1

Glycosyltransferases in the Differentiation of Slime Molds

HANS-JÖRG RISSE AND HELLMUTH HANS RÖSSLER

I. INTRODUCTION

The biochemistry of glycosyltransferases in slime molds and their function in the regulation of the biosynthesis of glycoconjugates are only fragmentarily understood. A review of this subject, accordingly, suffers from the same incompleteness. Moreover, the following topics have been excluded from this report or are touched on only by representative citations owing to space limitations: (a) the degradation of glycoconjugates, (b) the structure of glycoconjugates in slime molds, and (c) the mechanisms of energy production during the development of slime molds. Since reports on the biochemistry of glycosyltransferases in other species of slime molds are negligible, the results described here are restricted to those from *Dictyostelium discoideum*.

Slime molds have provoked the fascination of generations of developmental biologists, and their life cycles and cell differentiation have been reviewed extensively (Bonner, 1967; Gerisch, 1968; Killick and Wright, 1974; Loomis, 1975; Cappuccinelli and Ashworth, 1977; Newell, 1977; Gerisch and Guggenheim, 1980). The life cycle of the archetype of the family, *Dictyostelium discoideum,* can be divided into three major phases: (a) a vegetative phase, in which single ameboid cells grow and multiply; (b) an early differentiation phase, in which the cells are induced by nutrient deficiency to undergo differentiation and reach a state of specific aggregation; and (c) a late phase, in which the aggregated cells organize themselves and eventually form a fruiting body that is less sensitive to an unfavorable environment than the ameboid cell. The life cycle is completed by

135

the germination of spores after exposure of the sporangia to a fresh nutrient source. Slime molds are excellent models for the study of various developmental processes, such as the generation and utilization of chemotactic signals, the biochemical alterations of cell surfaces during a monocellular/multicellular transition, and intercellular control, important topics for the understanding of ontogenetic regulations in higher eukaryotes.

II. DEVELOPMENT AND GLYCOSYLTRANSFERASE ACTIVITY

Studies on glycosyltransferases have been performed within three functional domains in the life cycle of *Dictyostelium,* concerning (a) the modification of cell surface glycoconjugates during the "morphologically silent" phase between the induction of differentiation and aggregation, (b) the synthesis of the storage saccharides glycogen and trehalose (α-D-glucopyranosyl-α-D-glucopyranoside), and (c) the synthesis of polysaccharides typical of the late phases of culmination and sorocarp formation.

A. Glycosyltransferases in Early Periods of Development

1. Localization, Properties, and Developmental Regulation of Glycosyltransferases

From the work published mainly by Gerisch and his group it is obvious that, aside from the induction of the chemotactic system, the synthesis of new surface antigens, functioning as "contact sites," is the predominant process of the preaggregation phase (Riedel and Gerisch, 1969; Beug *et al.,* 1970; Müller and Gerisch, 1978; Müller *et al.,* 1979; Gerisch *et al.,* 1980). The "contact sites" are glycoproteins containing the sugars D-mannose, L-fucose, and *N*-acetyl-D-glucosamine. One primary event after the onset of differentiation, therefore, should be the stimulation of enzyme activities responsible for the generation of new surface structures.

Transferase activities for the three sugars are present in different subfractions of *Dictyostelium* cells. Fractions of the endoplasmic reticulum are rich in glycosyltransferase activity, but considerable activity has also been found in purified plasma membranes (Sievers *et al.,* 1978). There is no evidence, however, of kinetic or functional differences between glycosyltransferases in the endoplasmatic reticulum and those in the plasma membranes; the simplest explanation would be that the precursors for the plasma membrane enzymes are synthesized in the inner membranes.

The plasma membrane enzymes (as well as the endoplasmic enzymes) are developmentally regulated (Risse *et al.,* 1974, 1979; Rogge *et al.,* 1975).

Maximum activity, suppressible by cycloheximide, is detected for mannosyl- and fucosyltransferases during a 3-hour period preceding cell aggregation. It is tempting to assume that these enzymes are part of the machinery synthesizing the contact-site carbohydrates, but direct proof is lacking. N-Acetylglucosaminyl-transferase, in contrast, exhibits a different developmental course: After a decline in the first 3–4 hours of differentation (probably due to the degradation of enzymes from the vegetative period), a gradual rise of activity can be followed over the aggregation period without any significant maximum.

The plasma membrane glycosyltransferases are integral proteins. So far, solubilization by different detergents has not afforded active extracts. The enzymes in the membrane preparations depend strongly on the presence of divalent cations. Mannosyltransferase requires Mg^{2+} and is inhibited by Mn^{2+}, whereas N-acetylglucosaminyltransferase is better stimulated by Mn^{2+} and also by higher concentrations of Mg^{2+}. The cation concentrations are optimal at about 10 mM. Fucosyltransferase is activated by Mg^{2+} and Mn^{2+}, as well as by Ca^{2+} (H. H. Rössler and H. J. Risse, unpublished data, 1980).

The transfer of monosaccharides from ^{14}C-labeled nucleoside diphospho sugars to glycoprotein and glycolipid can be followed by measuring the radioactivity of the products. Mannose-containing cerebrosides and polyisoprenols were detected as the predominant products of the mannose transfer (see below), whereas fucose and N-acetylglucosamine are incorporated mainly into the protein portion.

Glycosyltransferases are also constituents of the nucleus. Mannosyl- and N-acetylglucosaminyltransferases were described in isolated nuclei (Rogge and Risse, 1974). The activities in nuclei are highest during the growth phase and decline gradually after the induction of differentiation (Rogge et al., 1977). The polymer acceptor in nuclei is not known, and there is as yet no evidence of any biological function of nuclear glycosyltransferases in *Dictyostelium*. From other systems, hypotheses on a possible function of nuclear glycoconjugates in the control of transcription have been derived (Stein *et al.*, 1975; Schaffrath *et al.*, 1976).

2. Mechanisms of Glycosyl Transfer

Since the early work of Leloir's group (Parodi and Leloir, 1979), many results have accumulated which prove that membrane-linked glycosyltransferases utilize polyisoprenylphospho- or polyisoprenyldiphosphosaccharides as hydrophobic substrates (Waechter and Lennarz, 1976; Shur and Roth, 1975). The sequence of reactions leading from a nucleoside diphospho sugar via a polyisoprenyl intermediate to the final glycoconjugate seems to be a very conservative and common type of mechanism found in prokaryotes as well as eukaryotes (Waechter and Lennarz, 1976). It is not surprising, therefore, that the membrane-linked glycosyltransferases in *Dictyostelium* follow the same pattern. Rössler *et al.*

extracted glycolipids from *in vitro* mixtures containing plasma membranes and ^{14}C-labeled GDP-mannose* and observed the synthesis of a set of different mannolipids containing various sugar moieties (Rössler *et al.*, 1978; Crean and Rossomando, 1977). Glycolipids were isolated after incubation with the corresponding nucleoside diphospho sugars of *N*-acetylglucosamine and fucose and were characterized by a relatively high hydrophilicity, as compared to the mannolipids, probably caused by a high carbohydrate content (Rössler *et al.*, 1980). *N*-Acetylglucosaminyltransferase in a particulate fraction was described by Crean *et al.* (1979), but the authors could not find glycolipids. Most of the products were polyisoprenylphospho derivatives, as shown by their kinetics of hydrolysis and, in one case, by mass spectroscopy. These findings, taken together with the result that direct incorporation of mannose into protein is rather poor, support the hypothesis that, in a first series of reactions, polyisoprenyl phosphate-linked oligosaccharides are synthesized which eventually are transferred *en bloc* to the cell surface glycoproteins.

Not only the sugar residues but also the polyisoprenol moieties play a role in the regulation of the sugar transfer reactions. Incubation of plasma membranes with GDP-mannose and exogenous polyisoprenyl phosphates gave evidence that the preferentially utilized polyisoprenol is an undecaprenol with a saturated terminal isoprenoic residue. Fully unsaturated polyisoprenyl phosphate of the same chain length is completely inactive; α-saturated chains of lengths other than 55 C atoms cause measurable but markedly reduced activities (Rössler *et al.*, 1980). The fully unsaturated undecaprenol is the typical lipid for sugar transfer in prokaryotes (Higashi *et al.*, 1967). On the other hand, the α-saturated polyisoprenols, the dolichols, in higher eukaryotic cells are characterized by longer chains (70–120 C atoms). The finding of a prokaryote type of chain length together with a eukaryote type of saturation might reflect the biochemical classification of *Dictyostelium* as a very archaic type of eukaryotic cell. It will certainly be of interest to expand the investigation of the glycosyl transfer mechanism to other primitive eukaryotes.

The synthesis of the glycolipids and glycoproteins is under developmental control and can be completely inhibited by blocking protein synthesis. Since it has not been possible thus far to dissociate enzyme activities responsible for the glycolipid synthesis from those catalyzing the final transfer to glycoprotein, the inhibition can be clearly assigned to the first but not necessarily to the second step of the reaction.

3. Glycosyltransferases in the Synthesis of Cerebrosides

Cerebrosides are also synthesized under developmental control. One of the mannolipids isolated by Rössler (Rössler *et al.*, 1978) turned out to be a man-

---

*Abbreviations: ATP, adenosine triphosphate; cAMP, 3′, 5′-cyclic adenosine monophosphate; EDTA, ethylenediaminetetraacetic acid; GDP, guanosine diphosphate; RNA, ribonucleic acid; UDP, uridine diphosphate.

nosylcerebroside. The synthesis of this product, however, is independent of the developmental course; it was not analyzed further. Possibly it is related to a mannose-containing cerebroside described by Wilhelms *et al.* (1974) as a constituent of a surface antigen present in both the vegetative and the differentiation periods of *Dictyostelium*. During the preaggregation phase, a glucosyltransferase is induced and participates in the synthesis of a glucosylcerebroside of molecular weight 700 daltons that contains palmitic acid as the fatty acid moiety (Rössler *et al.*, 1978). The enzyme activity is found mainly in the microsomal fraction and is abolished completely by treatment of the cells with cycloheximide before the period of maximum activity. In the absence of divalent cations, this lipid is the only product in which glucose is incorporated. After the addition of Mg^{2+}, glucose is also incorporated into another nonlipophilic product of as yet unexamined structure. A similar, or perhaps the same, cerebroside was found and described by Crean and Rossomando (1979).

It has not yet been possible to assign a physiological function to the glucocerebroside, but it is of interest that this compound is synthesized only during a 2-hour period immediately before cell aggregation.

B. Synthesis of Storage Products

1. Glycogen Synthase (UDP-Glucose : glycogen 4-α-D-Glucosyltransferase, EC 2.4.1.11)

When the activity of glycogen synthase is followed in the presence of a primer, no significant changes are detected during the course of differentiation (Wright and Dahlberg, 1967; Hames *et al.*, 1972). The developmental regulation, however, influences the localization and the dependence of the enzyme on activators. During cell growth, glycogen synthase exhibits maximum activity in the transition from the exponential to the stationary phase (Weeks and Ashworth, 1972). The enzyme is present in both the soluble supernatant and a 100,000 *g* sediment; the higher specific activity is found in the particulate fraction. The synthase is inhibited by ATP in physiological concentrations, and the inhibition can be reverted by glucose 6-phosphate. The regulation of the activity is governed by the UDP-glucose concentration and the ATP/glucose 6-phosphate ratio rather than by the enzyme concentration. Glycogen synthesis can be increased sevenfold by growing the cells in the presence of glucose. Wright *et al.* (1966) also described a soluble glycogen synthase in the early stages of the differentiation that was dependent on the presence of a primer and glucose 6-phosphate and activated *in vitro* by EDTA.

In later periods, beginning with the culmination stage, the enzyme activity gradually becomes insoluble and independent of a primer (Wright *et al.*, 1968). Using an enzymatic cycling procedure for the assay of liberated UDP, Harris and Rutherford (1976) corroborated and expanded these findings. Glycogen synthase becomes insoluble during the same period, when the degradation of soluble glycogen reaches a maximum, providing the cells with glucose units for the synthesis of trehalose, glycosaminoglycan, and cellulose, the carbohydrates typical of the final stage of sporangium formation (Jones and Wright, 1970; Firtel and Bonner, 1972). During the development of the stalk, a decreasing gradient of glycogen synthase activity runs from the apex to the base of the stalk. The same distribution is found in final sorocarps, in which the apical stalk cells contain activities similar to those in the spores, whereas in the cells at the base the enzyme is virtually absent. The glycogen synthesized in the late period is insoluble as well. It has been suggested that at the beginning of the culmination stage the synthase and the glycogen synthesized became entrapped in the cell wall, thus being withdrawn from degradation and saved for germination. Upon germination, the cell wall becomes degraded, and synthase as well as primer is liberated for the further glycogen synthesis in the myxamebas.

2. α,α-Trehalose-phosphate Synthase (UDP-Glucose : D-Glucose-6-phosphate 1-α-D-Glucosyltransferase, EC 2.4.1.15)

There is no doubt about the role of the disaccharide trehalose as an energy source for the germination of spores (Cotter and Raper, 1970). About 5% of the dry weight of *Dictyostelium* spores consists of this sugar (Clegg and Filosa, 1961). A glucosyltransferase using UDP-glucose as substrate catalyzes the synthesis of trehalose 6-phosphate, the immediate precursor of trehalose (Killick and Wright, 1975). The enzyme is present in a 33,000 g supernatant, and the activity assayed is sensitive to the conditions of preparation. Differences in activities reported by other authors may be caused by such influences (Roth *et al.*, 1968). The addition of trehalose during the preparation affords long-term protection to the enzyme (Killick and Wright, 1975).

The highest α,α-trehalose-phosphate synthase activities are found between the preculmination (18 hours) and the end of the sorocarp development (24 hours). This is in accord with the analysis of trehalose accumulation in the late stages (Rutherford and Jefferson, 1976). Surprisingly, trehalose was also detected in stalk cells, forming an apical–basal gradient similar to that found with glycogen synthase. The significance of this finding is not clear, since the stalk cells finally degenerate and do not take part in germination. Possibly the stalk trehalose reflects to some degree a diffusion of the sugar from the spores.

C. Galactosyltransferase in the Late Period of Development

The terminal stages of the *Dictyostelium* life cycle are the culmination and the formation of the sorocarp. As already reported, an insoluble type of glycogen is one of the carbohydrate products of the terminal stages. Cellulose, responsible for the rigidity of the stalk is the second,* and an acidic glycosaminoglycan, composed of the sugars N-acetylgalactosamine, galactose, and galacturonic acid, is the third glycoconjugate of interest (White and Sussman, 1963a,b). The enzyme responsible for the transfer of galactose to the polymer was the first glycosyltransferase described in *Dictyostelium* having a clear developmental regulation (Sussman and Osborn, 1964).

The enzyme is absent from growing cells and is induced after aggregation between 14 and 22 hours. The maximum peak is attained at the last stages of fruiting body construction. Afterward the enzyme becomes soluble and is eventually released into the extracellular space (Sussman and Lovgren, 1965). Inhibition of protein synthesis by cycloheximide prevents the synthesis of the enzyme from taking place (Sussman, 1965). Also, treatment with actinomycin abolishes the accumulation of the galactosyltransferase (Roth *et al.*, 1968; Sussman, 1967), supporting the suggestion that the process of enzyme stimulation is regulated at the level of transcription. This was confirmed for several other enzymes by Firtel *et al.* (1973) with the limitation that actinomycin D alone is not able to block the mRNA synthesis in *Dictyostelium* completely but only in combination with daunomycin. The aggregation of the cells is an essential factor for the development of the galactosyltransferase. This was shown by work with various mutants (Sussman, 1967) and by dissociation/reassociation experiments. After the dissociation of aggregated cells, the information for the developmental course remains conserved. Reaggregated cells reach the state of differentiation that they had before disaggregation in a much shorter time than do cells induced to differentiate for the first time. The accumulation of galactosyltransferase is then carried out in the normal way (Newell *et al.*, 1972). If the dissociated cells are prevented from aggregating, then the galactosyltransferase activity is destroyed (Takeuchi *et al.*, 1978). During the period of galactosyltransferase destruction, cAMP-phosphodiesterase accumulates dramatically. By treatment of the dissociated cells with cAMP, the phosphodiesterase activity can be completely abolished, and the breakdown of galactosyltransferase is partially prevented.

These experiments accentuate the significance of intercellular contacts and signals for the verification of the development of slime molds at the level of enzyme activities.

REFERENCES

Beug, H., Gerisch, G., Kempff, S., Riedel, V., and Cremer, G. (1970). *Exp. Cell Res.* **63**, 147–158.

*The biochemistry of cellulose synthase is not discussed in this review since there are too few data available.

Bonner, J. T. (1967). "The Cellular Slime Molds." Princeton Univ. Press, Princeton, New Jersey.

Cappuccinelli, P., and Ashworth, J. M., eds. (1977). "Development and Differentiation in the Cellular Slime Molds." Elsevier/North-Holland Biomedical Press, Amsterdam.

Clegg, J. S., and Filosa, M. F. (1961). *Nature (London)* **192,** 1077–1078.

Cotter, D. A., and Raper, K. B. (1970). *Dev. Biol.* **22,** 112–128.

Crean, E. V., and Rossomando, E. F. (1977). *Biochim. Biophys. Acta* **498,** 439–441.

Crean, E. V., and Rossomando, E. F. (1979). *Biochem. Biophys.* **196,** 186–191.

Crean, E. V., Lagerstedt, J. P., and Rossomando, E. F. (1979). *J. Bacteriol.* **140,** 188–196.

Firtel, R. A., and Bonner, J. (1972). *Dev. Biol.* **29,** 85–103.

Firtel, R. A., Baxter, L., and Lodish, H. F. (1973). *J. Mol. Biol.* **79,** 315–327.

Gerisch, G. (1968). *Curr. Top. Dev. Biol.* **3,** 159–197.

Gerisch, G., and Guggenheim, R. (1980). *Prog. Brain Res.* **51,** 3–15.

Gerisch, G., Krelle, H., Bozzaro, S., Eitle, E., and Guggenheim, R. (1980). *In* "Cell Adhesion and Motility" (A. S. G. Curtis and J. D. Pitts, eds.), pp. 293–307. Cambridge Univ. Press, London and New York.

Hames, B. D., Weeks, G., and Ashworth, J. M. (1972). *Biochem. J.* **126,** 627–633.

Harris, J. F., and Rutherforth, C. L. (1976). *J. Bacteriol.* **127,** 84–90.

Higashi, Y., Strominger, J. L., and Sweeley, C. C. (1967). *Proc. Natl. Acad. Sci. U.S.A.* **57,** 1878–1884.

Jones, T. D. H., and Wright, B. E. (1970). *J. Bacteriol.* **104,** 754–761.

Killick, K. A., and Wright, B. E. (1974). *Annu. Rev. Microbiol.* **28,** 139–166.

Killick, K. A., and Wright, B. E. (1975). *Arch. Biochem. Biophys.* **170,** 634–643.

Loomis, W. F. (1975). "Dictyostelium Discoideum: A Developmental System." Academic Press, New York.

Müller, K., and Gerisch, G. (1978). *Nature (London)* **274,** 445–449.

Müller, K., Gerisch, G., Fromme, I., Mayer, H., and Tsugita, A. (1979). *Eur. J. Biochem.* **99,** 419–426.

Newell, P. C. (1977). *Endeavour* [N. S.] **1,** 63–68.

Newell, P. C., Franke, J., and Sussman, M. (1972). *J. Mol. Biol.* **63,** 373–382.

Parodi, A. J., and Leloir, L. F. (1979). *Biochim. Biophys. Acta* **559,** 1–37.

Riedel, V., and Gerisch, G. (1969). *Wilhelm Roux' Arch. Entwicklungsmech. Org.* **162,** 268–285.

Risse, H. J., Rogge, H., Rath, M., and Platzek, F. (1974). *In* "Biomembranes: Architecture, Biogenesis, Bioenergetics, and Differentiation" (L. Packer, ed.), pp. 371–379. Academic Press, New York.

Risse, H. J., Rössler, H. H., and Malati, T. (1979). *Leather Sci. (Madras)* **26,** 91–97.

Rogge, H., and Risse, H. J. (1974). *Hoppe-Seyler's Z. Physiol. Chem.* **355,** 1467–1470.

Rogge, H., Neises, M., Passow, H., Grunz, H., and Risse, H. J. (1975). *In* "New Approaches to the Evaluation of Abnormal Embryonic Development (H. J. Merker and D. Neubert, eds.), pp. 772–791. Thieme, Stuttgart.

Rogge, H., Neises, M., and Risse, H. J. (1977). *Biochim. Biophys. Acta* **499,** 273–277.

Rössler, H. H., Peuckert, W., Risse, H. J., and Eibl, H. J. (1978). *Mol. Cell. Biochem.* **20,** 3–15.

Rössler, H. H., Schneider-Seelbach, E., Malati, T., and Risse, H. J. (1980). *Mol. Cell. Biochem.* **34,** 65–72.

Roth, R., Ashworth, J. M., and Sussman, M. (1968). *Proc. Natl. Acad. Sci. U.S.A.* **59,** 1235–1242.

Rutherford, C. L., and Jefferson, B. L. (1976). *Dev. Biol.* **52,** 52–60.

Schaffrath, D., Stuhlsatz, H. W., and Greiling, H. (1976). *Hoppe-Seyler's Z. Physiol. Chem.* **357,** 499–508.

Shur, B. D., and Roth, S. (1975). *Biochim. Biophys. Acta* **415,** 473–512.

Sievers, S., Risse, H. J., and Sekeri-Pataryas, K. H. (1978). *Mol. Cell. Biochem.* **20,** 103–110.

Stein, G. S., Roberts, R. M., Davis, J. L., Head, W. J., Stein, J. L., Thrall, C. L., VanVeen, J., and Welch, D. W. (1975). *Nature (London)* **258,** 639-641.

Sussman, M. (1965). *Biochem. Biophys. Res. Commun.* **18,** 763-767.

Sussman, M. (1967). *Fed. Proc., Fed. Am. Soc. Exp. Biol.* **26,** 77-83.

Sussman, M., and Lovgren, N. (1965). *Exp. Cell Res.* **38,** 97-105.

Sussman, M., and Osborn, M. J. (1964). *Proc. Natl. Acad. Sci. U.S.A.* **52,** 81-87.

Takeuchi, I., Okamoto, K., Tasaka, M., and Takemoto, S. (1978). *Bot. Mag., Spec. Issue* **1,** 47-60.

Waechter, C. J., and Lennarz, W. J. (1976). *Annu. Rev. Biochem.* **45,** 91-112.

Weeks, G., and Ashworth, J. M. (1972). *Biochem. J.* **126,** 617-626.

White, G., and Sussman, M. (1963a). *Biochim. Biophys. Acta* **74,** 173-178.

White, G., and Sussman, M. (1963b). *Biochim. Biophys. Acta* **74,** 179-188.

Wilhelms, O. H., Lüderitz, O., Westphal, O., and Gerisch, G. (1974). *Eur. J. Biochem.* **48,** 89-101.

Wright, B., and Dahlberg, D. (1967). *Biochemistry* **6,** 2074-2079.

Wright, B., Ward, C., and Dahlberg, D. (1966). *Biochem. Biophys. Res. Commun.* **22,** 352-356.

Wright, B., Ward, C., and Dahlberg, D. (1968). *Arch. Biochem. Biophys.* **124,** 380-385.

SECTION 2

Cell Surface Glycosyltransferase Activities during Fertilization and Early Embryogenesis

BARRY D. SHUR

THE GLYCOCONJUGATES, VOL. III

I. INTRODUCTION

Cell-surface and extracellular components are required for a variety of cellular interactions, including those that occur during fertilization, morphogenesis, and immune recognition. Complex glycoconjugates are likely to participate in many of these cellular interactions since, for example, the synthesis of specific glycosaminoglycans coincides with a number of cellular migrations (Toole, 1976; Van Roelen et al., 1980; Pratt et al., 1975) and tissue interactions (Kosher and Lash, 1975; Bernfield and Banerjee, 1972). Also, a variety of intercellular adhesions are dependent upon the presence of specific glycoside residues on adhering cell surfaces (see Roth et al., 1971; Lloyd and Cook, 1974; Marchase, 1977, for examples). Cells are thought to recognize these complex carbohydrates, either on adjacent cells or in the extracellular space, by displaying specific carbohydrate-binding proteins, or lectins, on their surfaces (Barondes and Rosen, 1976). In fact, lectin-like proteins have been isolated from a variety of tissues (Rosen et al., 1979), where they and their complementary carbohydrates may play some role in cellular interactions.

Over ten years ago, Roseman (1970) proposed that one class of these cell-surface, lectin-like proteins are the glycosyltransferases. These enzymes synthesize all the known complex carbohydrates by catalyzing the transfer of saccharides from sugar donors to growing polysaccharide chains. On the cell surface, glycosyltransferases could participate in cellular interactions by binding their specific carbohydrate substrates on adjacent cell surfaces or in the extracellular matrix. In so doing, surface glycosyltransferases could function during cellular adhesions (Roth, 1973) and cellular migrations (Shur, 1977a). Models for glycosyltransferase involvement in cellular interactions require that the enzymes be capable of binding their glycosyl acceptors in the absence of sugar donors. This has been shown to be true for at least some glycosyltransferases (Rearick et al., 1979). In the presence of the sugar donor, catalysis would separate this enzyme–substrate complex, thereby dissociating the cells from each other or from their matrix. Either the surface glycosyltransferases or their cell-bound glycosyl acceptors could be coupled to a "second messenger" (e.g., cAMP, Ca^{2+}), which could then influence transcriptional activity within the cell.

Assays for cell-surface glycosyltransferases are hampered by a variety of competing enzyme reactions, which must be carefully controlled for in each system. These controls, which are particular to assays of surface glycosyltransferase activity, are discussed in Section III. As a rule, when these obligatory controls have been performed, glycosyltransferases are detectable at the cell surface. Their presence on the surface does not mean, necessarily, that glycosyltransferases actively participate in cellular interactions. Therefore, systems are needed that can distinguish between an active and a passive role for surface glycosyltransferases in such interactions. Recent studies on surface glycosyltransferase involvement in fertilization and early embryogenesis represent efforts in this direction. It is the purpose of this section to focus on these studies, as well as on the role these surface enzymes may play during development. Surface transferase changes that accompany neoplastic transformation are not discussed, since they have been extensively reviewed elsewhere (Shur and Roth, 1975; Pierce *et al.*, 1980b).

Obviously, there are many fruitful approaches for analyzing the molecular basis of cell interactions. Collectively, these studies have identified a host of surface and extracellular components, which are likely to be important in morphogenesis. Many recent review articles discuss this subject (Frazier and Glaser, 1979; Culp *et al.*, 1979; Yamada and Olden, 1978; Hynes, 1979; Kraemer, 1979; Lilien *et al.*, 1979; see also Chapter 3 of this volume and Chapter 1, Volume IV). Each type of cellular interaction probably results from more than one underlying molecular mechanism. For example, the adhesion of many cells, including sponge (Jumblatt *et al.*, 1980) and cultured cells (Urushihara and Takeichi, 1980), is dependent on two distinct mechanisms, which can be distinguished by their Ca^{2+} requirements. The point is that surface glycosyltransferases are probably just one of many molecular species that enable cells to interact with their environment.

II. THE GLYCOSYLTRANSFERASES

A. Distribution

Glycosyltransferases are the enzymes that catalyze the synthesis of all the known complex polysaccharides, including glycoproteins, glycolipids, and glycosaminoglycans. Most glycosyltransferases are membrane bound (see Schachter and Roseman, 1980 and Chapter 1, this volume, for review) and have been found in nuclear (Richard *et al.*, 1975), mitochondrial (Bosmann, 1971b), endoplasmic reticulum, Golgi, and plasma membranes. Among these, the Golgi membranes usually demonstrate the highest specific activity. Some glycosyltransferase activities are found in soluble form in a variety of body fluids, including colostrum (McGuire *et al.*, 1965), milk (Magee *et al.*, 1974), serum (Schachter *et al.*, 1973), cerebrospinal fluid (Den *et al.*, 1970), vitreous humor (Den *et al.*, 1970), and amniotic fluid (Den *et al.*, 1970). The function of these soluble enzymes is unknown, with the exception of the milk enzyme, which synthesizes lactose.

B. Sugar Donors

Glycosyltransferases catalyze the addition of specific monosaccharide residues from sugar nucleotide donors (e.g., UDP-galactose) to the nonreducing terminus of growing polysaccharide chains (Fig. 1). Usually they require divalent cations

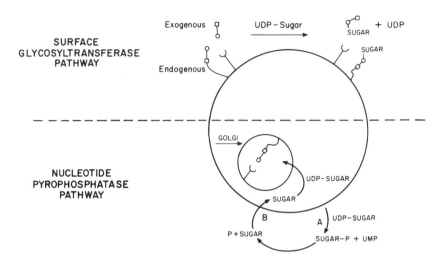

Figure 1 Two alternative pathways for sugar nucleotide metabolism. In the upper panel, surface glycosyltransferases transfer the sugar from the nucleotide donor (UDP-sugar) to either endogenous or exogenous disaccharide acceptors. This reaction produces trisaccharides and UDP. The availability of sugar nucleotide is dependent on the extent of sugar nucleotide hydrolysis, depicted in the lower panel. Here, nucleotide pyrophosphatase (reaction A) degrades the sugar nucleotide to sugar phosphate and UMP. Phosphatase activity then releases the free sugar, which can be transported into the cell (reaction B), be resynthesized into sugar nucleotide, and hypothetically participate in Golgi and endoplasmic reticulum glycosyltransferase reactions. Intracellular utilization of sugar nucleotide degradation products can be minimized by inhibiting reactions A and/or B, as described in the text.

for optimal activity. Each glycosyltransferase is named according to its sugar nucleotide substrate, so that there exist families of galactosyltransferases, glucosyltransferases, sialyltransferases, etc. It was long thought that the sole sugar donors for these enzymes were the sugar nucleotides. However, sugar nucleotides are not readily transported through plasma membranes, nor possibly any other membrane. It was unclear, therefore, how the cytosolic sugar nucleotides became accessible to the glycosyltransferases, which are situated outside of the cytoplasm. This dilemma was resolved by the characterization of lipid-soluble isoprenol–phosphate–sugar intermediates that shuttle mono- and oligosaccharide residues from sugar nucleotides to Golgi and endoplasmic reticulum glycoprotein glycosyltransferases (see Parodi and Leloir, 1979, for review). There is reason to believe that similar lipid–sugar intermediates function in the plasma membrane, transporting sugars from the cytoplasm to surface-associated glycosyltransferases (see Pierce *et al.*, 1980b, for review). However, it is not yet clear if lipid intermediates take part in all glycoprotein syntheses, or if they are involved in glycosaminoglycan metabolism other than keratan sulfate production (Hart and Lennarz, 1978). The role of lipid intermediates in glycosylation is discussed in detail in Chapter 1 of this volume.

C. Sugar Acceptors

Generally, the glycosyltransferase specificity for the sugar acceptor substrate is not nearly as stringent as it is for the sugar donor. For example, some purified transferases cannot distinguish between glycoprotein and glycolipid cores since they transfer their sugars to either substrate (Schwyzer and Hill, 1977; Rearick *et al.*, 1979; Beyer and Hill, 1980). On the other hand, within each family of glycosyltransferases (e.g., galactosyl) are enzymes that may show some specificity for the type of glycoconjugate being synthesized, such as the glycoprotein galactosyltransferases and the glycolipid galactosyltransferases (Paulson *et al.*, 1977). Even in these instances, however, the degree of specificity varies. For example, glycoprotein galactosyltransferase transfers glactose either to terminal *N*-acetylglucosamine residues in a variety of intact glycoproteins or to free *N*-acetylglucosamine. Here, binding to the intact glycoprotein is stronger than it is to the free monosaccharide, suggesting that the enzyme recognizes portions of the molecule in addition to the terminal sugar (Schanbacher and Ebner, 1970).

Finally, it is now known that the sequence of monosaccharides within the oligosaccharide results, in part, from the availability of particular glycosyltransferases (Beyer *et al.*, 1979). For example, some sialyltransferases and fucosyltransferases can compete for the same substrate. This may also be true for an *N*-acetylglucosaminyltransferase and sialyltransferase, which may compete for galactosyl residues in growing keratan oligosaccharides (Choi and Meyer, 1975). The reader is referred to Chapter 1 of this volume for a more complete discussion of glycosyltransferase specificities.

D. Substrate Modifier Proteins

The sugar acceptor specificity for one glycosyltransferase has been shown to be altered by the presence of a substrate modifier protein. Specifically, milk lactose synthetase is associated with two proteins (Brew *et al.*, 1968). One is an *N*-acetylglucosamine galactosyltransferase, which is apparently identical to other glycoprotein *N*-acetylglucosamine galactosyltransferases from a variety of sources. The second protein is α-lactalbumin, which inhibits the transfer of galactose to *N*-acetylglucosamine and stimulates the transfer to glucose, a substrate of normally low affinity for the galactosyltransferase. In this way, lactose (i.e., galactose-glucose) is synthesized by an enzyme normally accustomed to synthesizing *N*-acetyllactosamine. Although α-lactalbumin is the only known example of a glycosyltransferase substrate modifier protein, it is conceivable that they occur elsewhere, dictating the formation of specific enzyme–substrate complexes.

From the brief introduction above, it should be clear that the final composition of complex carbohydrates is dependent on many factors, including enzyme availability, acceptor specificity, and possibly substrate modifier proteins. Since approximately ten monosaccharides and three major categories of complex carbohydrates are commonly encountered in eukaryotic cells, the total number of glycosyltransferases is most likely very large, and the carbohydrate heterogeneity that they can produce must be enormous.

E. Glycosyltransferase Assays

Most glycosyltransferase assays are simple and are based on the isolation of radiolabeled glycosylated product from unused, low molecular weight sugar nucleotide and its degradation products. In quantitative assays, this can be accomplished by (a) high-voltage borate electrophoresis (Roth *et al.*, 1971), during which low molecular weight sugar compounds migrate away from the glycosylated product, (b) acid precipitation of glycoprotein products, (c) chloroform-methanol extraction of glycolipid products, (d) ion-exchange and molecular sieve chromatography, or (e) repeated washing of an immobilized substrate (Turley and Roth, 1979). It is possible to localize radiolabeled glycosyltransferase reaction products endogenous to the cells and tissues by autoradiography (Shur, 1977a). Here the unused low molecular weight compounds are simply washed away from the tissue after fixation. Since autoradiography localizes the glycosylated reaction product, not the enzyme itself, appropriate controls (see Section III) must be included to demonstrate enzyme localization as well. It may be possible to radiolabel the enzyme directly with irreversibly bound substrate analogs. For example, periodate-oxidized sugar nucleotides form Schiff bases within the sugar nucleotide-binding site of the enzyme (Powell and Brew, 1976), which can then be reduced with borohydride. Finally, an assay has been developed to identify N-acetylglucosamine galactosyltransferase within nondenaturing polyacrylamide gels and presumably tissues (Pierce *et al.*, 1980a). By this procedure, the production of UDP from UDP-galactose is enzymatically coupled to the formation of fluorescent NADH.

III. NECESSARY CONTROLS FOR ASSAY OF SURFACE-ASSOCIATED GLYCOSYLTRANSFERASES

A. Competing Sugar Nucleotide Pyrophosphatase and Glycosidase Activities

With two exceptions (Yogeeswaran *et al.*, 1974; Turley and Roth, 1979), all reports documenting surface-associated glycosyltransferase activities have relied on the addition of radiolabeled sugar nucleotides to the assay. Since virtually all cells examined have nucleotide pyrophosphatase and sugar phosphatase activities (Deppert and Walter, 1976), the successive actions of these enzymes on sugar nucleotides yield nucleotide monophosphates, sugar phosphates, and free sugars (see Fig. 1). Although there is no evidence for sugar nucleotide transport across the plasma membrane, free sugars gain easy access to the cell interior via the hexose transport systems. In some systems, the cells are incapable of using these specific monosaccharides resulting from sugar nucleotide hydrolysis (Shur, 1977a,b; Shur and Bennett, 1979; Durr *et al.*, 1977). In these instances, intracellular glycosyltransferases probably do not contribute to the final glycosylated product. However, most cells do incorporate free sugars into high molecular weight products. Therefore, cell-surface glycosyltransferase assays using radiolabeled sugar nucleotides must control for the potential intracellular utilization of free labeled sugars resulting from sugar nucleotide hydrolysis. To do this, one can assay surface transferase activities in the presence of either (a) inhibitors of sugar nucleotide hydrolysis (reaction A, Fig. 1) or (b) inhibitors of free labeled sugar utilization (reaction B, Fig. 1).

In the first instance, sugar nucleotide degradation can be competitively inhibited by the addition of vast molar excesses of some other unlabeled sugar nucleotide or nucleotide (Shur, 1977b; Patt *et al.*, 1976). Since both glycosyltransferases and nucleotide pyrophosphatases act on similar substrates, one must be sure to avoid inhibiting the glycosyltransferase as well. Alternatively, reagents are available that selectively inhibit free labeled sugar utilization (reaction B, Fig. 1). One can either inhibit the transport of free sugar into the cell by phloretin, phloridzin, or cytochalasin B (Shur, 1977b), or dilute out the specific activity of the free radiolabeled sugar with an unlabeled excess of the identical sugar (Roth *et al.*, 1971; Shur, 1977b; Cervén, 1977). However, an increase in the molar concentration of sugar may elevate the rate of sugar transport into the cell, offsetting any dilution in specific activity.

As in any enzymological study, cell-surface glycosyltransferase assays must employ saturating quantities of all enzyme substrates (i.e., sugar nucleotide, sugar acceptor, cation). This is particularly relevant when considering competing nucleotide pyrophosphatase activities, since the percentage of sugar nucleotide degradation decreases with increasing sugar nucleotide concentrations.

Just as there exist phosphatases which are active on the sugar nucleotide substrates, glycosidases may exist which degrade either the polysaccharide sugar acceptor or the glycosylated reaction product. The consequences of glycosidase action, if any, in surface glycosyltransferase assays should be determined, especially when one is comparing activities among different cell types. If necessary, glycosidases can be readily inhibited by the appropriate lactone (Levvy and Conchie, 1966).

B. Leakage of Intracellular Glycosyltransferases

The second main consideration when assaying for surface glycosyltransferases is to control for the possible secretion, or "leakage," of intracellular enzymes into the incubation medium. This soluble activity can be monitored by assaying incubation supernatants collected from cells that have been incubated under standard conditions but without sugar nucleotide. Any soluble, "leaked" transferase activity can then be assayed, free of the cell's enzymes, in the presence of labeled sugar nucleotide and a sugar acceptor source.

Almost all investigators have controlled for the presence of "leaked" transferases acting on cell-surface acceptors during surface-localized transferase assays (see Shur and Roth, 1975; Pierce *et al.*, 1980b). Never have significant levels of the surface-associated activity been attributed to soluble enzymes.

Lysed, or dead, cells may expose intracellular glycosyltransferases to the incubation medium. Therefore, one must carefully monitor the viability of the cell population by dye exclusion tests, rates of DNA synthesis, leakage of cytosolic enzymes, or other standard techniques. In some systems, a small background level of dead cells may be unavoidable. Any contribution that their intracellular glycosyltransferases may make to the final radiolabeled product can be determined if the activity present in suspensions of totally lysed cells is known. When this has been analyzed, the residual level of dead cells fails to contribute significant levels of intracellular activity (Struck and Lennarz, 1976; Hoflack *et al.*, 1979). In one instance (Hoflack *et al.*, 1979), lysed cells were shown to release unlabeled sugar nucleotide, which markedly diluted the specific activity of the radioisotope in the assay.

In order to circumvent some of the problems discussed above, workers have turned to a variety of other techniques. For example, isolated plasma membrane (Weiser *et al.*, 1978; Merritt *et al.*, 1977; Cummings *et al.*, 1979) and cell coat fractions (Graham *et al.*, 1978) both demonstrate glycosyltransferase activities. Two studies assayed surface transferase activity by utilizing the cell's endogenous pool of sugar nucleotide (Yogeeswaran *et al.*, 1974; Turley and Roth, 1979), avoiding the problems inherent in the use of exogenously added labeled sugar nucleotides. Other studies have involved the use of (a) large molecular weight sugar acceptors that do not enter cells (Verbert *et al.*, 1976), (b) enzyme substrate modifiers that also fail to enter cells (LaMont *et al.*, 1977), and (c) conditions for affinity labeling of surface glycosyltransferases (Cummings *et al.*, 1979).

Not all investigators have taken into consideration the consequences of competing enzyme reactions discussed above. However, those controls that have been reported for each analysis are briefly mentioned in the appropriate section.

IV. CELL SURFACE GLYCOSYLTRANSFERASES DURING FERTILIZATION

It has been nearly 70 years since Lillie first suggested that gametes adhere to each other via complementary molecules on adjacent cell surfaces (Lillie, 1913). Lillie's receptor–ligand hypothesis has been a central theme throughout studies on fertilization ranging from yeast (Burke *et al.*, 1980) to mammals (Bleil and Wassarman, 1980). Many observations support this underlying premise for gamete recognition, including the characterization of egg receptors from sea urchin (Vacquier and Moy, 1977) and porcine sperm (Peterson *et al.*, 1980) and of sperm ligands from mouse (Bleil and Wassarman, 1980) and sea urchin (Glabe and Vacquier, 1978) eggs. The isolated egg receptor from sea urchin sperm behaves as a lectin toward specific egg coat carbohydrates (Moy and Vacquier, 1979; Glabe and Lennarz, 1979), suggesting an involvement of complementary proteins and carbohydrates in fertilization. These egg receptors on sperm may actually be glycosyltransferases that function during fertilization by binding their appropriate glycosyl substrates in the overlying egg coat (Roth, 1973; Durr *et al.*, 1977). This is an attractive hypothesis because it is consistent with many different aspects of gamete recognition within a wide range of species. For example, in some species, gamete membranes separate after mating, suggestive of a catalytic means for membrane dissociation (Snell and Roseman, 1979). Also, the sea urchin sperm receptor for eggs is housed in the acrosome, which develops from the glycosyltransferase-laden Golgi complex. Finally, most higher eukaryote sperm initially must recognize and adhere to the outer acellular coat enveloping the egg, which is abundant in complex carbohydrates (SeGall and Lennarz, 1979).

To date, cell surface glycosyltransferases have been examined during two mating reactions, that between + and − alga gametes and that between mouse sperm and eggs. They will be discussed individually.

A. Algal Mating

In the green alga *Chlamydomonas*, species-specific gamete recognition first occurs between + and − gamete flagellar membranes. This eventually leads to cytoplasmic continuity between the cell bodies and subsequent zygote formation. The adhering flagella spontaneously separate, after which they are no longer adhesive to one another. Ideally, molecular models for *Chlamydomonas* mating should explain this flagellar de-adhesion. These nonadhesive membranes serve as an ideal "sexually incompetent" control, which some workers have used to test their working hypotheses.

Six individual glycosyltransferase activities toward endogenous acceptors have been identified on the surfaces of + and − gametes (McLean and Bosmann, 1975). When these gametes are mixed, as occurs during mating, the resultant glycosyltransferase activity is two to three times greater than the expected intermediate value. This elevation in transferase activity suggests enzyme–substrate complementarity between gametes. When flagellar membrane vesicles are isolated from these cells, not only is their glycosyltransferase specific activity higher than that on intact cells, but mixing + and − vesicles results in an even greater enzymatic stimulation (four to eight times) over the intermediate values.

When similar assays are conducted (Bosmann and McLean, 1975) using sexually incompetent, vegetative cells, no glycosyltransferase stimulation is seen upon mixing + and − cells. Similarly, mixing sexually competent + and − gametes from two different sexually incompatible species of *Chlamydomonas* fails to elevate the transferase activity. These results show that glycosylation between gametes not only requires sexually competent cells, but also demonstrates the appropriate species specificity.

A few of the relevant controls have been performed to determine whether the transferase activity is surface-associated (Colombino *et al.*, 1978). For example, the CMP-sialic acid product radioactivity can be removed by mild trypsinization of intact cells. Also, a vast molar excess of free sialic acid fails to affect sialyltransferase stimulation when + and − gametes are mixed, thus eliminating intracellular utilization of free sialic acid. These and other observations speak collectively for a surface-localized transferase activity, which may be involved in gamete adhesion. Not until these enzymes and their acceptors have been purified, antibodies against them raised, and/or appropriate gamete mutations analyzed will we be better able to determine the role of surface transferases in algal mating.

B. Mouse Fertilization

Fertilization is more difficult to analyze in higher eukaryotes than in algae, not only because it often occurs internally, but also because the gametes are invested in a variety of extracellular coats and membranes. Despite these handicaps, investigators have been able to reconstruct the normal sequence of events leading to zygote formation. In marine invertebrates, the sperm-bound egg receptor is housed in a Golgi-derived acrosome granule, which ruptures and polymerizes into a filamentous protein that binds egg vitelline membrane residues (Moy and Vacquier, 1979). In mammals (see Gwatkin, 1976; Shapiro and Eddy, 1980, for reviews) sperm are covered with "coating" factors, which must be removed within the female in order for sperm to bind to the outer egg coat, or zona pellucida (Chang, 1957). This zona binding may stimulate the acrosome to rupture, exposing hydrolytic enzymes, which enable the sperm to propel through the zona pellucida. Recent studies (Saling and Storey, 1979) indicate that only sperm with intact acrosomes bind to the zona, implying that the zona pellucida receptor is normally exposed on the outer plasma membrane covering the intact acrosome. The possibility that exposed hydrolases in acrosome-reacted sperm have destroyed their egg receptors has not yet been eliminated. Once the sperm reaches the egg plasma membrane, the sperm postacrosomal and egg plasma membranes fuse, allowing the sperm nucleus to enter. This scheme implies that there are two distinct sperm receptor sites, one for the zona pellucida and the other for the egg plasma membrane. It is noteworthy that the species specificity of binding resides in the zona receptor, since zona-free eggs can be fertilized by heterologous sperm (Thadani, 1980).

The contribution that sperm motility makes to this process is not clear. Although amotile sperm are incapable of fertilizing eggs (Bennett, 1975), it is not known whether amotility affects the sperm's ability to reach the egg or affects the sperm's ability to propel through the zona pellucida. Sperm motility, which becomes apparent after "capacitation," requires the presence of specific sperm surface components and cAMP (Hoskins *et al.*, 1978). In this regard, a sperm cAMP-dependent protein kinase has been identified and is thought to function during sperm motility (Garbers *et al.*, 1973; Hoskins *et al.*, 1972).

The involvement of cell surface glycosyltransferases in mammalian fertilization is being explored (Durr *et al.*, 1977; Shur and Bennett, 1979). Extensively washed mature mouse sperm contain surface-associated sialyltransferase, galactosyltransferase, and *N*-acetylglucosaminyltransferases. These enzymes are either inactive, or show moderate activity, when assayed against endogenous sperm acceptors but are highly active when assayed against soluble exogenous acceptors, including zona pellucida digests. In one instance, the low glycosyl acceptor levels on sperm have been shown not to result from glycosidase degradation of the product. However, the effect of other glycosidases on glycosyl acceptors and on transferase products in general has not been adequately examined (Shur and Bennett, 1979).

Three criteria attest to the surface localization of the sperm glycosyltransferase activity. First, mature sperm do not possess any of the traditional sources of intracellular transferases, such as the endoplasmic reticulum or the Golgi complex. Their only remnants occur as the acrosomal membranes. However, the transferase activity probably is not contributed by ruptured acrosomes since only sperm with intact acrosomes are being assayed (Durr *et al.*, 1977). Second, the degree of sugar nucleotide hydrolysis that occurs during these assays is very low, producing negligible quantities of free sugar available for intrasperm transport. Third, and most significant, sperm are totally incapable of synthesizing carbohydrates from free sugar precursors, including sialic acid (Durr *et al.*, 1977) and galactose (Shur and Bennett, 1979). Therefore, the radiolabeled sugar nucleotides are the direct precursors of the glycosylated products, which are catalyzed by plasma membrane-associated enzymes.

That sperm have surface glycosyltransferase activities toward egg acceptors, but have only moderate levels of acceptors themselves, suggests a transferase involvement in egg binding since sperm have been designed specifically for egg fusion. Three observations are consistent with this possibility. First, the temperature optimum for the sperm sialyltransferase activity has been examined (Durr *et al.*, 1977), since sperm function is temperature sensitive. The sialyltransferase shows optimal activity at 35°, the internal temperature of the mouse testis, above which enzyme activity decreases. On the other hand, sperm glycosidase and phosphatase both show an increase in activity coincident with increasing temperature, up until 45°. Thus, temperature sensitivity is a characteristic, at least partly specific, of the sialyltransferase.

The second observation that supports sperm glycosyltransferase involvement in egg binding comes from an autoradiographic analysis of mouse egg sialyltransferase activity (Durr *et al.*, 1977). In this study, sialyltransferase–sialyl acceptor complexes were visualized by using radiolabeled CMP-sialic acid. The glycosylated product could then be identified by autoradiography. Assayed in the absence of sperm, eggs and/or cumulus cells show a moderate level of diffuse sialytransferase activity covering their surfaces. With sperm present, beside the basal level of activity seen before, localized patches of intense enzyme activity are also seen on the egg surface, which correspond to the sites of sperm penetration. These results imply that sperm sialyltransferases are capable of binding egg sialyl acceptors at the site of gamete recognition.

Finally, galactosyltransferase activity is much higher than normal on sperm that have a genetic predisposition for increased fertilizing capability (Shur and Bennett, 1979). Of nine sperm enzymes assayed, only one (i.e., galactosyltransferase) shows this elevation in activity coincident with elevated sperm transmission. These studies are discussed in detail later (Section VII,B) but are mentioned here as another observation that argues for glycosyltransferase involvement in fertilization.

How many different glycosyltransferases exist on the mammalian sperm surface, or whether any of them have specific spatial distributions within the plasma membrane, is not known. These surface enzymes may not be directly involved in gamete binding, but rather exist solely to synthesize complex carbohydrates that are subsequently required for cellular interactions. The low endogenous acceptor activity associated with washed sperm speaks against this possibility. These points can be clarified only through the use of isolated enzymes, their acceptors, specific inhibitors, competitive haptenes, and serological and genetic probes. By using some of these techniques, the egg receptor from sea urchin sperm (Vacquier and Moy, 1977) and the sperm ligand from mouse zona pellucida (Bleil and Wassarman, 1980) have been isolated. It would be interesting to determine whether they show glycosyltransferase or glycosyl acceptor activity, respectively.

V. CELL SURFACE GLYCOSYLTRANSFERASES DURING EMBRYOGENESIS

A. Chick Gastrulation

Some of the most intriguing cellular interactions known are those that occur during, and soon after, gastrulation. It is during this time that migratory and inductive events lay down the organ rudiments from what was initially a double layer of epithelium. The catalog of cellular and tissue interactions that occur during these first few crucial days of development is enormous, as must be the developmental heterogeneity within the embryo at any one time. Although this inherent heterogeneity limits the amount of information one can obtain from gastrulating embryos, it allows one to survey the molecular nature of numerous morphogenetic interactions at a single time. For this reason, one of the earliest studies on the surface localization and potential morphogenetic function of glycosyltransferases was an analysis of gastrulating chick embryos (Shur and Roth, 1973; Shur, 1977a,b). In this autoradiographic study, cell-surface transferase activities toward endogenous extracellular acceptors were visualized within the embryo by labeled sugar nucleotide incorporation.

Collectively, these assays show that embryonic chick cells display four different surface glycosyltransferase activities during gastrulation. Galactosyl-, N-acetylglucosaminyl-, sialyl-, and fucosyltransferases are present. Many controls attest to the surface localization of the activity. First, sugar nucleotide hydrolytic products (i.e., free sugar, sugar phosphate) contribute negligibly to the product radioactivity (Shur, 1977b). For example, total inhibition of the sugar nucleotide that does occur has no effect on the amount of product formed, nor do effective inhibitors of free sugar transport (see Fig. 1). Assays in which only the free sugar pools, resulting from sugar nucleotide hydrolysis, are labeled produce low levels of incorporation. Taken together, these results and others (Shur, 1977b) show that the sugar nucleotides, rather than their degradation products, are directly responsible for the transferase activities seen. Second, no soluble enzyme is found in the supernatant, and no intracellular transferase activities contributed by lysed cells can be detected. Finally, three different monosaccharides (i.e., glucose, glucuronic acid, N-acetylgalactosamine) produce very high activity when administered as the free sugar but are virtually inactive when given as the sugar nucleotide. This implies that the sugar nucleotides are excluded from the intracellular pathways, which are otherwise capable of synthesizing complex carbohydrates containing these monosaccharides. Even those sugar nucleotides that are highly active fail to label some of the tissues that are capable of incorporating the same free sugars.

More interesting than a simple demonstration of glycosyltransferase–acceptor complexes at the cell surface is the cellular and temporal specificity of the activities themselves (Shur, 1977a). Each of the four active glycosyltransferases shows a spatially and temporally characteristic distribution of activity. These distributions of surface glycosyltransferase activities can be classified into two broad categories. First, surface glycosyltransferases are highly active on migrating cells, such as primary mesenchyme, primordial germ cells, and neural crest cells. Second, enzyme activities are present at the interface of a number of "inductive" interactions, including medullary plate–presomite mesoderm, optic cup–skin ectoderm, and notochord–somite. When the four different surface glycosyltransferase activities are quantitated by high-voltage borate electrophoresis, galactosyltransferase is by far the most active of those transferases present on migrating cells and during inductive interactions. It was suggested

(Shur, 1977a) that surface transferases might be required for the migratory and inductive behavior of these tissues by binding their glycosyl acceptors in the extracellular matrix.

Embryonic cells most likely migrate on complex carbohydrates during development. For example, specific glycosaminoglycans are synthesized during the migration of cranial neural crest cells (Pratt *et al.*, 1975), a variety of mesenchymal cells (Toole, 1976), and tissue culture cells (Culp *et al.*, 1979). Similarly, extracellular collagen and/or glycosaminoglycans are specifically synthesized by, and can substitute for, a variety of inducing tissues (Kosher and Lash, 1975; Kosher and Church, 1975). It is conceivable that migrating cells and inducible tissues utilize their surface glycosyltransferases to recognize these complex carbohydrates in their environment. The presence of the required sugar nucleotide and/or lipid intermediate would catalyze the reaction allowing migration to advance or would signal an inductive message. If cells do utilize surface glycosyltransferases to migrate on glycosaminoglycans, then their surface transferase reaction products are likely to be these same glycosaminoglycans, which are glycosylated during migration. The surface glycosyltransferase reaction products have been characterized in early mouse embryo assays (Shur, 1981c) and have been shown to be large keratan-like extracellular glycoconjugates (see Section VII,E). Studies to be discussed below (Section V,B) have more directly examined surface transferase involvement in cellular migration.

To probe surface glycosyltransferase function in morphogenesis further, three independent studies examined the effects of sugar nucleotides on cell behavior. First (Shur, 1977b), high levels (5 mM) of UDP-galactose and UDP-N-acetylglucosamine are teratogenic to developing chick embryos because they interfere with the development of those migratory cell types that autoradiographically display the highest transferase activity. Identical concentrations of substrates for inactive transferases (i.e., UDP-glucuronic acid) or free sugars have little effect on subsequent development. Second, 10 mM UDP-galactose markedly inhibits the *in vitro* transformation of mouse morulae into blastocysts (Shur *et al.*, 1979). Identical levels of UDP-glucose and/or galactose have no effect. Third, much lower levels of sugar nucleotides (0.1 mM) affect the *in vitro* migratory behavior of embryonic cells and neurite axon projections (Karfunkel *et al.*, 1977). In each of these three studies, the inhibitory effect of exogenously added sugar nucleotides was thought to result from some perturbation of the normal transferase–acceptor relationship. However, there is no direct evidence for this.

B. Cell Migration *in Vitro*

An *in vitro* assay has been designed to test more directly surface glycosyltransferase involvement in cellular migration (Turley and Roth, 1979). In these studies, SV40-transformed mouse fibroblasts migrated away from explants most rapidly when cultured on a hyaluronate substrate, migrated less when cultured on chondroitin 6-sulfate, and failed to leave the explant site when cultured on polygalacturonic acid substrates. To assess glycosyltransferase involvement, these cells were prelabeled with free sugars from which endogenous sugar donors were synthesized. When applied to the glycosaminoglycan substrates, these prelabeled cells deposited sugars identical to the constituent monosaccharides of the underlying substrate, that is, glucuronic acid and N-acetylgalactosamine sulfate on chondroitin sulfate, and glucuronic acid and N-acetylglucosamine on hyaluronate. No radioactivity was deposited by these cells onto polygalacturonic acid substrates.

Two important results emerge from these studies (Turley and Roth, 1979). First, the monosaccharides deposited on these glycosaminoglycan substrates were covalently linked to the matrix, since the radioactivity could not be displaced by SDS–urea treatment. This result implies that extracellular glycosyltransferases catalyzed these sugar additions. In controls, no product radioactivity resulted when either culture supernatants or lysed prelabeled cells were added to the glycosaminoglycan substrates, eliminating any contribution by soluble or intracellular enzymes. Second, the degree of glycosylation was reciprocal to the degree of migration. That is, hyaluronate was glysosylated less than chondroitin sulfate, but it supported more extensive cell migration. These workers suggest (Turley and Roth, 1979) that the rate of migration may be inversely proportional to the number of oligosaccharide acceptors available for binding. As the density of enzyme–substrate complexes increases, so may cell adhesion to the substrate, diminishing the migration rate. It would be interesting to see if nontransformed, contact-inhibited fibroblasts also display this inverse relationship between glycosylation and migration. It would also be interesting to determine if this adhesion to the substrates, and migration upon them, can be perturbed by irreversible glycosyltransferase inhibitors, competitive haptenes, etc.

C. Limb Bud Chondrogenesis

When limb bud mesenchymal cells are removed from the influence of the overlying apical ectodermal ridge, either *in vivo* or *in vitro*, they undergo a cellular condensation phase prior to synthesizing cartilage (Thorogood and Hinchliffe, 1975; Kosher *et al.*, 1979a). That is, the mesenchymal cells, which were initially separated from one another by an extensive hyaluronate matrix, become closely packed together, possibly due to decreased hyaluronate accumulation (Kosher *et al.*, 1981). Since elevated cAMP levels cause precocious cartilage differentiation by these mesenchymal cells (Kosher *et al.*, 1979b), one function of cellular condensation may be to elevate cAMP levels.

Recent studies examined whether an interaction between cell surface glycosyltransferases and acceptors on adjacent cell surfaces occurs during the cellular condensation phase of *in vitro* limb bud chondrogenesis (Shur *et al.*, 1981). These studies were motivated by reports of contact-dependent glycosylation between normal mouse fibroblasts in culture (see Section VI,C) (Roth and White, 1972). Therefore, three surface glycosyltransferases were assayed on limb bud mesenchymal cells against their endogenous extracellular acceptors. Two, glucosyl- and *N*-acetylglucosaminyltransferases, are relatively inactive by both quantitative and autoradiographic techniques. On the other hand, surface galactosyltransferase activity is high and shows a twofold elevation in endogenous activity coincident with the onset of cellular condensation. At later stages, although the cells are still morphologically condensed, endogenous activity decreases. During subsequent cartilage synthesis, enzyme activity declines below its initial precondensation level. Control experiments have established that the observed galactosyltransferase activity is due to surface-associated enzymes. Specifically, intracellular utilization of sugar nucleotide degradation products cannot account for the activity, nor can soluble enzyme be found in the supernatant.

On the basis of these observations, it was suggested (Shur *et al.*, 1981) that subridge mesenchymal cells possess surface galactosyltransferases that are capable of binding glycosyl acceptors on adjacent cell surfaces. However, because of the extensive extracellular matrix, cell–cell contact ·is minimal, surface transferases are unable to interact with their substrates, and activity toward endogenous acceptors is low. As the cells begin to condense, surface galactosyltransferases gain access to acceptors on adjacent cells, producing high transferase activity. Galactosylation of surface acceptors occurs *in situ,* decreasing the availability of acceptors and resulting in diminished levels of transferase activity. There are, of course, other possible explanations for the decrease in surface transferase activity associated with late condensation, including turnover of the enzymes, the acceptors, or both. In any case, this galactosyltransferase–acceptor binding and/or subsequent glycosylation may result in an elevation of cAMP, which triggers chondrogenesis.

To test this model, it would be worthwhile to observe the effects of specific glycosyltransferase inhibitors, etc., on limb bud cAMP content and chondrogenesis. It may be possible to initiate precocious chondrogenesis in sparse cell cultures by the addition of solubilized galactosyl acceptors. Analyses of limb bud development and of cell migration are appealing because they are amenable to *in vitro* perturbations. Both are better defined, simpler systems than intact gastrulating embryos and are ideal for more sophisticated tests of the involvement of surface transferases in cellular interactions.

VI. *IN VITRO* ASSAYS OF ADHESIVE RECOGNITION IN WHICH THE ROLE OF SURFACE GLYCOSYLTRANSFERASES HAS BEEN EXPLORED

In his pioneering studies (see Townes and Holtfreter, 1955), Holtfreter showed that cells dissociated from amphibian embryos and then reaggregated would redistribute within the aggregate to recreate their relative positions within the embryo. He suggested that cells were guided to their final positions, during both *in vitro* reaggregation and embryonic development, by specific cellular affinities. Holtfreter's work stimulated the development of a variety of assays by which selective cell affinities, or adhesions, could be quantitated. The underlying principle for all of these "adhesion" assays is that they reflect cell affinities required for morphogenesis (see Marchase *et al.*, 1976, for review). By using these assays, the molecular mechanisms underlying morphogenetically crucial cellular affinities are being studied. In this context, the role of surface glycosyltransferases in specific *in vitro* cellular adhesions is being explored.

A. Embryonic Neural Retina

The first report that implicated surface glycosyltransferases in cellular interactions came from an analysis of embryonic neural retina cells (Roth *et al.*, 1971). Intact neural retina cells show surface galactosyltransferase activity toward both endogenous and exogenous acceptors. Control experiments minimized any contribution to the final product by either (a) intracellular utilization of UDP-galactose hydrolytic products or (b) leakage of enzyme into the medium.

Known galactosyltransferase acceptors serve as competitive haptenes when added to neural retina cell adhesion assays, since they inhibit adhesion from 20 to 50%. As a control, oligosaccharides that do not act as galactosyl acceptors have no inhibitory effect on neural retina adhesion. Pretreatment of cells with β-galactosidase increases their adhesion rate, presumably by creating additional binding sites for surface galactosyltransferases.

In more recent studies (Porzig, 1978), surface galactosyltransferase activity toward exogenous N-acetylglucosamine was analyzed during neural retina development. Under optimal assay conditions, activity is highest on the surfaces of mitotically active marginal cells. Cells removed from increasingly older neural retinas show a continual decline in enzymatic activity. No differences in activity could be found among the four neural retina quadrants.

Collectively, these two studies (Roth *et al.*, 1971; Porzig, 1978) clearly show that neural retina cells contain surface galactosyltransferase activity toward endogenous and exogenous acceptors. Whether these enzymes participate in neural retina adhesion by binding glycosyl acceptors on adjacent cell surfaces has yet to be determined, although the ability of exogenous galactosyl acceptors to inhibit neural retina cell adhesion is consistent with this possibility.

Embryonic neural retina cells, like many other cells, secrete a class of molecules into the medium that promote cellular aggregation when added back to the culture (see Moscona, 1976, for review). One such neural retina aggregation-promoting factor is thought to require terminal N-acetylgalactosamine residues for activity (McDonough and Lilien, 1978; Rutz and Lilien, 1979). Interestingly, N-acetylgalactosaminyltransferase activity is present at the neural retina cell surface, where it is able to glycosylate this aggregation-promoting factor (Balsamo and Lilien, 1980). There is as yet no way of knowing whether (a) this surface N-acetylgalactosaminyltransferase functions simply to add the required N-acetylgalactosamine residue to the aggregation-promoting factor, which then binds to another lectin-like protein, or (b) the enzyme directly serves as the surface receptor for the aggregation factor. One noteworthy result of these studies is that colchicine prevents enzyme–substrate complexes from being mobilized to the cell surface from intracellular pools.

Other workers (Garfield *et al.*, 1974) assayed a similar, if not identical, retinal aggregation factor for galactosyltransferase and galactosyl acceptor activity. Within the limitations of this study, no enzyme or acceptor activity was detectable. This result implies that the neural retina factor demonstrates some acceptor specificity for particular glycosyltransferases.

B. Retinotectal Adhesions

Axonal projections from the chick neural retina innervate the optic tectum of the brain. The topographical orientation of the retinal dorsal–ventral axis is inverted onto the tectum, such that ventral retinal cells innervate the dorsal aspect of the tectum, and the dorsal retina projects to the ventral tectum. Experiments that beautifully demonstrate the morphogenetic relevance of *in vitro* adhesive specificities have shown that dissociated dorsal and ventral retina cells, respectively, adhere to ventral and dorsal tectal surfaces (Barbera, 1975). Protease and glycosidase treatments indicate that retinotectal adhesions may be specified by complementary interactions between proteins and carbohydrates (Barbera, 1975; Marchase, 1977). Furthermore, these proteins and carbohydrates are thought to be distributed in gradients of opposite polarity on both retina and tectum. The proteinaceous receptor would be concentrated on ventral retina and ventral tectum, whereas its glycosidic ligand would be concentrated on dorsal retina and tectum. Retinotectal adhesions would then result by maximizing the number of protein receptor–carbohydrate ligand complexes.

Results show (Marchase, 1977) that the dorsally located carbohydrate ligand requires terminal N-acetylgalactosamine residues for its activity. The hypothesis was made that the ganglioside GM2 may be this ligand since it possesses terminal

N-acetylgalactosamines and since GM2-containing lecithin vesicles preferentially adhere to receptor-rich ventral tecta. However, there is no detectable difference in GM2 levels isolated from dorsal versus ventral retinal sonicates.

Since studies cited above (Roth *et al.*, 1971) suggested that neural retina cell-surface galactosyltransferases might participate in retina cell adhesions, the galactosyltransferase activities in ventral versus dorsal retina were compared (Marchase, 1977). As found by other workers (Porzig, 1978), ventral and dorsal retinas possess similar levels of galactosyltransferase activity toward N-acetylglucosamine. On the other hand, galactosyltransferase activity toward the ganglioside GM2 is 30% greater in ventral retina than in dorsal. Also, there is a marked rise in retinal GM2 galactosyltransferase activity at the time when ventral retina adhesive specificity toward dorsal tectum becomes apparent. From these results, it was suggested (Roth and Marchase, 1976; Barbera, 1975) that the ventral retina proteinaceous receptor may be GM2 galactosyltransferase, which participates in retinotectal adhesions by binding its dorsally concentrated substrate, GM2. However, the inability to find a dorsal–ventral gradient of GM2 within the retina reduces the likelihood of GM2 involvement. Glycosyltransferases may be involved, however, since synaptosomal preparations from embryonic chick brain show high transferase activities (Den and Kaufman, 1968; Bosmann, 1973). At the tips of growth cones, these enzymes may partially mediate neuronal specificity.

C. Fibroblasts

Many reports have compared the surface glycosyltransferase activities of normal and transformed cells. Collectively, these studies have shown that no one change in surface activity is diagnostic for neoplasia. Surface transferase activity can increase, decrease, or show no change after transformation. The same surface activity may even require different incubation conditions on normal and transformed cells for optimal activity. This subject has been extensively reviewed elsewhere (Shur and Roth, 1975; Pierce *et al.*, 1980b) and will not be discussed here. However, in the context of this review, some reports have explored surface glycosyltransferase involvement in cultured cell adhesions and social behavior. These studies will be briefly mentioned.

One of the most provocative findings to come from surface transferase studies deals with their potential role in contact inhibition of growth (Roth and White, 1972). Upon contact with one another, normal cells cease dividing; malignant cells do not. A variety of normal cells, including 3T3 mouse fibroblasts, show decreasing levels of surface transferase activity toward endogenous acceptors coincident with increasing cell density; malignant cells, including 3T12 cells, do not. Assayed in sparse suspensions, 3T3 cells show less endogenous surface galactosyltransferase activity than when these cells are assayed in a pellet. Malignant 3T12 cells, as well as synchronized populations of mitotic 3T3 cells (Webb and Roth, 1974), show similar levels of activity in both suspension and pellet incubations. These and other results have been incorporated into a model for growth control (Cebula and Roth, 1976). Reduced to its simplest form, it suggests that normal 3T3 cells glycosylate one another upon contact (trans-glycosylation) but are unable to glycosylate themselves (cis-glycosylation) because of a segregation of enzymes and acceptors within a relatively amobile plasma membrane. The glycosylated cell responds to this signal by reducing its growth rate. Transformed cells (i.e., 3T12) and normal mitotic cells show high levels of cis binding as a result of increased plasma membrane fluidity. Defects in either glycosylation (e.g., in lipid intermediates) or in the ability to respond to glycosylation result in higher growth rates and possibly malignancy.

Not all normal cells show a decrease in endogenous surface acceptors with increasing cell contact, and not all investigators have been able to document contact-dependent glycosylation (Patt and Grimes, 1974). The failure to assay contact-dependent glycosylation may have resulted from an inability to minimize cell-cell contact effectively during the transferase incubation. On the other hand, studies from a number of laboratories (see Shur and Roth, 1975; Pierce *et al.*, 1980b) do show that surface glycosyltransferases can glycosylate acceptors on adjacent cell surfaces and may therefore participate in growth control.

The role of surface glycosyltransferases in cultured cell adhesions has also been examined. Neuraminidase pretreatment of many different cell types often accelerates their rate of reaggregation. With rat dermal fibroblasts (Lloyd and Cook, 1974), this neuraminidase-produced increased adhesivity seems to result from an exposure of sialyl acceptors, which can then act as binding sites for adjacent cell-surface sialyltransferases. By using lectins as probes, it can be shown that specific monosaccharide residues are exposed by neuraminidase treatment. Glycoproteins possessing these same terminal monosaccharides inhibit the neuraminidase-produced increase in adhesivity and serve as substrate for intact cell-surface sialyltransferases. Isolated plasma membranes from these cells also display this sialyltransferase activity toward the haptene inhibitors of aggregation. These results imply that surface sialyltransferase–sialyl acceptor complexes bridge adhering cells together after neuraminidase treatment. Neuraminidase pretreatment also stimulates L-1210 (Porter and Bernacki, 1975), hamster (Sasaki and Robbins, 1974), and chick embryo (Spataro *et al.*, 1975) surface sialyltransferases toward endogenous acceptors.

D. Physiological and Lower Eukaryotic Adhesive Interactions Correlated with Surface Glycosyltransferase Activities

Adhesive interactions during physiological regulation and in lower eukaryotes have also been correlated with surface glycosyltransferase activities. Even though these studies do not pertain directly to fertilization and early embryogenesis, their results are nevertheless germane to surface glycosyltransferase involvement in development. Therefore, they will be briefly mentioned. The reader is referred to earlier reviews for a more detailed discussion (Pierce *et al.*, 1980b; Shur and Roth, 1975).

1. Immune Recognition

In the mature animal, the most intricate examples of cellular interactions presently known are those that characterize the immune system. A number of these require intimate cell contact. Some may involve specific receptor–ligand complementarity. For example, foreign antigens must complex with appropriate surface receptors during cytotoxic T-lymphocyte generation. Others may involve lectin-like interactions, and in this regard it has been shown that lymphocytes possess numerous surface glycosyltransferases (Hoflack *et al.*, 1979; LaMont *et al.*, 1974; Painter and White, 1976; Patt *et al.*, 1976; Baker *et al.*, 1980). A mechanism has been proposed whereby these surface enzymes may specify and subsequently recognize self from nonself (Parish, 1977). There are no direct data yet to support this provocative idea. There are data, however, which suggest that cytotoxic T-lymphocyte surface glycosyltransferases may participate in T-cell cytolysis of their target cells (Kurt *et al.*, 1981). T-Cell surface galactosyltransferase activity increases as the ability of T cells to bind target cells increases. Additionally, the surface galactosyltransferase activity that results when T lymphocytes are mixed with their target cells is greater than the additive product of the individual cell type activities. These results suggest that cytotoxic T-lymphocyte surface glycosyltransferases recognize, bind, and glycosylate specific target cell acceptors.

2. Hemostasis

Surface glycosyltransferases have also been investigated during the adhesion of platelets to vascular endothelium. Wounding of the endothelium exposes collagen fibrils to which circulating platelets adhere, initiating clot formation. Platelets then release ADP, which stimulates subsequent platelet–platelet aggregation. A variety of results, including those obtained with purified membrane fractions, demonstrate that platelet surfaces contain glycosyltransferase activities. One, sialyltransferase, may participate in the ADP-initiated platelet–platelet aggregation (Bosmann, 1972). Another platelet enzyme, collagen glucosyltransferase, is thought to be involved in the initial binding of platelets to collagen fibrils (Barber and Jamieson, 1971; Jamieson *et al.*, 1971; Bosmann, 1971a). Both carbohydrate and peptide residues of collagen are required for platelet adhesion (Puett *et al.*, 1973), and, interestingly, the platelet glucosyltransferase recognizes both the terminal galactosyl residue of collagen and adjacent peptide sequences (Smith *et al.*, 1977). Approximately 5–10% of the total cell glucosyltransferase activity resides in the outer membrane. In more recent studies (Leunis *et al.*, 1980), the platelet collagen glucosyltransferase has been purified and kinetically compared with the same enzymatic activity found in plasma, granulocytes, and lymphocytes. Within the limits of this analysis, the enzymes seem identical. These results, the authors suggest (Leunis *et al.*, 1980), speak against a unique role for collagen glucosyltransferases in platelet–collagen interaction. The ubiquity of plasma enzyme may likely interfere with platelet enzyme binding to collagen residues. However, higher degrees of specificity may exist either within the enzymes themselves, or within the platelet, that dictate specific platelet glucosyltransferase–collagen binding despite the presence of plasma enzyme.

3. Intestinal Mucosa Development

The intestinal epithelium is continually being replenished by a pool of mitotically active, undifferentiated cells located in the base of the intestinal crypts. From here, cells migrate up the crypt wall and mature upon the villus, which projects into the lumen. By sequential EDTA dissociations, villus and crypt epithelial cells can be isolated free of cross-contamination (Weiser, 1973). Furthermore, by sucrose density gradient sedimentation, Golgi and surface membranes can be separated out of villus and crypt cell homogenates (Weiser *et al.*, 1978). Enzymatic markers, electron microscopy, and cell-surface antisera can be used to identify the various microvilli, lateral–basal, and Golgi membrane fractions. With these preparations, glycosyltransferase activities have been examined on internal and surface membranes. Contrary to previous results (Weiser, 1973), Golgi and surface membranes from both villus and crypt cells show high glycosyltransferase activities (Weiser *et al.*, 1978). These surface enzymes might participate either in adhesion to the underlying basal lamina or in the synthesis of brush border glycoconjugates. These studies are discussed in detail in Chapter 2, Section 3 of this volume and Chapter 4, Section 3, of Volume IV.

4. Slime Mold Aggregation

The development of the slime mold *Dictyostelium discoideum* has been used as a model system by which to understand the nature of, and necessity for, cellular interactions. Cellular aggregation during the development of this organism is thought to be mediated by specific carbohydrate-binding proteins, or lectins, adhering to complex oligosaccharides on adjacent cells (Barondes and Rosen, 1976). The molecular weights of these lectins are less than those expected for glycosyltransferases (Rosen *et al.*, 1979). Nevertheless, three glycosyltransferases (mannosyl, glucosyl, and *N*-acetylglucosaminyl) are present on the slime mold cell surface (Sievers *et al.*, 1978). Two of these activities are higher on aggregation-competent cells than on incompetent cells. Surface glycosyltransferases could participate in cellular adhesion by either (a) synthesizing the complex carbohydrate lectin receptors, as the authors suggest, or (b) serving as a cell-bound lectin-like receptor for glycosyl acceptors on adjacent cells. The reader will find a more complete discussion of these results in Chapter 2, Section 1, of this volume.

5. Sponge Aggregation-Promoting Factor

The first reported isolation of a soluble aggregation-promoting factor was from sponges (Moscona, 1963). In the species *Geodia cydonium*, the aggregation factor is associated with three extracellular glycosyltransferases: sialyl (Müller *et al.*, 1977), galactosyl, and glucuronyl (Müller *et al.*, 1979). On the cell surface are the receptors for the aggregation factor, along with two glycosidases: galactosidase and glucuronidase. It has been proposed (Müller *et al.*, 1979) that the aggregation factor recognizes and binds cell-surface glucuronic acid residues for its aggregation-promoting activity. The availability of surface glucuronic acid may be regulated by cell-surface glucuronidase and extracellular glucuronyl-transferase activities. It has not yet been shown, however, whether purified aggregation factor receptors actually serve as substrate for these two enzymes. These workers further speculated (Müller *et al.*, 1979) that surface galactosidase and extracellular galactosyltransferase may be involved in the ''sorting-out'' that occurs after single-cell aggregation.

VII. USE OF THE *T/t* COMPLEX MORPHOGENETIC MUTANTS TO TEST FURTHER THE INVOLVEMENT OF SURFACE GLYCOSYLTRANSFERASES IN CELLULAR INTERACTIONS

A. The *T/t* Complex

The *T/t* complex of the mouse is located near the centromeric end of chromosome 17 and contains over 100 known dominant (*T*) and recessive (*t*) mutations (see Bennett, 1975, for review). The effects of the dominant (*T*) alleles are generally confined to embryonic development and tail formation. On the other hand, the effects of the recessive (*t*) alleles are very plieotropic. Many of these recessive *t* alleles have been extracted from wild-mouse populations, in which they exist as natural polymorphisms at this locus. The effects of *t* alleles can most easily be understood by discussing the consequences of recombination within the *T/t* complex. Even though the presence of recessive *t* alleles suppresses recombination along a large segment of the chromosome, rare recombinational events do occur with a frequency of 1/500 to 1/1000. When crossing-over occurs (Fig. 2), *t* alleles usually are subdivided along the chromosome into two (Bennett *et al.*, 1976; Lyon and Mason, 1977), functional components. First is a proximal, centromeric segment, which contains the ''tail-interaction'' factor of the *T/t* complex. By themselves, *t* alleles (+/*t*) do not affect tail development, whereas *T/+* mice have short tails. However, *t* alleles interact with *T* alleles to

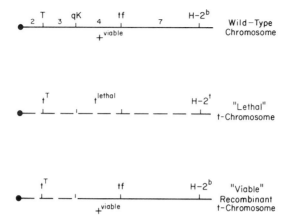

Figure 2 Three seventeenth chromosomal segments adjacent to the centromere: a wild-type chromosome with some of the known markers and the relative map distances between them, a "lethal" t chromosome, and a recombinant t chromosome resulting from crossing-over between the qk and tf markers. The recombinant chromosome has retained the tail-interaction favor (t^T) adjacent to the centromere but has lost the distal segments associated with embryonic lethality (t^{lethal}) and gametic segregation–distortion. Sperm galactosyltransferase activity is elevated only when the lethal/segregation–distorting portions of the T/t complex are present (Shur and Bennett, 1979; Shur, 1981a).

produce complete taillessness in T/t mice. A testicular protein of 63,000 daltons, pl6.9, that is possibly allelic to this tail-interaction factor has been identified by 2-D gel fluorography (Silver *et al.*, 1979). This protein not only is more acidic when synthesized by t-bearing cells, but is a direct transcriptional product of genes located in this segment of the T/t complex (Danska and Silver, 1980). Using this gene product as a probe, one can then begin to clone and isolate adjacent T/t complex genes.

The second, distal segment of the T/t complex produced by recombination is associated with embryonic lethality in t/t homozygotes and with segregation–distortion of t-bearing sperm. That is, males heterozygous ($+/t$) for homozygous-lethal t alleles transmit their t sperm much more frequently than their wild-type gamete. This segregation–distortion does not result from a deficiency of normal sperm or an overabundance of t sperm (Hammerberg and Klein, 1975). It does help explain why t alleles are maintained in wild populations despite t/t lethality. When this distal component has been lost by recombination, the resultant animals are viable when homozygous and transmit their t allele normally. However, they are still tailless when T/t. There is reason to believe that the mutant T/t complex has resulted from a substantial rearrangement of the interstitial heterochromatin in this portion of the chromosome (Lyon and Mason, 1977; Silver *et al.*, 1980).

The recessive t alleles, which cause embryonic lethality when homozygous, fall into six partially overlapping complementation groups, each one interfering with embryonic development at characteristic times and places (Bennett, 1975). The nature and mode of action of the biochemical lesion responsible for this mutant morphogenesis, and possibly segregation–distortion as well, are being actively investigated. At present, there are two major avenues of research.

First are studies that show a defect in intermediary metabolism in t mutant cells (Ginsberg and Hillman, 1975; Nadijcka and Hillman, 1975). Excessive intracellular lipid accumulation, possibly resulting from defects in ATP metabolism, distinguishes some t/t homozygotes from the wild type. These metabolic alterations may be responsible, in part, for the detrimental effects of some, but clearly not all, mutant t alleles (Wudl et al., 1977). Small intermediary metabolic defects are also seen in t-bearing sperm (Ginsberg and Hillman, 1974), which is consistent with the notion that intermediary metabolic defects are responsible for some t-dependent effects on embryonic development and fertilization. Differences in intermediary metabolism could result in differences in sperm motility, which then produce t-sperm segregation–distortion (Katz et al., 1980; Erickson et al., 1979). It is noteworthy that preliminary studies with in vitro fertilization of t sperm (McGrath and Hillman, 1980a,b) and t sperm injected into oviducts (Olds-Clarke and Becker, 1978) suggest that differences in sperm transport through the female may partially account for altered transmission frequencies. The possibility also exists that t sperm penetrate the zone pellucida faster than do wild-type sperm.

The second line of investigation into the mode of action of T/t alleles suggests that these mutant genes interfere with a variety of cellular interactions during development and fertilization (Bennett, 1975). There is histological, cytological, serological, and biochemical evidence supporting such an hypothesis. For example, transmission electron micrographs of mesenchymal cells in homozygous mutant T/t complex embryos show more extensive areas of intercellular contact than between wild-type cells (Spiegelman and Bennett, 1974). As shown by rotation-mediated aggregation assays, T/T cells have different adhesion kinetics than wild-type cells (Yanagisawa and Fujimoto, 1977). Somites isolated from T/T mutant embryos fail to respond to normal inductive signals from either wild-type or mutant neural tubes (Bennett, 1958). These studies imply that the cell surface is the site of action of the T/t-complex alleles. More directly, specific antisera have been raised against T sperm from congenic mice (Bennett et al., 1972). Antisera also have been produced following immunization with t sperm; these antisera contain some specificities common to many t sperm and other specificities unique to each t allele assayed (Artzt and Bennett, 1977).

Anti-t sperm antisera must be absorbed with wild-type, or $T/+$, sperm to remove non-t antibodies, which results in an antiserum of low sensitivity, or titer. Also, some investigators have been unable to raise antisera against t sperm using similar, but not identical assays (Gable et al., 1979). The inability to detect anti-t specificities and the low titer associated with positive anti-t antisera may result from the occurrence of hydrolytic acrosomal enzymes, which may partially degrade the antigens and/or antisera. To circumvent these problems, *t* testicular cells, rather than *t* sperm, have been used to generate anti-*t* antisera (Cheng and Bennett, 1980). Data show that these antisera are directed against carbohydrates on *t* testis cells, with *t* antisera against different complementation groups directed against different monosaccharide residues (Cheng and Bennett, 1980). However, there have been no reports of *t* specific antisera, against either sperm or testicular cells, using mice congenic for appropriate *t* alleles.

Most studies involving *T/t* complex tissues, fail to carefully control for biochemical variations independent of the presence of *t* alleles, i.e., strain and colony variation. In this regard, wild-type sperm from different strains of mice show quantitative variations in a number of biochemical parameters (Erickson *et al.*, 1979; Shur and Bennett, 1979). Future studies aimed at understanding the mode of action of the *T/t* alleles must compare either littermates segregating for the *t* allele of interest, or better yet, mice congenic for the appropriate *t* allele.

Independent of the primary molecular mechanism underlying mutant morphogenesis, these *T/t* genes clearly interfere with the same processes in which surface glycosyltransferases are thought to participate. For example, *t* alleles dramatically affect fertilization, while sperm surface transferases glycosylate egg carbohydrates (Durr *et al.*, 1977). Also, the spatial and temporal distribution of surface galactosyltransferase activity in normal chick embryos (Shur, 1977a,b) coincides well with the distribution of pathology seen in *t/t* and *T/T* mouse embryos (Bennett, 1975). Consequently, the *T/t* complex mutations can be used as probes to define further any involvement of surface glycosyltransferases during fertilization and development. Any specific glycosyltransferase defect found in mutant tissues may be instrumental in producing the mutant phenotype. When an appropriate *in vitro* system that accurately reflects a *T/t*-dependent defect, one should be able to phenocopy normal and mutant tissues by perturbing surface glycosyltransferase activities.

B. *t*-Sperm Glycosyltransferases

With this overall approach in mind, glycosyltransferase activities were assayed on normal and *t*-bearing sperm (Shur and Bennett, 1979). Sperm were chosen for this analysis because (a) normal sperm have enzymatic activity toward egg surfaces, (b) they are convenient cells for biochemical analysis (i.e., highly specialized for egg binding, large numbers, a relatively pure population, single cells), (c) they show *t*-dependent alterations in fertilizing ability, and (d) they can be functionally assayed *in vitro*.

Nine different enzymes were assayed on sperm from segregation-distorting +/*t* males versus wild-type (+/+) controls: three glycosyltransferases, three glycosidases, and three phosphatases. Only one of these enzymatic activities proved to be significantly different. Under optimal enzymatic conditions, galactosyltransferase activity toward N-acetylglucosamine is twice as high in heterozygote $(+/t)$ preparations relative to normal $(+/+)$ littermates. Endogenous acceptor activity is low on washed sperm but high in the initial sperm supernatant. Egg zona pellucida digests serve as a good acceptor source. Elevated t-sperm galactosyltransferase activity does not result from (a) differences in contaminating somatic cells, (b) differences in competing enzyme activities, or (c) differences in acrosome integrity.

Subsequent studies showed that $+/t$ sperm have a specific increase in galactosyltransferase activity due, in part, to a deficiency of some inhibitory component normally made by a $+/+$ sperm (Shur and Bennett, 1979). For example, when equal aliquots of $+/+$ and $+/t$ sperm are mixed and then assayed, galactosyltransferase activity is only 20% of the theoretical intermediate level. Because t transferases are susceptible to $+$-sperm inhibition, $+/t$-sperm enzyme assays underestimate the actual level of galactosyltransferase activity on t sperm. To circumvent this problem, sperm from males homozygous for a semilethal allele (t^{s1}/t^{s1}) were assayed relative to control littermates (Shur, 1981a). In this way, segregation-distorting t sperm were shown to have four times the activity of wild type. The enzyme levels of t^{s1}/t^{s1} sperm were nearly twice those expected, as calculated from heterozygote $+/t^{s1}$ assays. Thus, these t alleles, in homozygous form, act synergistically on sperm galactosyltransferase activity. This result supports the presence of a normal sperm galactosyltransferase inhibitor deficient on t sperm. An identical situation is found when sperm bearing two different complementing lethal t alleles (t^x/t^y) are assayed relative to $+/t^x$ and $+/t^y$ littermates (Shur, 1981a). The t^x/t^y activity is identical to t^{s1}/t^{s1} levels and almost double the additive levels of t^x and t^y. This result shows that two complementation group members (i.e., t^{12}, t^{w5}) and presumably others behave as if they are homozygous for this particular biochemical lesion.

The t^x/t^y males are sterile, but this is not due to deficiencies of egg receptors on the sperm. If given the opportunity *in vitro*, t^x/t^y sperm are fully capable of binding the egg (McGrath and Hillman, 1980a). The sterility of t^x/t^y likely results from defective spermatogenesis, which in turn, results in defective reproductive efficiency (Tucker, 1980), sperm motility (Bennett and Dunn, 1967) and so on.

Sperm bearing various recombinant t-chromosomes (see Section VII,A) were analyzed for galactosyltransferase activity in order to associate this enzyme stimulation with a particular function of the T/t complex (Shur, 1981a) (see Fig. 2). In neither a single dose $(+/t)$ nor a double dose (t/t) do viable, non-segregation-distorting, recombinant t alleles affect sperm galactosyltransferase activity. Additionally, sperm bearing any one of four dominant T mutations analyzed show enzyme levels identical to wild type. All four T alleles map to the proximal, centromeric segments of the T/t complex. These results show the absolute necessity of exclusively analyzing littermates segregating for the T/t allele of interest, since one of these four dominant T mutations reportedly affected sperm transferase activity when compared to mostly nonlittermate controls (Shur and Bennett, 1979). Collectively, these results show that only when the distal portions of the T/t complex are present, which are associated with embryonic lethality and increased t-sperm transmission frequency, is sperm galactosyltransferase activity likewise elevated.

A variety of results suggest that sperm galactosyltransferases participate during fertilization by binding zona pellucida N-acetyglucosamine residues. As mentioned previously, sperm galactosyltransferase activity is elevated on segregation-distorting t sperm, and sperm galactosyltransferase can glycosylate zona pellucida digests. Additionally, freshly isolated epididymal sperm are coated with high levels of sperm galactosyltransferase acceptors, which are shed from the sperm surface prior to sperm–zona pellucida binding (Shur and Hall, 1982). These sperm-bound galactosyl acceptors inhibit sperm-zona binding when added back to *in vitro* fertilization assays. Similarly, sperm preincubated in Ca^{2+}-containing medium (Saling and Storey, 1979), loose their endogenous galactosyl acceptors, show increased activity toward exogenous N-acetylglucosamine, and simultaneously, show increased binding to eggs (Shur and Hall, 1982). Washing the sperm in the absence of Ca^{2+} results in a loss of endogenous acceptors and a coincident rise in sperm–zona binding.

Results show (Shur and Hall, 1982) that the sperm galactosyl acceptors present in epididymal fluids are immunoprecipitated by antiserum raised against F9 teratocarcinoma cells (Artzt *et al.*, 1973), and are sensitive to endo-β-galactosidase digestion. Anti-F9 antiserum stimulates sperm–zona binding, coincident with decreased galactosylation of sperm substrates and increased galactosylation of N-acetylglucosamine. Results have eliminated antisera effects on competing enzymes, and galactosyltransferases are not absorbed from the antiserum onto the sperm surface. Normal mouse serum has no effect on enzymatic activity nor on sperm–zona binding (Shur and Hall, 1982).

The effects of anti-F9 antiserum on sperm galactosyltransferases prompted an investigation of F9 teratocarcinoma cell surface galactosyltransferases (Shur, 1981b). As on sperm, poly N-acetyllactosamine substrates are associated with galactosyltransferase on the F9 cell surface. A discussion of some salient features of F9 teratocarcinoma cells will precede a discussion of F9 cell surface galactosyltransferases.

C. F9 Teratocarcinoma Cells

Teratomas can occur either spontaneously from the abnormal proliferation of primordial germ cells or from the implantation of early embryos into extrauterine sites (see Martin, 1980, for review). In addition to a variety of differentiated cells, these heterogeneous masses often contain a malignant, proliferating stem cell population called embryonal carcinoma (EC). When established *in vitro,* EC cell lines can be either nullipotential, i.e., have lost the capacity to differentiate, or multipotential. Some multipotential EC cells can differentiate into derivatives of all three embryonic germ layers, similar to that which is found in teratomas. This suggests that EC cells are developmentally similar to early, cleavage-stage blastomeres (Martin and Evans, 1975). Specifically, EC cells are thought to be comparable to the pluripotential cells of the inner cell mass (ICM), which go on to form the embryo. Morphological, histochemical, and biochemical data suggest that ICM cells, like EC cells, give rise to primary endoderm as one of their first differentiated products (Hogan, 1980; Lo and Gilula, 1980). The degree and diversity of EC cell differentiation into endoderm can be modulated by culture conditions, such as increased cell densities (Sherman and Miller, 1978). The frequency of EC cell differentiation (Solter *et al.,* 1979) and the induction of hyaluronate synthesis (Prehm, 1980) are enhanced by the addition of retinoic acid to cultures of F9 EC cells. When coupled with cyclic AMP, retinoic acid induces F9 cultures to acquire some neural characteristics (Kuff and Fewell, 1980).

Recently, there has been much interest in the characterization of EC cell-surface glycopeptides. Embryonal carcinoma surface glycopeptides can be resolved by Sephadex G-50 chromatography into two broad categories (Muramatsu *et al.,* 1979a,b). The minor constituents are the low molecular weight, high-mannose and complex-type glycopeptides, which are found on most, possibly all, cells. Second, an unexpected, large molecular weight (>6000) fraction, low in mannose but high in fucose, galactose, and glucosamine, is excluded from the column. These large fucosylglycopeptides are characteristic of all teratocarcinoma cells examined and also of early mouse embryos (Jacob, 1979; Muramatsu *et al.,* 1980; Buc-Caron *et al.,* 1978). Upon differentiation of EC cells in culture, and by 9–10 days of mouse embryo gestation, synthesis of large fucosylglycopeptides is minimal and replaced by elevated levels of traditional, low molecular weight glycopeptides. The terminal structure of some of these large fucosylglycopeptides is now known (Gooi *et al.,* 1981), and they have internal linkages (sensitive to endo-β-galactosidase) characteristic of keratan-like (Rodén, 1980) and poly N-acetyllactosamine glycoconjugates (Gooi *et al.,* 1981).

Specific immunoprecipitation has shown that these fucosylglycopeptides contain some of the F9 antigenic specificity (Muramatsu *et al.,* 1979a). This is consistent with the parallel expression of fucosylglycopeptides and F9 antiserum reactivity during embryonic development and EC cell differentiation. The anti-F9 antiserum is directed against the glycosidic residues of these EC surface glyco-peptides (Buc-Caron and Duponey, 1980; Gooi *et al.,* 1981), as anti-*t* antisera are directed against specific monosaccharide residues on *t* testicular cells (Cheng and Bennett, 1980). The F9 fucosylglycopeptides may be crucial for normal cellular interactions, since some anti-F9 antiserum inhibits morula compaction (Kemler *et al.*, 1977; Ducibella, 1980) and decreases the adhesion and cyto-differentiation of F9 cells in culture (Jacob, 1979).

D. F9 Teratocarcinoma Cell Surface Glycosyltransferases

Surface galactosyltransferases on F9 cells have been assayed utilizing saturating levels of sugar donor, sugar acceptor, and cation, (Shur, 1981b). The results are noteworthy for two reasons. First, galactosyltransferase activity toward endogenous acceptors is significantly greater than any other surface transferase activity. Second, endogenous galactosyltransferase activity is often as great as that toward saturating concentrations (20 mM) of exogenous N-acetylgucosamine,

The galactosyltransferase activity is cell surface localized by many criteria. For example, UDP-galactose degradation products cannot account for any significant amount of product formed in UDP-galactose incubations. Also, no leakage of galactosyltransferases into the incubation medium is detectable when either endogenous or exogenous acceptors are used.

Similar to the case with sperm, when F9 cells are pretreated with anti-F9 antiserum, washed, and assayed for N-acetylglucosamine galactosyltransferase activity, exogenous activity is stimulated twofold (Shur, 1981b). Nonimmune, normal mouse serum has no effect on enzyme activity. Likewise, heat inactivating the galactosyltransferases endogenous to the antisera does not affect the stimulation of F9 cell N-acetylglucosamine galactosyltransferases. When anti-F9 antiserum is present in excess throughout the assay, galactosyltransferase activity is inhibited toward endogenous acceptors and stimulated toward exogenous N-acetylglucosamine. When F9 cells are induced to differentiate into endoderm with retinoic acid, they no longer react with F9 antisera nor show any antisera effects on surface galactosyltransferase activity. Taken together, these results suggest that F9 antisera recognizes the endogenous substrate for F9 cell surface galactosyltransferases. In the presence of anti-F9 antisera, the F9 endogenous acceptors are displaced, increasing the availability of enzymes for exogenous acceptor glycosylation.

The F9 surface galactosyltransferase reaction product has been identified as a poly N-acetyllactosamine glycoconjugate (Shur, 1981b). About 20% of the product radioactivity can be extracted by glycolipid solvents. After extensive pronase digestion, the resultant glycopeptides are excluded near the void volume of a Sephadex G-50 column, unlike the profile expected for the traditional, low molecular weight oligosaccharides. Finally, the reaction product is sensitive to purified keratanase (i.e., endo-β-galactosidase) (Nakazawa and Suzuki, 1975) and is preferentially precipitated with anti-F9 antiserum, relative to preimmune, normal mouse serum.

Detergent extracts of F9 cells were used as a source of poly N-acetyllactosamine glycoconjugate since detergent-extracted material is precipitable with anti-F9 antiserum, is excluded from a Sephadex G-50 column after pronase digestion, and is sensitive to endo-β-galactosidase digestion (M. Muramatsu *et al.*, 1979a; Shur, 1981b). These dialyzed extracts compete with exogenous N-acetylglucosamine for F9 surface galactosyltransferases (Shur, 1981b). Pretreatment with β-glucosaminidase decreases both the exogenous acceptor capabilities of the F9 extract and the capacity of the extract to compete for N-acetylglucosamine galactosyltransferases. Pretreatment with either endo-β-galactosidase or β-galactosidase produces the reciprocal effect.

Finally, the galactosyltransferase reaction product is solubilized into the assay medium during incubation with UDP-galactose (Shur, 1981b). Control incubations without UDP-galactose fail to solubilize acceptor activity. Incubations containing the galactosyltransferase inhibitor UDP-dialdehyde (Powell and Brew, 1976) do not release any glycosylated product into the incubation medium. This suggests that the cell dissociates from its substrate after catalysis, which supports the possibility that one of the surface receptors for poly N-acetyllactosamine glycoconjugates is a galactosyltransferase.

E. Normal and *T/T* Embryonic Surface Glycosyltransferases

Early mouse embryos also synthesize large poly N-acetyllactosamine glycoconjugates (Buc-Caron *et al.*, 1978; Muramatsu *et al.*, 1980; Jacob, 1979). Since surface galactosyltransferases may be one of the receptors for these glycosides, surface glycosyltransferase activities were assayed during mouse embryogenesis. This study (Shur, 1981c) used the identical quantitative and autoradiographic assays that were used to identify surface transferase activities during chick embryogenesis (Shur, 1977a,b). However, three sugar nucleotides (UDP-galactose, UDP-glucose, UDP-N-acetylglucosamine) were added at saturating concentrations. CMP-Sialic acid, GDP-fucose, and UDP-N-acetylgalactosamine are not available in an unlabeled form, so these assays contained only micromolar amounts of labeled sugar nucleotides. The necessary controls were performed and showed the activity to be surface-associated. In these mammalian tissues, free sugar pools from hydrolyzed labeled sugar nucleotides range from only 0 to 1% of the input radioactivity, minimizing any contribution by sugar nucleotide degradation products.

Eight-day wild-type embryos show high levels of surface galactosyltransferase activity both autoradiographically and per microgram DNA (Shur, 1981c). The activity is localized primarily on the newly formed primitive streak and the resultant mesenchymal tissues. This is generally consistent with the galactosyltransferase distribution in early chick embryos (Shur, 1977a). Surprisingly, incubations containing the five other sugar nucleotides show background levels of radioactivity by both assay criteria.

Nine-day wild-type embryos continue to show high levels of surface galactosyltransferase activity, which is localized to primitive neural and mesenchymal tissues (Shur, 1981c). Low levels of some other surface glycosyltransferases become detectable, but are too inactive to be detected autoradiographically. By 10 days of gestation, galactosyltransferase levels are very low, with the exception of head and early limb bud mesenchyme.

The galactosyltransferase product resulting from 8- and 9-day embryo assays was partially characterized (Shur, 1981c). It is not extractable by chloroform/methanol or chloroform/methanol/H_2O solvents. After extensive pronase digestion, the radioactivity is excluded in the void volume of a Sephadex G-50 column. Keratanase (endo-β-galactosidase) digestion shifts more than 60% of the radioactivity into the included volume. Finally, the galactosylated reaction product is preferentially precipitated with anti-F9 antiserum. As before, these results imply that the galactosyltransferase product is a large, poly N-acetyllactosamine extracellular glycoconjugate.

An identical quantitative and autoradiographic analysis was conducted on morphogenetically arrested T/T embryos to explore further the function of surface galactosyltransferases during development (Shur, 1981c). T/T embryos are characterized by a generalized mesenchymal cell defect, which overtakes the embryo as a pathological wave, arresting the development of some tissues before others. At the light microscopic level, some areas of the embryo appear to be unaffected, while in other portions the advancing pathology serves to distinguish T/T embryos from wild type.

By 9 days of development, T/T embryos are clearly recognizable (Bennett, 1975) and have 1.8–2.0 times the surface galactosyltransferase activity of normal controls (Shur, 1981c). Histologically normal mesenchymal tissues in T/T embryos show this elevated activity. Ten-day limb buds from T/T embryos, although morphologically abnormal, show more than six times the endogenous galactosyltransferase activity of wild-type limb buds of similar size.

In contrast to galactosyltransferase activity, sialyltransferase activity in 9-day T/T embryos is only 70% of the already low activity detectable in 9-day wild-type embryos (Shur, 1981c). This result eliminates the possibility that the elevated galactosyltransferase activity seen in intact T/T embryos is due to greater accessibility of UDP-galactose to T/T tissues relative to wild type. If this were the case, CMP-sialic acid would also show elevated incorporation in T/T, which is the opposite of what is seen.

The reason for elevated galactosyl acceptor levels in T/T embryos is not known. One possible explanation, among many, is that normal galactosylation of extracellular glycoconjugates did not occur in T/T due to defective mesenchymal cell migration. This would result in higher residual levels of acceptors in

T/T than in wild type. If some of these galactosyl residues would normally be terminated by sialic acid moieties, then a decrease in galactosylation would likewise produce a decrease in sialyl acceptors.

This transferase analysis has attempted to use T/T embryos as probes to explore surface galactosyltransferase function during wild-type morphogenesis. These results are consistent with those from sperm (Shur and Bennett, 1979; Shur, 1981a; Shur and Hall, 1982) and F9 teratocarcinoma (Shur, 1981b) assays, which focus attention on one specific model for glycosyltransferase–acceptor complexes functioning during fertilization and early development.

F. A Specific Model for Murine Fertilization, Teratocarcinoma Cell Adhesion, and Mesenchymal Cell Migration

Observations compiled from sperm, teratocarcinoma cells, and mouse embryos can be incorporated into a specific, tentative model for cellular interactions during fertilization and early development (Fig. 3). The model is testable, and therefore its primary usefulness lies in the design of future experiments.

According to the model, teratocarcinoma cells, early embryonic cells, and sperm are all characterized by large, poly N-acetyllactosamine glycoconjugates, which react with F9 antiserum, in their extracellular environment. These are bound to the cell surface via terminal N-acetylglucosamine binding to surface galactosyltransferases. During embryonic development, this material can be recognized (Jacob, 1979; Muramatsu *et al.*, 1980; Buc-Caron *et al.*, 1978) biochemically (large fucosylglycopeptides) and serologically (anti-F9 antisera) until around the ninth or tenth gestation day, which coincides with the decline in embryonic surface galactosyltransferase activity (Shur, 1981c). The cells may recognize, adhere to, and migrate on this matrix, which they could dissociate from by catalysis of enzyme–substrate complexes. Consistent with this is the solubilization of F9 antigenic material from whole F9 cells after UDP-galactose incubation (Shur, 1981b). Similar keratan-like extracellular material has been identified in frog embryos during mesenchyme formation and migration (Johnson, 1977).

A similar poly N-acetyllactosamine galactosyl acceptor may affect the availability of egg-binding sites on sperm (i.e., galactosyltransferases). The fertilizing capability of the sperm would be proportional to the number of exposed transferases, which is regulated by the degree of competing endogenous acceptors adsorbed to the surface. These competing endogenous acceptors may be one class of "sperm coating factors" (Chang, 1957), which are removable by washing the sperm. The sperm receptor on eggs is also likely to be a glycoprotein with terminal N-acetylglucosamine residues but it has not yet been fully characterized (Bleil and Wassarman, 1980). This model predicts that the

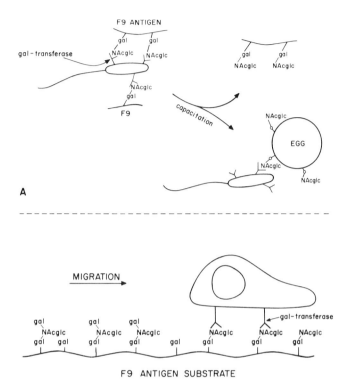

Figure 3 A specific model for murine fertilization and mesenchymal cell migration. In both panels, poly N-acetyllactosamine glycoconjugate interacts with the cell surface via terminal N-acetylglucosamine (NAcglc) binding to surface galactosyltransferases. Only small portions of the oligosaccharide side chains are shown and consist partly of the traditional repeating keratan disaccharide, galactose-N-acetylglucosamine. Some of these disaccharides may be fucosylated (Gooi *et al.*, 1981). On sperm, this glycoconjugate may be one class of "coating" factor, most of which is removable by washing the sperm and which inhibits sperm–egg binding. Terminal N-acetylglucosamine residues also occur in the sperm receptor on eggs, but this receptor has not yet been characterized. During early development, mesenchymal cells may migrate on keratan-like matrices via surface galactosyltransferase binding to terminal N-acetylglucosamine residues. Galactosylation would dissociate the cell from its substrate, similar to the effects of UDP-galactose *in vitro* (Shur, 1981b), and allow the cell to advance to unglycosylated N-acetylglucosamine binding sites.

sperm N-acetylglucosame galactosyltransferase may glycosylate N-acetylglucosamine residues in a variety of oligosaccharides. N-Acetylglucosamine galactosyltransferases purified from other sources show little specificity for the N-acetylglucosamine terminating substrate (Schanbacher and Ebner, 1970). Thus, the relative nonspecificity of the sperm transferase could explain why sperm readily fuse with a wide range of somatic cells, including vaginal (Bendich *et al.*, 1976).

In mutant cells, the model predicts that t alleles interfere with this galactosyltransferase–acceptor association, but in ways not presently clear. We know only that segregation-distorting t sperm show a specific increase in galactosyltransferase activity, and that mesenchymally-defective T/T embryos show defects in surface galactosylation of extracellular substrates. On t sperm, increased galactosyltransferase availability results, hypothetically, in a greater number of egg-binding sites, partially explaining segregation–distortion.

VIII. CONCLUSIONS AND FUTURE DIRECTIONS

From the work discussed here, as well as that discussed in other review articles (Pierce *et al.*, 1980b; Shur and Roth, 1975), it is clear that glycosyltransferases are components of the plasma membrane. The inability to assay surface transferase activities stems in large part from the use of assay conditions that are optimal for sugar nucleotide degradation rather than for glycosyltransferase activity (Deppert *et al.*, 1974; Deppert and Walter, 1978). In retrospect, it is not surprising that transferases are surface constituents, since the plasma membrane is likely formed, in part, from Golgi vesicle fusion. On the contrary, it would be interesting to find cells with glycosyltransferase-negative surfaces, not only because they may selectively delete Golgi membrane proteins, but also because these cells may show interesting social behavior.

During the past 10 years, surface glycosyltransferase activities have been correlated with a host of cellular interactions, including fertilization, cell migration, embryonic induction, chondrogenesis, contact inhibition of growth, cell adhesion, hemostasis, intestinal cell differentiation, and immune recognition. Rather than adding to this list of correlations, investigators should turn toward a more definitive analysis of surface glycosyltransferase function in any one of these cellular interactions. Some of the available systems are obviously preferable to others. For example, in many studies dissociative enzymes are required to prepare single cells for adhesion and glycosyltransferase assays. Similarly, in some assays glycosidase pretreatment is necessary to expose adhesion sites on cell surfaces. In both of these instances, the use of exogenous enzymes to prepare the cells may create binding sites not normally involved in intercellular adhesion.

A more conservative approach would be to investigate a well-defined cellular interaction, which one could perturb *in vitro* with a defined probe. In this way, one could assay surface glycosyltransferase activities in the presence and absence of the perturbation. At least two classes of probes are available for this kind of analysis: genetic and biochemical.

First, a mutation could be introduced into the cells, which would interfere with the cellular interaction being studied. There are many morphogenetic mutants available in the mouse, which most likely can be studied *in vitro* and for which wild-type littermates serve as control. This genetic approach is analogous to the use of temperature-sensitive mutants, the effects of which can be turned on and off by temperature shifts. Alternatively, one could introduce biochemicals into a defined, *in vitro* system which would inhibit specific cellular interactions. The choice of biochemicals would obviously depend on the molecular hypothesis being tested. In the case of surface glycosyltransferase studies, these reagents could include purified enzymes, their haptene acceptors, antisera raised against them, and specific enzyme inhibitors. In this regard, the use of periodate-oxidized analogs of sugar nucleotide substrates has been shown to inhibit some glycosyltransferase activities irreversibly (Cummings *et al.*, 1979). These nucleotide dialdehydes, which are discussed in more detail in Chapter 4, Section 1, Volume IV, may prove invaluable in studying the function of surface glycosyltransferases.

Ideally, both genetic and biochemical probes could be incorporated into the same analysis. The *T/t*-complex mutations and the F9 teratocarcinoma cells represent systems for which genetic and biochemical probes are available. First, *T/t*-mutant alleles interfere with a variety of cellular interactions, including fertilization and morphogenesis. Both of these processes can be studied *in vitro* without the use of dissociative enzymes. In the case of fertilization, sperm and eggs are readily available as single cells for *in vitro* assays. Embryonic morphogenesis is more difficult to study *in vitro,* but teratocarcinoma cells have been shown to be analogous to early, pluripotential embryonic cells. Consequently, teratocarcinoma cells, which are easily dissociated with EDTA, can be used as a model system to study embryonic cellular adhesions. Second, antisera that interfere with teratocarcinoma and early embryonic cellular interactions are available. Finally, solubilized endogenous acceptors can be used as competitive haptene substrates for surface galactosyltransferases. Using these three probes, among others, a more definitive analysis of surface glycosyltransferases in fertilization and morphogenesis should be feasible. Results to date are consistent with such a role for surface glycosyltransferases in development.

ACKNOWLEDGMENTS

I am indebted to Dr. S. Cooperstein, Dr. J. Grasso, Dr. G. Maxwell, and Dr. R. Kosher for their suggestions regarding this manuscript. Original work conducted in my laboratory was supported by grants from the University of Connecticut Research Foundation, the Anna Fuller Fund, the American Cancer Society, the National Institutes of Health, and by a Basil O'Connor Starter Research grant from the March of Dimes Birth Defects Foundation.

REFERENCES

Artzt, K., and Bennett, D. (1977). *Immunogenetics (N.Y.)* **5,** 97–107.
Artzt, K., Dubois, P., Bennett, D., Condamine, H., Babinet, C., and Jacob, F. (1973). *Proc. Natl. Acad. Sci. U.S.A.* **70,** 2988–2992.

Baker, A. P., Smith, W. J., and Holden, D. A. (1980). *Cell. Immunol.* **51,** 186–191.

Balsamo, J., and Lilien, J. (1980). *Biochemistry* **19,** 2479–2484.

Barbera, A. J. (1975). *Dev. Biol.* **46,** 167–191.

Barber, A. J., and Jamieson, G. A. (1971). *Biochim. Biophys. Acta* **252,** 533–545.

Barondes, S. H., and Rosen, S. D. (1976). *In* "Neuronal Recognition" (S. H. Barondes, ed.), pp. 331–356. Plenum, New York.

Bendich, A., Borenfreund, E., Witkin, S. S., Beju, D., and Higgins, P. J. (1976). *Prog. Nucleic Acid Res. Mol. Biol.* **17,** 43–75.

Bennett, D. (1958). *Nature (London)* **181,** 1286.

Bennett, D. (1975). *Cell* **6,** 441–454.

Bennett, D., and Dunn, L. C. (1967). *J. Reprod. Fertil.* **13,** 421–428.

Bennett, D., Goldberg, E., Dunn, L. C., and Boyse, E. A. (1972). *Proc. Natl. Acad. Sci. U.S.A.* **69,** 2076–2080.

Bennett, D., Dunn, L. C., and Artzt, K. (1976). *Genetics* **83,** 361–372.

Bernfield, M. R., and Banerjee, S. D. (1972). *J. Cell Biol.* **52,** 664–673.

Beyer, T. A., and Hill, R. L. (1980). *J. Biol. Chem.* **255,** 5373–5379.

Beyer, T. A., Rearick, J. I., Paulson, J. C., Prieels, J.-P., Sadler, J. E., and Hill, R. L. (1979). *J. Biol. Chem.* **254,** 12531–12541.

Bleil, J. O., and Wassarman, P. M. (1980). *Cell* **20,** 873–882.

Bosmann, H. B. (1971a). *Biochem. Biophys. Res. Commun.* **43,** 1118–1124.

Bosmann, H. B. (1971b). *Nature (London) New Biol.* **234,** 54–56.

Bosmann, H. B. (1972). *Biochim. Biophys. Acta* **279,** 456–474.

Bosmann, H. B. (1973). *J. Neurochem.* **20,** 1037–1049.

Bosmann, H. B., and McLean, R. J. (1975). *Biochem. Biophys. Res. Commun.* **63,** 323–327.

Brew, K., Vanaman, T., and Hill, R. L. (1968). *Proc. Natl. Acad. Sci. U.S.A.* **59,** 491–497.

Buc-Caron, M. H., and Duponey, P. (1980). *Mol. Immunol.* **17,** 655–664.

Buc-Caron, M. H., Condamine, H., and Jacob, F. (1978). *J. Embryol. Exp. Morphol.* **47,** 149–160.

Burke, D., Mendonca-Previato, L., and Ballou, C. E. (1980). *Proc. Natl. Acad. Sci. U.S.A.* **77,** 318–322.

Cebula, T. A., and Roth, S. (1976). *In* "Biogenesis and Turnover of Membrane Macromolecules" (T. S. Cook, ed.), pp. 235–250. Raven, New York.

Cervén, E. (1977). *Biochim. Biophys. Acta* **467,** 72–85.

Chang, M. C. (1957). *Nature (London)* **175,** 258–259.

Cheng, C., and Bennett, D. (1980). *Cell* **19,** 537–543.

Choi, H. U., and Meyer, K. (1975). *Biochem. J.* **151,** 543–553.

Colombino, L. F., Bosmann, H. B., and McLean, R. J. (1978). *Exp. Cell Res.* **112,** 25–30.

Culp, L. A., Murray, B. A., and Rollins, B. J. (1979). *J. Supramol. Struct.* **11,** 401–427.

Cummings, R. D., Cebula, T. A., and Roth, S. (1979). *J. Biol. Chem.* **254,** 1233–1240.

Danska, J. S., and Silver, L. M. (1980). *Cell* **22,** 901–904.

Den, H., and Kaufman, B. (1968). *Fed. Proc., Fed. Am. Soc. Exp. Biol.* **27,** 346.

Den, H., Kaufman, B., and Roseman, S. (1970). *J. Biol. Chem.* **245,** 6607–6615.

Deppert, W., and Walter, G. (1976). *J. Cell. Physiol.* **90,** 41–52.

Deppert, W., and Walter, G. (1978). *J. Supramol. Struct.* **8,** 19–37.

Deppert, W., Werchaw, H., and Walter, G. (1974). *Proc. Natl. Acad. Sci. U.S.A.* **71,** 3068–3072.

Dooher, G. B., and Bennett, D. (1974). *J. Embryol. Exp. Morphol.* **32,** 749–761.

Ducibella, T. (1980). *Dev. Biol.* **79,** 356–366.

Durr, R., Shur, B., and Roth, S. (1977). *Nature (London)* **265,** 547–548.

Erickson, R. P., Butley, M. S., Martin, S. R., and Betlach, C. J. (1979). *Genet. Res.* **33,** 129–136.

Frazier, W., and Glaser, L. (1979). *Annu. Rev. Biochem.* **48,** 491–523.

Gable, R. J., Levinson, J. R., McDevitt, H. O., and Goodfellow, P. N. (1979). *Tissue Antigens* **13,** 177–185.

Garbers, D. L., First, N. L., and Lardy, H. A. (1973). *J. Biol. Chem.* **248,** 875–879.

Garfield, S., Hausman, R. E., and Moscona, A. A. (1974). *Cell Differ.* **3,** 215–219.

Ginsberg, L., and Hillman, N. (1974). *J. Reprod. Fertil.* **38,** 157–163.

Ginsberg, L., and Hillman, N. (1975). *J. Embryol. Exp. Morphol.* **33,** 715–723.

Glabe, C. G., and Lennarz, W. J. (1979). *J. Cell Biol.* **83,** 595–604.

Glabe, C. G., and Vacquier, V. D. (1978). *Proc. Natl. Acad. Sci. U.S.A.* **75,** 881–885.

Gooi, H. C., Feizi, T., Kapadia, A., Knowles, B. B., Solter, D., and Evans, M. J. (1981). *Nature (London)* **292,** 156–158.

Graham, J. M., Hynes, R. O., Rowlatt, C., and Sandall, K. (1978). *Ann. N.Y. Acad. Sci.* **312,** 221–239.

Gwatkin, R. B. L. (1976). *In* ''The Cell Surface in Animal Embryogenesis and Development'' (G. Poste and G. L. Nicolson, eds.), pp. 1–43. Elsevier/North-Holland Biomedical Press, Amsterdam.

Hammerberg, C., and Klein, J. (1975). *Nature (London)* **253,** 137–138.

Hart, G. W., and Lennarz, W. J. (1978). *J. Biol. Chem.* **253,** 5795–5801.

Hoflack, B., Cacan, R., Montreuil, J., and Verbert, A. (1979). *Biochim. Biophys. Acta* **568,** 348–356.

Hogan, B. L. M. (1980). *Dev. Biol.* **76,** 275–285.

Hoskins, D. D., Casillas, E. R., and Stephens, D. T. (1972). *Biochem. Biophys. Res. Commun.* **48,** 1331–1338.

Hoskins, D. D., Brandt, H., and Acott, T. S. (1978). *Fed. Proc., Fed. Am. Soc. Exp. Biol.* **37,** 2534–2542.

Hynes, R. O. (1979). *In* ''Surfaces of Normal and Malignant Cells'' (R. O. Hynes, ed.), pp. 103–148. Wiley, New York.

Jacob, F. (1979). *Curr. Top. Dev. Biol.* **13,** 89–115.

Jamieson, G. A., Urban, C. L., and Barber, A. J. (1971). *Nature (London) New Biol.* **243,** 5–7.

Johnson, K. E. (1977). *J. Cell Sci.* **25,** 335–354.

Jumblatt, J. E., Schlup, V., and Burger, M. M. (1980). *Biochemistry* **19,** 1038–1042.

Karfunkel, P., Hoffman, M., Phillips, M., and Black, J. (1977). *In* ''Formshaping Movements in Neurogenesis'' (C. Jacobson and T. Eberdahl, eds.), pp. 23–31. Almqvist & Wiksell, Stockholm.

Katz, D., Erickson, R. P., and Nathanson, M. (1980). *J. Exp. Zool.* **210,** 529–535.

Kemler, R., Babinet, C., Eisen, H., and Jacob, F. (1977). *Proc. Natl. Acad. Sci. U.S.A.* **74,** 4449–4452.

Kosher, R. A., and Church, R. L. (1975). *Nature (London)* **258,** 327–330.

Kosher, R. A., and Lash, J. W. (1975). *Dev. Biol.* **42,** 362–378.

Kosher, R. A., Savage, M. P., and Chan, S.-C. (1979a). *J. Embryol. Exp. Morphol.* **50,** 75–97.

Kosher, R. A., Savage, M. P., and Chan, S.-C. (1979b). *J. Exp. Zool.* **209,** 221–228.

Kosher, R. A., Savage, M. P. and Walker, K. H. (1981). *J. Embryol. Exp. Morphol.* **63,** 85–98.

Kraemer, P. M. (1979). *In* ''Surfaces of Normal and Malignant Cells'' (R. O. Hynes, ed.), pp. 149–198. Wiley, New York.

Kuff, E. L., and Fewell, J. W. (1980). *Dev. Biol.* **77,** 103–115.

Kurt, E. A., Shur, B. D., and Lindquist, R. R. (1981). *Fed. Proc., Fed. Am. Soc. Exp. Biol.* (abstr.) **40,** 1150.

LaMont, J. T., Perrotto, J. L., Weiser, M. M., and Isselbacher, K. J. (1974). *Proc. Natl. Acad. Sci. U.S.A.* **71,** 3726–3730.

LaMont, J. T., Gammon, M. T., and Isselbacher, K. J. (1977). *Proc. Natl. Acad. Sci. U.S.A.* **74,** 1086–1090.

Leunis, J. C., Smith, D. F., Nwokozo, N., Fishback, B. L., Wu, C., and Jamieson, G. A. (1980). *Biochim. Biophys. Acta* **611,** 79–86.

Levvy, G. A., and Conchie, J. (1966). *In* "Methods in Enzymology" (E. F. Neufeld and V. Ginsburg, eds.), Vol. 8, 571–584. Academic Press, New York.

Lilien, J., Balsamo, J., McDonough, J., Hermolin, J., Cook, J., and Rutz, R. (1979). *In* "Surfaces of Normal and Malignant Cell" (R. O. Hynes, ed.) pp. 389–419. Wiley, New York.

Lillie, F. R. (1913). *Science* **38**, 524–528.

Lloyd, C. W., and Cook, G. M. W. (1974). *J. Cell Sci.* **15**, 575–590.

Lo, C., and Gilula, N. (1980). *Dev. Biol.* **75**, 78–92.

Lyon, M. F., and Mason, I. (1977). *Genet. Res.* **29**, 255–266.

McDonough, J., and Lilien, J. (1978). *J. Supramol. Struct.* **7**, 409–418.

McGrath, J., and Hillman, N. (1980a). *J. Embryol. Exp. Morphol.* **59**, 46–58.

McGrath, J., and Hillman, N. (1980b). *Nature (London)* **283**, 479–481.

McGuire, E. J., Jourdian, G. W., Carlson, D. M., and Roseman, S. (1965). *J. Biol. Chem.* **240**, PC4112–PC4115.

McLean, R. J., and Bosmann, H. B. (1975). *Proc. Natl. Acad. Sci. U.S.A.* **72**, 310–313.

Magee, S. C., Mawal, R., and Ebner, K. E. (1974). *Biochemistry* **13**, 99–102.

Marchase, R. B. (1977). *J. Cell Biol.* **75**, 237–257.

Marchase, R. B., Vosbeck, K., and Roth, S. (1976). *Biochim. Biophys. Acta* **457**, 385–416.

Martin, G. (1980). *Science* **209**, 768–775.

Martin, G., and Evans, M. J. (1975). *Proc. Natl. Acad. Sci. U.S.A.* **72**, 1441–1445.

Merritt, W. D., Morre, D. J., Franke, W. W., and Keenan, T. W. (1977). *Biochim. Biophys. Acta* **497**, 820–824.

Moscona, A. A. (1963). *Proc. Natl. Acad. Sci. U.S.A.* **49**, 742–747.

Moscona, A. A. (1976). *In* "Neuronal Recognition" (S. H. Barondes, ed.), pp. 205–226. Plenum, New York.

Moy, G. W., and Vacquier, V. D. (1979). *Curr. Top. Dev. Biol.* **13**, 31–44.

Müller, W. E. G., Arendes, J., Kurelec, B., Zahn, R. K., and Müller, I. (1977). *J. Biol. Chem.* **252**, 3836–3842.

Müller, W. E. G., Zahn, R. K., Kurelec, B., Müller, I., Uhlenbruck, G., and Vaith, P. (1979). *J. Biol. Chem.* **254**, 1280–1287.

Muramatsu, T., Gachelin, G., Damonneville, M., Delarbre, C., and Jacob, F. (1979a). *Cell* **18**, 183–191.

Muramatsu, T., Gachelin, G., and Jacob, F. (1979b). *Biochim. Biophys. Acta* **587**, 392–406.

Muramatsu, T., Condamine, H., Gachelin, G. and Jacob, F. (1980). *J. Embryol. Exp. Morphol.* **57**, 25–36.

Nadijcka, M., and Hillman, N. (1975). *J. Embryol. Exp. Morphol.* **33**, 725–730.

Nakazawa, K., and Suzuki, S. (1975). *J. Biol. Chem.* **250**, 912–917.

Olds-Clarke, P., and Becker, A. (1978). *Biol. Reprod.* **18**, 132–140.

Painter, R. C., and White, A. (1976). *Proc. Natl. Acad. Sci. U.S.A.* **73**, 837–841.

Parish, C. R. (1977). *Nature (London)* **267**, 711–713.

Parodi, A., and Leloir, L. F. (1979). *Biochim. Biophys. Acta* **559**, 1–37.

Patt, L. M., and Grimes, W. J. (1974). *J. Biol. Chem.* **249**, 4157–4165.

Patt, L. M., Endres, R. O., Lucas, D. O., and Grimes, W. J. (1976). *J. Cell Biol.* **68**, 799–802.

Paulson, J. C., Rearick, J. I., and Hill, R. L. (1977). *J. Biol. Chem.* **252**, 2363–2371.

Peterson, R. N., Russell, L., Bundman, D., and Freund, M. (1980). *Science* **207**, 73–74.

Pierce, M., Cummings, R. D., and Roth, S. (1980a). *Anal. Biochem.* **102**, 441–449.

Pierce, M., Turley, E. A., and Roth, S. (1980b). *Int. Rev. Cytol.* **65**, 1–47.

Porter, C. W., and Bernacki, R. J. (1975). *Nature (London)* **256**, 648–650.

Porzig, E. F. (1978). *Dev. Biol.* **67**, 114–136.

Powell, J. T., and Brew, K. (1976). *Biochemistry* **15**, 3499–3505.

Pratt, R. M., Larsen, M. A., and Johnston, M. C. (1975). *Dev. Biol.* **44**, 298–305.

Prehm, P. (1980). *FEBS Lett.* **111**, 295–298.

Puett, D., Wasserman, B. K., Ford, J. D., and Cunningham, L. W. (1973). *J. Clin. Invest.* **52**, 2495–2506.

Rearick, J. I., Sadler, J. E., Paulson, J. C., and Hill, R. L. (1979). *J. Biol. Chem.* **254**, 4444–4451.

Richard, M., Marin, A., and Louisot, P. (1975). *Biochem. Biophys. Res. Commun.* **64**, 108–114.

Rodén, L. (1980). *In* "The Biochemistry of Glycoproteins and Proteoglycans" (W. J. Lennarz, ed.), pp. 267–371. Plenum, New York.

Roseman, S. (1970). *Chem. Phys. Lipids* **5**, 270–297.

Rosen, S. D., Kaur, J., Clark, D. L., Pardos, B. T., and Frazier, W. A. (1979). *J. Biol. Chem.* **254**, 9408–9445.

Roth, S. (1973). *Q. Rev. Biol.* **48**, 541–563.

Roth, S., and Marchase, R. B. (1976). *In* "Neuronal Recognition" (S. H. Barondes, ed.), pp. 227–248. Plenum, New York.

Roth, S., and White, D. (1972). *Proc. Natl. Acad. Sci. U.S.A.* **69**, 485–489.

Roth, S., McGuire, E. J., and Roseman, S. (1971). *J. Cell Biol.* **51**, 536–547.

Rutz, R., and Lilien, J. (1979). *J. Cell Sci.* **36**, 323–342.

Saling, P. M., and Storey, B. T. (1979). *J. Cell Biol.* **83**, 544–555.

Sasaki, T., and Robbins, P. W. (1974). *In* "Biology and Chemistry of Eucaryotic Cell Surfaces" (E. Y. C. Lee and E. E. Smith, eds.), pp. 125–157. Academic Press, New York.

Schachter, H., and Roseman, S. (1980). *In* "Biochemistry of Glycoproteins and Proteoglycans" (W. J. Lennarz, ed.), pp. 85–160. Plenum, New York.

Schachter, H., Michaels, M., Tilley, C. A., Crookston, M. D., and Crookston, J. H. (1973). *Proc. Natl. Acad. Sci. U.S.A.* **70**, 220–224.

Schanbacher, F. L., and Ebner, K. E. (1970). *J. Biol. Chem.* **245**, 5057–5061.

Schwyzer, M., and Hill, R. (1977). *J. Biol. Chem.* **252**, 2346–2355.

SeGall, G. K., and Lennarz, W. J. (1979). *Dev. Biol.* **71**, 33–48.

Shapiro, B. M., and Eddy, E. M. (1980). *Int. Rev. Cytol.* **66**, 257–302.

Sherman, M. I., and Miller, R. A. (1978). *Dev. Biol.* **63**, 27–34.

Shur, B. D. (1977a). *Dev. Biol.* **58**, 23–39.

Shur, B. D. (1977b). *Dev. Biol.* **58**, 40–55.

Shur, B. D. (1981a). *Genet. Res.* (in press).

Shur, B. D. (1981b). Submitted for publication.

Shur, B. D. (1981c). *Dev. Biol.*, in press.

Shur, B. D., and Bennett, D. (1979). *Dev. Biol.* **71**, 243–259.

Shur, B. D., and Roth, S. (1973). *Am. Zool.* **13**, 1129–1135.

Shur, B. D., and Roth, S. (1975). *Biochim. Biophys. Acta* **415**, 473–512.

Shur, B. D., and Hall, N. G. (1982). Submitted for publication.

Shur, B. D., Oettgen, P., and Bennett, D. (1979). *Dev. Biol.* **73**, 178–181.

Shur, B. D., Vogler, M., and Kosher, R. A. (1981). *Exp. Cell Res.*, in press.

Sievers, S., Risse, H.-J., and Sekeri-Pataryas, K. H. (1978). *Mol. Cell. Biochem.* **20**, 103–110.

Silver, L. M., Artzt, K., and Bennett, D. (1979). *Cell* **17**, 275–284.

Silver, L. M., White, M., and Artzt, K. (1980). *Proc. Natl. Acad. Sci. U.S.A.* **77**, 6077–6080.

Smith, D. F., Kosow, D. P., Wu, C., and Jamieson, G. A. (1977). *Biochim. Biophys. Acta* **483**, 263–278.

Snell, W. J., and Roseman, S. (1979). *J. Biol. Chem.* **254**, 10820–10829.

Solter, D., Shevinsky, L., Knowles, B., and Strickland, S. (1979). *Dev. Biol.* **70**, 515–521.

Spataro, A. C., Morgan, H. R., and Bosmann, H. B. (1975). *Proc. Soc. Exp. Biol. Med.* **149**, 486–490.

Spiegelman, M., and Bennett, D. (1974). *J. Embryol. Exp. Morphol.* **32**, 723–738.

Struck, D. K., and Lennarz, W. J. (1976). *J. Biol. Chem.* **251**, 2511–2519.

Thadani, V. M. (1980). *J. Exp. Zool.* **212,** 435–453.

Thorogood, P. V., and Hinchliffe, J. R. (1975). *J. Embryol. Exp. Morphol.* **33,** 581–606.

Toole, B. P. (1976). *In* "Neuronal Recognition" (S. H. Barondes, ed.), pp. 275–329. Plenum, New York.

Townes, P. L., and Holtfreter (1955). *J. Exp. Zool.* **128,** 53–118.

Tucker, M. J. (1980). *Nature (London)* **288,** 367–368.

Turley, E. A., and Roth, S. (1979). *Cell* **17,** 109–115.

Urushihara, H., and Takeichi, M. (1980). *Cell* **20,** 363–371.

Vacquier, V. D., and Moy, G. W. (1977). *Proc. Natl. Acad. Sci. U.S.A.* **74,** 2456–2460.

Van Roelen, C., Vakoet, L., and Andries, L. (1980). *J. Embryol. Exp. Morphol.* **56,** 169–178.

Verbert, A., Cacan, R., and Montreuil, J. (1976). *Eur. J. Biochem.* **70,** 49–53.

Webb, G. C., and Roth, S. (1974). *J. Cell Biol.* **63,** 796–805.

Weiser, M. M. (1973). *J. Biol. Chem.* **248,** 2542–2548.

Weiser, M. M., Neumeier, M. M., Quaroni, A., and Kirsch, K. (1978). *J. Cell Biol.* **77,** 722–734.

Wudl, L. R., Sherman, M. I., and Hillman, N. (1977). *Nature (London)* **270,** 137–140.

Yamada, K. M., and Olden, K. (1978). *Nature (London)* **275,** 179–184.

Yanagisawa, K. O., and Fujimoto, H. (1977). *J. Embryol. Exp. Morphol.* **40,** 277–283.

Yogeeswaran, G., Laine, R. A., and Hakamori, S. (1974). *Biochem. Biophys. Res. Commun.* **59,** 591–599.

SECTION 3

Glycosyltransferases in Fetal, Neonatal, and Adult Colon: Relationship to Differentiation

J. THOMAS LAMONT

I. INTRODUCTION

The mammalian colonic epithelium is less highly specialized than other parts of the gastrointestinal tract, having as its primary function the absorption of water and sodium from dietary waste products. Compared to the stomach and small intestine, the colon is richly invested with specialized goblet cells, which store mucin glycoproteins for secretion into the lumen. It appears that mucin provides a physical protective barrier against the highly toxic fecal contents, including proteases and glycosidases, microorganisms, ammonia, bile salts, and waste products (Forstner, 1978). Colonic epithelial cells, therefore, contain a relatively large amount of glycosyltransferases, which synthesize the carbohydrate side chains of mucin and membrane glycoproteins and glycolipids. These transferases are located in the Golgi apparatus, endoplasmic reticulum, and plasma membrane, and their level of activity appears to correlate with the functional differentiation of the colon, especially the capacity to synthesize mucus glycopro-

THE GLYCOCONJUGATES, VOL. III

TABLE I
Relationship between Glycoprotein Metabolism and Differentiation in
Colonic Epithelial Cells

	Cell type		
	Early fetal	Late fetal and adult	Neoplastic
Level of maturation	Undifferentiated	Differentiated	Dedifferentiated
Histology	Stratified epithelial cells	Goblet, absorptive, and undifferentiated columnar cells	Loss of structural organization, reduction in number of goblet or absorptive cells, and increase in undifferentiated cells
Glycosyltransferases	Low	High	Low
Carcinoembryonic antigen	Present	Absent or low levels	Increased

teins (LaMont and Ventola, 1978). The relationship between glycosyltransferases, cell-surface glycoproteins, and differentiation is of particular interest in the study of colonic cancer, the commonest form of intestinal malignancy in man. In this section, the biochemical characteristics of colonic glycoprotein glycosyltransferases are correlated with functional, morphological, and clinical aspects of the colon at various developmental levels (see Table I).

II. EMBRYONIC DEVELOPMENT AND DIFFERENTIATION OF COLON

A. Embryology of Colon

The embryonic development of mammalian colon has been most intensively studied in experimental animals, especially rats and mice. The proximal colon arises from the embryonal midgut, whereas the distal colon is an outgrowth of the primitive hindgut. Maturation of fetal rat colon occurs during the final 4 days of the 22-day gestational period (Helander, 1973). Up to 18 days, the fetal colon is lined by relatively undifferentiated stratified epithelium 3–10 cells thick. These cells contain a large nucleus, but their capacity to synthesize secretory proteins is quite limited, as evidenced by a sparse endoplasmic reticulum and minute Golgi apparatus. Goblet cells are absent. Between 18 and 22 days, the organ undergoes a period of dramatic differentiation. The stratified epithelium at 18 days is replaced by a single layer of columnar epithelium, which invaginates to form crypts surrounding a well-developed lumen. Two major cell types evolve: (a) surface absorptive cells lining the lumen and (b) rudimentary goblet cells containing mucin granules. Autoradiographic studies by Rampal *et al.* (1978) indicate that secretion of mucin into the lumen occurs between 20 and 22 days of gestation.

B. Differentiation of Glycoprotein Synthesis

The colon 15 days after birth is morphologically and functionally similar to adult colon in that it is capable of reabsorption of water and sodium chloride and secretion of mucus. The proximal and distal adult colon differ with regard to the type of mucin secreted, perhaps reflecting their different embryonic origin. Histochemical studies reveal that the predominant mucin in the proximal colon is a sialomucin. In the distal colon, sialomucin predominates in the lower half of the crypt, whereas sulfated mucins predominate in the upper half (Filipe, 1969, 1972). The functional significance of these histochemical differences is unknown. Recent studies from our laboratory indicate that separate sialo- and sulfomucins cannot be separated by gel filtration or ion-exchange chromatography and that both sulfate and sialic acid are contained on the same mucin molecule (LaMont and Ventola, 1980).

In addition to regional differences in the colon, glycoprotein synthesis may vary quantitatively and qualitatively in cells as they migrate from crypts to surface. *In vitro* autoradiographic studies reveal that fucose, glucosamine, *N*-acetylmannosamine, and sulfate are preferentially incorporated into surface columnar and goblet cells compared to deeper crypt cells (Neutra *et al.*, 1977). A similar gradient of glucosamine incorporation into small intestinal villus versus crypt cells was demonstrated after *in vivo* injection of precursor (Weiser, 1973a).

C. Neoplastic Dedifferentiation

As in other tissues, neoplasia of the colon is accompanied by histological and biochemical dedifferentiation. Although many colonic neoplasms in animals and man retain the ability to synthesize mucus, a reduction in the number of goblet cells as well as the loss of a well-defined brush border and glycocalyx are hallmarks of colonic neoplasia. Considerable variability among tumors occurs, ranging from well-differentiated benign tumors (adenomas) containing mature-appearing goblet cells to invasive, undifferentiated cancers containing cells similar to those seen in the primitive embryonic colon (Morson and Dawson, 1972). Qualitative alterations in glycoproteins in colonic neoplasia have been documented by histochemical and biochemical techniques. In both experimental animals and man, striking differences in mucin histochemistry are observed in colon cancers, as well as in the normal-appearing mucosa (transitional epithelium) surrounding tumors (Filipe, 1979). Alterations in mucin histochemistry may actually precede histological evidence of cellular atypia (Filipe, 1972). The most common finding is a generalized increase in sialomucin with the goblet cell and a decrease in the amount of sulfomucin. It is not clear whether these changes are primary to the process of neoplastic change or are a secondary response to injury by chemical or physical carcinogens. The availability of a potent colonic carcinogen, 1,2-dimethylhydrazine, should make it possible to undertake sequential studies of the biochemical changes in colonic glycoproteins during the development of colon cancer (LaMont and O'Gorman, 1978).

III. COLONIC GLYCOPROTEIN BIOSYNTHESIS

A. Autoradiographic and Biochemical Studies

Much of our knowledge of colonic glycoprotein biosynthesis is derived from autoradiographic studies in experimental animals injected *in vivo* with tritiated monosaccharide precursors before being killed. The advantage of using this experimental approach is that it demonstrates the type of cells involved in glycoprotein synthesis and the subcellular organelles in which glycoproteins are produced. Both goblet and columnar cells of rat colon synthesize and secrete glycoproteins continuously throughout their 5- to 6-day migration from crypt to surface (Neutra and Leblond, 1966a,b). Monosaccharide precursors are first incorporated into nascent glycoproteins in the supranuclear Golgi complex, from which they migrate apically to endoplasmic reticulum and then to mucus granules. After 4–8 hours, newly synthesized mucus is slowly discharged into the lumen. In absorbtive (nongoblet) cells, the intracellular migration of newly synthesized glycoproteins is much more rapid, reaching the glycocalyx, or fuzzy coat of the microvillus membrane, within approximately 30 minutes after injection of precursor. Intracellular migration of glycoprotein precursors in colonic epithelium of man has been studied *in vitro* and follows a similar but slower pattern than that described for rat colon (Neutra *et al.*, 1977). As noted above, an increase in glycoprotein synthesis in surface compared to crypt cells has been noted in human rectal epithelial biopsies incubated with monosaccharide precursors. It has been suggested that small intestinal epithelial glycoproteins undergo a similar gradient of synthesis, with the innermost core sugars (hexosamines and neutral sugars) being attached in the crypt region but the terminal sialic acid being attached as the cells migrate to the differentiated villous region (Weiser, 1973a,b).

Biochemical characterization of colonic glycoproteins has focused primarily on mucin, the major secretory product of the colon. Precursor labeling studies with [14C]threonine and [14C]glucose in sheep colonic scrapings indicated that the synthesis of carbohydrate side chains of mucin is strongly dependent on prior formation of the peptide portion of the molecule (Draper and Kent, 1963; Allen and Kent, 1968). Puromycin inhibited the incorporation of threonine into colonic mucin to a much greater degree than the incorporation of glucose or sulfate. Similar results were obtained with cycloheximide in rabbit colon by MacDermott *et al.* (1974). The compositional and structural features of colonic mucin are quite similar to those of other epithelial mucins (see Volume I of this series). For example, pig colonic mucin is a very large macromolecule (MW 15×10^6) consisting of 13.3% protein and 73.3% carbohydrate (Marshall and Allen, 1978). A smaller subunit of MW 670,000 is obtained by pronase digestion followed by mercaptoethanol reduction. Because of the very large size of the mucin molecule, it seems quite likely that polypeptide subunits of much lower molecular weight are first synthesized on membrane-bound ribosomes and then glycosylated in the Golgi apparatus. The glycosylated subunits then aggregate to form larger polymers within the membrane-bound thecae of the goblet cell (Forstner, 1978).

B. Glycoprotein Synthesis in Colonic Neoplasia

Major alterations of both membrane and mucin glycoprotein synthesis occur in both experimental and human colonic neoplasms. Gold and his colleagues (1965; Gold and Freedman, 1965) first reported the occurrence of a tumor-associated antigen, carcinoembryonic antigen (CEA), in human colon cancers and fetal colon but not in normal human colon. The CEA was subsequently shown to be a glycoprotein of MW 180,000 (Slater and Coligan, 1975) occurring predominantly on the glycocalyx of the malignant cell and to a much lesser extent on the normal colonic epithelial cell (Gold *et al.*, 1970). The serum of colon cancer patients contains elevated amounts of CEA, which can be used as a diagnostic and prognostic test in the management of this disease (Thompson *et al.*, 1969). It is thought that CEA is a normally occurring embryonic glycoprotein that disappears in normally differentiated adult colon but reappears in dedifferentiated tumor cells. This process may involve defective glycosylation of normal glycoprotein precursors in the tumor cells. Evidence for a diminished number of carbohydrate constituents in human colon cancers has been provided by compositional analysis of glycoproteins from tumors and normal colon tissue. Kim and Isaacs (1975) showed that tumor glycoporteins contained reduced amounts of sialic acid, fucose, and hexosamines. These differences were not secondary to enhanced activity of glycosidase enzymes, but were correlated with diminished glycosyltransferase activities (see below).

Studies from our laboratory confirm that glycoprotein synthesis is diminished in human colon cancers (LaMont and O'Gorman, 1978). We measured glycoprotein synthesis and secretion in organ culture of human colon mucosa obtained at colectomy for colon cancer. Incorporation of [³H]glucosamine into neoplastic tissue, particularly the membrane fraction, was significantly diminished compared to normal tissue. Colon cancer membranes have also been shown to have diminished blood group A, H, and Le^a activity but higher levels of blood group H precursor activity, again suggesting incomplete glycosylation of normally occurring glycoproteins and glycolipids in colon tumors (Kim *et al.*, 1974).

IV. COLONIC EPITHELIAL GLYCOSYLTRANSFERASES

A. Enzyme Characteristics

The glycosyltransferase enzymes in intestinal epithelial cells appear to have pH optima, cation requirements, and substrate specificities quite similar to those described in other tissues. The pH optima for small intestinal (Kim *et al.*, 1971) and colonic (Kim and Isaacs, 1975) transferases is between 6 and 7, and the enzymes show a pronounced dependence on manganese. Addition of Triton X-100 stimulate galactosyltransferase activity from small intestine (Kim *et al.*, 1971) and colon (LaMont and Ventola, 1978). Intestinal glycosyltransferase activity has been measured with endogenous (Weiser, 1973b) as well as exogenous (Kim *et al.*, 1971) receptors. The endogenous receptor in rat small intestine appears to be incompletely glycosylated mucin glycoprotein still attached at its site of glycosylation in the smooth membranes. Interestingly, rat colon galactosyltransferase activity with endogenous acceptors·cannot be measured (LaMont and Ventola, 1978). Exogenous acceptors for intestinal glycosyltransferases can be either oligosaccharides or proteins. For example, galactosyltransferase of rat small intestine can transfer galactose to *N*-acetylglucosamine, forming lactosamine, and sialyltransferase can add sialic acid to lactose, producing sialyllactose (Weiser, 1973b). Exogenous protein acceptors are serum glycoproteins such as fetuin or orosomucoid from which sialic acid and galactose have been removed.

Weiser (1973b) demonstrated that the specific activity of cell-surface glycosyltransferases in small intestinal epithelial cells is related to their level of differentiation. Intestinal crypt cells are undifferentiated and mitotically active, in contrast to villus cells, which are differentiated with regard to transport and digestive functions and do not divide. The levels of galactosyl-, *N*-acetylglucosaminyl-, and fucosyltransferases were approximately 10-fold greater on crypt versus villus cells, whereas sialyltransferase was comparatively higher on villus cells. High levels of certain cell-surface transferases has been correlated with mitotic activity in other cell systems, including lymphocytes (LaMont *et al.*, 1974a), tissue culture fibroblasts (LaMont *et al.*, 1977), and tumor cells (Bosmann *et al.*, 1974). This correlation has not been confirmed in colonic epithelial cells owing to the difficulty of preparing isolated cell fractions from deep crypt (undifferentiated) and surface (differentiated) cells.

B. Glycosyltransferases in Fetal Intestine

During fetal development, the level of glycosyltransferases increases in certain tissues but declines in others. Glycosyltransferase levels in rat fetal pancreas increase markedly between 11 and 12 days *in utero* to birth (Carlson *et al.,* 1973) and decrease in chick brain (Garfield and Ilan, 1976) and rat liver (Jato-Rodriguez and Mookerjea, 1974). We studied the level of galactosyltransferase in developing rat colon and correlated this with morphological differentiation and autoradiographic studies with tritiated monosaccharide precursors (LaMont and Ventola, 1978; Rampal *et al.,* 1978). Galactosyltransferase activity in fetal colon homogenates increased four- to sevenfold between 18 and 22 days, the last 4 days of gestation. Enzyme activity then increased rather slowly after birth to reach adult level at about day 15. Fetal galactosyltransferase activity was identical to adult activity with regard to pH optimum (6.8–7.2), K_m for UDP-galactose ($6.1 \times 10^{-5}M$), and stimulatory effect of manganese and Triton X-100. CDP-Choline had no stimulatory effect on fetal galactosyltransferase but did stimulate adult enzyme approximately twofold at 2 mM (final concentration CDP-choline in assay). This effect was shown to be secondary in part to the protection of substrate UDP-galactose against enzymatic hydrolysis. Coincident with the increase in galactosyltransferase activity in the last 4 days of fetal life was a marked acceleration of incorporation of [³H]fucose and [³H]galactose but not [³H]glucosamine into fetal colonic glycoproteins following injection of these precursors into the mothers. Sepharose 4B chromatography of the radiolabeled glycoproteins in colonic cytosol revealed that [³H]galactose was incorporated into two major glycoproteins including a high molecular weight (>2,000,000) fraction in the void volume and a lower molecular weight included peak. Interestingly, autoradiographic studies revealed selective incorporation of [³H]galactose into goblet cells but not columnar absorptive cells. Similar studies in rat stomach have indicated that galactosyltransferase activity is located primarily in mucus-secreting cells (Poort, 1977). These results indicate that maturation of fetal rat intestine during the last 4 days of gestation is accompanied by the appearance of goblet cells, enhanced synthesis of mucin glycoproteins, and elevation of membrane-bound galactosyltransferase enzyme activity.

C. Glycosyltransferases in Adult Colon

Adult colonic epithelial cells are rich in glycosyltransferases owing to the large number of mucin goblet cells in this organ. Regional differences between the proximal and distal colon of rats with regard to glycosyltransferase activity were noted (Freeman *et al.,* 1978). Galactosyltransferase and sialyltransferase activity were significantly greater in proximal than in distal colonic mucosa. The injection of rats with 1,2-dimethylhydrazine, a potent colonic carcinogen, results in increased activity of galactosyltransferases in normal-appearing mucosa from tumor-bearing rats (Freeman *et al.*, 1978).

Compared to normal colonic epithelia, human colonic adenocarcinomas have reduced levels of various glycosyltransferases, including galactosyltransferase, N-acetylglucosaminyltransferase, fucosyltransferase, and sialyltransferase (Kim *et al.*, 1974; LaMont and Isselbacher, 1975). We showed that colon cancer homogenates had a reduced capacity to transfer galactose or fucose to partially purified CEA when compared to normal tissue, suggesting that CEA may accumulate in tumor tissue because of deficient glycosylation of a normal glycoprotein precursor (LaMont and Isselbacher, 1975). Kim and Isaacs (1975) defined three distinct galactosyltransferase enzymes in colonic epithelium on the basis of acceptor (substrate) specificities. Transfer of galactose to protein acceptors with terminal N-acetylglucosamine residues was significantly reduced in tumor homogenates compared to normal, whereas transfer to protein acceptors with terminal N-acetylgalactosamine residues was unimpaired. Reduced levels of glycosyltransferase enzymes are also observed in experimental colon cancers induced by the injection of 1,2-dimethylhydrazine (LaMont *et al.*, 1974b; Freeman *et al.*, 1978).

Human fetal and colon cancer cell lines maintained in organ culture release a significant amount of soluble galactosyltransferase into the medium during growth in culture (Whitehead *et al.*, 1979). This secretion appears to be an active process rather than a result of cell breakdown or death. Podolsky and Weiser (1975) and Podolsky *et al.* (1977) demonstrated that the sera of experimental animals and human beings with malignancies of various organs contain high levels of a cancer-associated galactosyltransferase isoenzyme not present in the sera of non-tumor patients (see Chapter 4, Section 3, Volume IV). These observations raise the possibility that a colon tumor-specific galactosyltransferase might be released from colon tumor cells into the circulation and be useful as a diagnostic test in the management of this disease.

V. SUMMARY

Colonic epithelial cells are rich in glycosyltransferase enzymes, which synthesize the oligosaccharide portion of mucin and membrane glycoproteins. The development of these enzymes in late fetal colonic epithelium parallels the appearance of goblet cells and the capacity to synthesize high molecular weight mucin (see Table I). Similarly, the reduction of glycosyltransferases in neoplastic colon cells correlates with the reduction in the number of goblet cells, impaired glycosylation of membrane glycoproteins, and appearance of "carcinofetal" glycoproteins with incomplete carbohydrate side chains like CEA. Thus, mucin glycosyltransferase enzymes reflect the level of differentiation of the colonic epithelial cell.

REFERENCES

Allen, A., and Kent, P. W. (1968). *Biochem. J.* **106,** 301–309.

Bosmann, H. B., Case, K. R., and Morgan, H. R. (1974). *Exp. Cell Res.* **83,** 15–24.

Carlson, D. M., David, J., and Rutter, W. J. (1973). *Arch. Biochem. Biophys.* **157,** 605–612.

Draper, P., and Kent, P. W. (1963). *Biochem. J.* **86,** 248–254.

Filipe, M. I. (1969). *Gut* **10,** 577–586.

Filipe, M. I. (1972). *J. Clin. Pathol.* **25,** 123–128.

Filipe, M. I. (1979). *Invest. Cell Pathol.* **2,** 195–216.

Forstner, J. F. (1978). *Digestion* **17,** 234–263.

Freeman, F. J., Kim, Y., and Kim, Y. S. (1978). *Cancer Res.* **38,** 3385–3390.

Garfield, S., and Ilan, J. (1976). *Biochim. Biophys. Acta* **444,** 154–163.

Gold, P., and Freedman, S. O. (1965). *J. Exp. Med.* **121,** 439–462.

Gold, P., Gold, M., and Freedman, S. O. (1968). *Cancer Res.* **28,** 1331–1333.

Gold, P., Krupey, J., and Ansari, H. (1970). *JNCL, J. Natl. Cancer Inst.* **45,** 219–225.

Helander, H. F. (1973). *Acta Anat.* **85,** 153–176.

Jato-Rodriguez, J. J., and Mookerjea, S. (1974). *Arch. Biochem. Biophys.* **162,** 281–292.

Kim, Y. S., and Isaacs, R. (1975). *Cancer Res.* **35,** 2092–2097.

Kim, Y. S., Perdomo, J., and Nordberg, J. (1971). *J. Biol. Chem.* **246,** 5466–5476.

Kim, Y. S., Isaacs, R., and Perdomo, J. S. (1974). *Proc. Natl. Acad. Sci. U.S.A.* **71,** 4869–4875.

LaMont, J. T., and Isselbacher, K. J. (1975). *JNCL, J. Natl. Cancer Inst.* **54,** 53–56.

LaMont, J. T., and O'Gorman, T. A. (1978). *Gastroenterology* **75,** 1157–1162.

LaMont, J. T., and Ventola, A. S. (1978). *Am. J. Physiol.* **235,** E213–E217.

LaMont, J. T., and Ventola, A. S. (1980). *Biochim. Biophys. Acta* **626,** 234–243.

LaMont, J. T., Perrotto, J. L., Weiser, M. M., and Isselbacher, K. J. (1974a). *Proc. Natl. Acad. Sci. U.S.A.* **71,** 3726–3730.

LaMont, J. T., Weiser, M. M., and Isselbacher, K. J. (1974b). *Cancer Res.* **34,** 3225–3228.

LaMont, J. T., Gammon, M. T., and Isselbacher, K. J. (1977). *Proc. Natl. Acad. Sci. U.S.A.* **74,** 1086–1090.

MacDermott, R. P., Donaldson, R. M., and Trier, J. S. (1974). *J. Clin. Invest.* **54,** 545–554.

Marshall, T., and Allen, A. (1978). *Biochem. J.* **173,** 569–578.

Morson, B. C., and Dawson, I. M. (1972). "Gastrointestinal Pathology." Blackwell, Oxford.

Neutra, M. R., and Leblond, C. P. (1966a). *J. Cell Biol.* **30,** 119–136.

Neutra, M. R., and Leblond, C. P. (1966b). *J. Cell Biol.* **30,** 137–150.

Neutra, M. R., Grand, R. J., and Trier, J. S. (1977). *Lab. Invest.* **36,** 535–546.

Podolsky, D. K. and Weiser, M. M. (1975). *Biochem. Biophys. Res. Comm.* **65,** 545–551.

Podolsky, D. K., Weiser, M. M., Westwood, J. C., and Gammon, M. (1977). *J. Biol. Chem.* **252,** 1807–1813.

Poort, C. (1977). *J. Histochem. Cytochem.* **25,** 57–60.

Rampal, P. J., LaMont, J. T., and Trier, J. S. (1978). *Am. J. Physiol.* **235,** E207–E212.

Slater, H. S., and Coligan, J. E. (1975). *Biochemistry* **14,** 2323–2330.

Thompson, D. M., Krupey, J., Freedman, S. O., and Gold, P. (1969). *Proc. Natl. Acad. Sci. U.S.A.* **64,** 161–167.

Weiser, M. M. (1973a). *J. Biol. Chem.* **248,** 2536–2541.

Weiser, M. M. (1973b). *J. Biol. Chem.* **248,** 2542–2548.

Whitehead, J. S., Fearney, F. J., and Kim, Y. S. (1979). *Cancer Res.* **39,** 1259–1263.

SECTION 4

Proteoglycans in Developing Embryonic Cartilage

PAUL F. GOETINCK

I. INTRODUCTION

Differentiation is a developmental process that involves the appearance of functionally specialized cells from less specialized precursor cells. The terminal differentiated state of a cell represents the culmination of a series of restrictions that are imposed on precursor cells in terms of their developmental potentialities. The events associated with the differentiation of chondrocytes (chondrogenesis) have been studied in a number of embryonic organ systems. The vertebral cartilages and the cartilage of the appendicular skeleton in particular have been the subject of intense investigations. These studies have been concerned with the characterization of the specific morphological and biochemical properties of chondrocytes as well as with the regulatory mechanisms that lead to their differentiated state.

Studies on the differentiation of cartilage cells differ from studies concerned with the maintenance of the already differentiated state of chondrocytes. Chondrocytes can, under certain conditions, alter the expression of their differentiated state in response to certain external stimuli. This phenomenon, often termed "dedifferentiation," is a reversible process and often mimics certain aspects of normal developmental processes (Abbott and Holtzer, 1968; Holtzer and Abbot,

197

1968; Cahn, 1968; Abbott *et al.,* 1972; Holtzer *et al.,* 1972; Palmoski and Goetinck, 1972; Levitt and Dorfman, 1973, 1974; Mayne *et al.,* 1975, 1976; Okayama *et al.,* 1977; Pacifici *et al.,* 1977; Hassell *et al.,* 1978; Lewis *et al.,* 1978; Gross and Rifkin, 1979; Pennypacker and Goetinck, 1979). Although these studies are directly related to the understanding of chondrogenesis, they are not reviewed here. This section is concerned strictly with changes associated directly with the process of differentiation of chondrocytes.

A number of hypotheses have been advanced with respect to the observed changes in the biochemical parameters used in evaluating chondrogenesis. Early studies were concerned with the changes in synthesis of glycosaminoglycans, particularly chondroitin sulfate, during cartilage differentiation. Using this parameter, studies revealed that chondroitin sulfate was synthesized at very low levels in precartilaginous tissues and that the synthesis increased dramatically during chondrogenesis. These studies led to a hypothesis which stated that chondrogenic expression, as measured by chondroitin sulfate synthesis, involves a mere increase in the synthesis of a preexisting molecule (Zwilling, 1968). A similar hypothesis was advanced as attention turned from the synthesis of glycosaminoglycans to the synthesis of the parent molecule, the proteoglycan, during chondrogenesis (Goetinck *et al.,* 1974; Vasan and Lash, 1977; Lash and Vasan, 1978). Now, however, the overwhelming body of evidence indicates that new proteoglycans appear during chondrogenesis (Levitt and Dorfman, 1973; Okayama *et al.,* 1976; Kitamura and Yamagata, 1976; DeLuca *et al.,* 1977, 1978; McKeown and Goetinck, 1979; Sawyer and Goetinck, 1981; Goetinck *et al.,* 1981). The molecular basis for these changes is reviewed in this section. Experimental evidence is presented which indicates that changes occur in both the glycosaminoglycans and the core protein of the proteoglycans and that changes in the core protein may represent the activation of different genes during chondrogenesis.

The following reviews will provide the reader with a thorough background on the many experimental studies aimed at the understanding of chondrogenesis: Thorp and Dorfman (1967), Holtzer (1968), Lash (1968), Zwilling (1968), Holtzer *et al.* (1972), Levitt and Dorfman (1974), Manasek (1975), and Ede (1977).

II. CELLULAR ASPECTS OF CARTILAGE DEVELOPMENT

Vertebral cartilage differentiates from the somites. The latter are formed as a result of the condensation of loosely bound mesodermal cells into segmentally arranged blocks of tissues on either side of the developing neural tube and notochord. The differentiation of the somite cells into chondrocytes depends on interactions with the notochord and neural tube. This interaction is a classical example of a secondary embryonic induction, which in the case of somite chondrogenesis can be demonstrated *in vivo* and *in vitro* (Holtzer, 1968; Lash, 1968).

The general features leading to cartilage formation in the limb of the avian embryo are as follows. The limb buds arise from the somatopleural thickening lateral to the somites at stage 16–17 (51–64 hours of incubation) of Hamburger and Hamilton (1951). Up to stage 22 (3.5–4 days) the mesodermal core of the limb bud is histologically homogeneous. At about stage 22–23 (4 days), primordia of the cartilaginous skeletal elements can be recognized as condensations in the core of the limb (Fell and Canti, 1934). At stage 25 (4.5–5 days) overtly differentiated chondrocytes can be identified microscopically.

Chondrogenesis in the limb can be examined either in intact limb buds at various stages of development or in cultures derived from stage-24 limb bud cells. When such stage-24 limb bud cells are cultured at high density (5×10^6 cells per 35-mm culture dish), chondrogenesis takes place *in vitro* (Caplan, 1970; Schacter, 1970; Goetinck *et al.*, 1974; Levitt *et al.*, 1975). A micromass adaptation of the cell culture system has been devised (Solursh *et al.*, 1978) in which 2×10^5 cells in 10 μl of medium are allowed to attach to a culture dish. This method allows one to investigate chondrogenesis in cells derived from limb buds of a single 4-day embryo (Sawyer and Goetinck, 1981). The chondrogenesis seen in the somite and the limb systems is a normally occurring developmental event that can be studied either *in vivo* or *in vitro*. An artificial induction system has also been used to study chondrogenesis. This method involves the subcutaneous implantation of demineralized bone matrix into young rats or the culturing of minced embryonic rat muscle on demineralized bone matrix (Nogami and Urist, 1970, 1974; Reddi and Huggins, 1972; Reddi and Anderson, 1976; Reddi *et al.*, 1978; Nathanson *et al.*, 1978; Nathanson and Hay, 1980a,b; Weiss and Reddi, 1981).

In the chick embryo, the sternum is a particularly good organ from which to obtain a large number of relatively pure chondrocytes. Developmentally, the sternum of birds first appears as a pair of mesodermal condensations, which differentiate into two separated cartilaginous plates. These sternal plates move toward the midline and fuse to form the embryonic sternum (Fell, 1939). Although the sternum is a good organ from which to obtain pure chondrocytes, it is not as suitable an organ for studying the entire process of cartilage differentiation because of its complex developmental pattern.

The somite, limb, sternal, and matrix-induced chondrocytes all differentiate from mesenchymal precursor cells. However, chondrocytes of Meckel's cartilage and the hypobranchial skeleton are derived from the neural crest, which originates in the ectoderm, and thereby have a different developmental history. Meckel's cartilage serves as the embryonic mandible, the proximal portion of which gives rise to the articular bone of the adult mandible (Johnston, 1966; Le Lièvre, 1974; Le Lièvre and Le Douarin, 1975).

III. SULFATED PROTEOGLYCAN STRUCTURE

The current model of sulfated proteoglycans of cartilage has been derived mainly from studies on proteoglycan from bovine nasal septum, pig laryngeal cartilage, Swarm rat chondrosarcoma, and chick limb bud chondrocytes (Hascall, 1977). The proteoglycan monomer consists of a core protein of about 2.0×10^5 daltons to which are attached two types of sulfated side chains and two types of oligosaccharides. The sulfated carbohydrate chains are chondroitin sulfate and keratan sulfate (Rodén and Horowitz, 1978).

Chondroitin sulfate, a repeating unit of glucuronic acid and sulfated N-acetylgalactosamine, is linked to core proteins via a linkage region, which consists of glucuronic acid–galactose–galactose–xylose. The linkage to the core protein is through an O-glycosidic bond between xylose and the hydroxyl group of serine. Sulfation can be at either the 4 or the 6 position of the N-acetylgalactosamine. Typically there are 80–100 chondroitin sulfate chains of about 20,000 daltons per protein backbone.

Keratan sulfate, a repeating disaccharide of N-acetylglucosamine and galactose, is linked, in cartilage, to core protein by an O-glycosidic bond between N-acetylgalactosamine and hydroxyl groups of serine and threonine. Keratan sulfate occupies the 6 position of the linkage N-acetylgalactosamine, and position 3 of this linkage amino sugar is occupied by a disaccharide, galactose–N-acetylneuraminic acid. The N-acetylglucosamine and variably the galactose residues are sulfated. There are typically 20–30 keratan sulfate residues of about 6,000–10,000 daltons per protein backbone. There exists a third type of sulfated glycosaminoglycan, dermatan sulfate. This glycosaminoglycan is a chondroitin sulfate (formerly called chondroitin sulfate B) in which the principal uronic acid is L-iduronic acid. The formation L-iduronic acid residues is catalyzed by an epimerase, which inverts the C-5 in D-glucuronic acid after the uronic acid has been incorporated in the polymer (Malmström *et al.*, 1975; Lindahl *et al.*, 1972). Dermatan sulfate is not found in the major cartilage proteoglycan, although its presence has been reported in minor proteoglycan fractions (Kimata *et al.*, 1978) and in proteoglycans synthesized by dedifferentiated human chondrocyte cultures (Oegema and Thompson, 1981).

In addition to the chondroitin sulfate and keratan sulfate side chains, two classes of oligosaccharides have been described in proteoglycans from chick limb bud chondrocytes and Swarm rat chondrosarcoma (DeLuca *et al.*, 1980; Lohmander *et al.*, 1980). One class, made up of three structurally related oligosaccharides of different size, is a neuraminic acid-rich oligosaccharide that is linked to core protein via the same linkage as keratan sulfate, i.e., an O-glycosidic bond between N-acetylgalactosamine and hydroxyl group of serine and threonine. The second class, a neuraminic acid-containing, mannose-rich oligosaccharide, appears to be linked to asparagine residues through N-glycosylamine bonds to N-acetylglucosamine.

The proposed generalized model of proteoglycan separates the molecule into three domains: A hyaluronic acid-binding domain, which is relatively free of carbohydrate, represents one end of the core protein through which the proteoglycan interacts with hyaluronic acid to form aggregates; adjacent to the hyaluronic acid-binding domain is the keratan sulfate-attachment domain of the

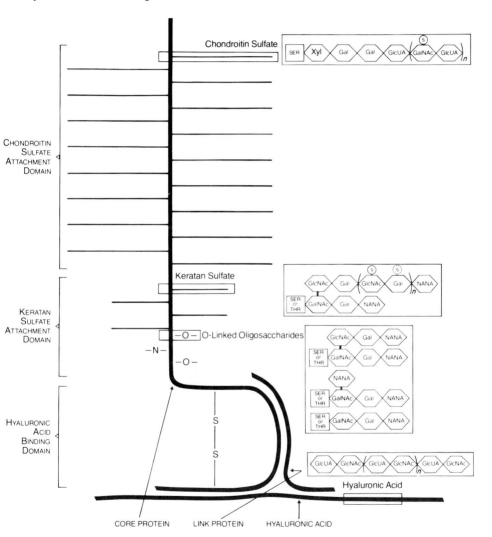

Figure 1 Schematic representation of a proteoglycan monomer interacting with hyaluronic acid and link protein. A compositional diagrammatic representation of chondroitin sulfate, keratan sulfate, the O-linked oligosaccharides, and hyaluronic acid is presented in the insets. Abbreviations: Gal, galactose; GalNAc, N-acetygalactosamine; GlcNAc, N-acetyglucosamine; GlcUA, glucaronic acid; NANA, N-acetyl neuraminic acid; S, sulfate; SER, sereine; THR, threonine; Xyl, xylose.

core protein; and finally, the other end of the core protein represents the chon-droitin sulfate-attachment domain. A diagrammatic representation of a cartilage proteoglycan molecule is given in Figure 1. The hyaluronic acid-binding domain can also interact with link protein. The interaction of link protein with the proteoglycan and the hyaluronic acid serves to stabilize the aggregate structure.

IV. PROTEOGLYCAN SYNTHESIS

The synthesis of a complex macromolecule such as a proteoglycan clearly involves a large number of biosynthetic steps. The sequence of biosynthetic events begins with the synthesis of a core protein, which, as a secreted protein, must be synthesized on ribosomes bound to the endoplasmic reticulum. The cell-free translation product of chick embryonic chondrocyte core protein is estimated to be about 340,000 daltons (Upholt et al., 1979). A considerable amount of protein processing must take place between the initial translation product and the final proteoglycan core protein, which is estimated to be about 200,000 daltons (Zimmerman et al., 1980). One of these steps is likely to be the removal of a signal peptide, but this modification may not be enough to account for the entire difference between the molecular weight of the cell-free translation product and the final core protein as determined by biophysical methods. Next are a series of posttranslational modifications of the core protein involving the synthesis of polysaccharides and oligosaccharides. The initiation of chon-droitin sulfate synthesis begins with the transfer of xylose from the uridine nucleotide donor to serine residues of the core. The synthesis of the linkage region takes place through the sequential addition of galactose, galactose, and glucuronic acid residues followed by the synthesis of the repeating disaccharide, which is finally further modified by the addition of sulfate groups. The synthesis of chondroitin sulfate as a component of proteoglycan therefore requires the activity of six different glycosyltransferases as well as the three enzymes that are involved in the activation and the transfer of sulfate groups to the N-acetylgalactosamine residues of the repeating disaccharide (Rodén and Horowitz, 1978).

The keratan sulfate and the O-linked oligosaccharides, which are structurally similar to keratan sulfate, are also the result of the activity of glycosyltransferases resulting in the transfer of N-acetylgalactosamine, galactose, and neuraminic acid. The N-linked oligosaccharides, which are rich in mannose, would require the enzymatic machinery for the synthesis of lipid intermediates and the synthesis of the mannose complex found in the synthesis of glycoproteins (Narasimhan et al., 1980; Harpaz and Schachter, 1980a,b). The general sequence of the events leading to the complete synthesis of a proteoglycan can be summarized as fol-lows:

I. *Synthesis of core protein*
  A. Initial translation of core protein mRNA
  B. Posttranslational processing of core proteins
II. *Posttranslational modification of core protein*
  A. Synthesis of polysaccharide chains
    1. Chondroitin sulfate
      a. Chain initiation
      b. Completion of carbohydrate–protein linkage region
      c. Synthesis of repeating disaccharide
      d. Modification of polysaccharide chain, *i.e.,* sulfation
    2. Keratan sulfate
      a. Chain initiation
      b. Synthesis of repeating disaccharide at the 6 position of the linkage amino sugar
      c. Capping of repeating disaccharide with neuraminic acid
      d. Capping of galactose with neuraminic acid
      e. Modification of polysaccharide chains, *i.e.,* sulfation
  B. Synthesis of oligosaccharide chains
    1. O-Linked oligosaccharide
      a. Chain initiation and various additions of sugars at the 6 position
      b. Capping of the 6 position on sugars linked to the 6 position of linkage amino sugar with neuraminic acid
      c. Addition of galactose to the 3 position of linkage amino sugar
      d. Capping of galactose with neuraminic acid
    2. N-Linked oligosaccharide

The order in which the steps are outlined above is not meant to imply a biosynthetic sequence. The list of events may not be complete, and some of the steps are speculative.

It is clear that many biosynthetic events must take place in a temporally and spatially regulated fashion in order for a fully functional proteoglycan to be synthesized. The synthesis of such a molecule, therefore, requires the activity of a large number of genes. For example, the synthesis of a hypothetical proteoglycan consisting of a core protein and only chondroitin sulfate side chains would require a minimum of ten different proteins and therefore at least as many genes coding for the polypeptides involved. This number does not include the regulatory genes involved in the synthesis of a particular proteoglycan. For the synthesis of the proteoglycan, one would add the number of genes coding for the polypeptides involved in the synthesis of keratan sulfate and the O- and N-linked oligosaccharides. Any one of the genes involved in the synthesis of a proteoglycan is a potential site of regulation during development. Changes in the enzymatic activity could, for example, alter the number, size, kind, and composition of the polysaccharide or oligosaccharide of the proteoglycan and thereby affect the functional properties of this macromolecule. Furthermore, mutations affecting either the structure or the regulation of the genes involved could also lead to a macromolecule with altered properties. For example, mice that are homozygous for the recessive gene, brachymorphy, have a proteoglycan with undersulfated glycosaminoglycans. This results from a defective enzyme in the synthesis of the sulfation cofactor $3',5'$-phosphoadenosine phosphate sulfate (PAPS) (Orkin *et al.*, 1976; Schwartz *et al.*, 1978). In tissues of active proteoglycan synthesis, PAPS becomes rate limiting, and as a result glycosaminoglycans are undersulfated. Phenotypically, such mice have shortened limbs. Nanomelia, a recessive mutation in the chicken that acts at the level of the core protein, also has dramatic developmental consequences (Goetinck *et al.*, 1981).

V. EXTRACTION AND TERMINOLOGY OF PROTEOGLYCANS

Proteoglycans are usually obtained from cartilage by means of the nondisruptive extraction method introduced by Sajdera and Hascall (1969) and Hascall and Sajdera (1969). The general isolation procedure is shown in Figure 2. In this method, cartilage is extracted in a dissociative solvent, usually $4\ M$ guanidine hydrochloride (GuHCl). At this relatively high ionic strength, the proteoglycan aggregates are dissociated. Upon cesium chloride (CsCl) density equilibrium centrifugation under associative conditions, the proteoglycan aggregates are found in the bottom fraction. According to the nomenclature of Heinegård (1972) this is the A_1 fraction. The aggregate-containing fraction can then be brought to dissociative conditions ($4\ M$ GuHCl) and again subjected to a CsCl density equilibrium centrifugation. Under these conditions the proteoglycan aggregate

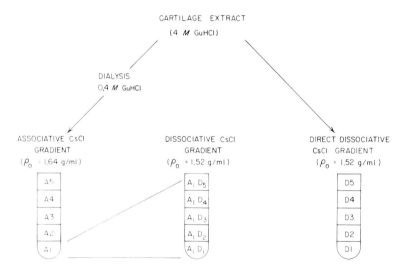

Figure 2 General method of purification and designation of proteoglycan fractions obtained by CsCl density gradient centrifugation. The nomenclature is that introduced by Heinegård (1972).

dissociates, and, after centrifugation, the bottom fraction (A_1D_1) contains monomer and the top fraction (A_1D_5) contains link protein. Hyaluronic acid is found in the middle of this gradient (Gregory, 1973). This procedure results in fractions enriched in particular components and does not represent an absolute purification scheme. This is particularly true for the A_1D_5 fraction, which contains proteoglycans of low buoyant density in addition to link protein (Heinegård, 1972; Baker and Caterson, 1977; Oegema et al., 1977; McKeown-Longo et al., 1981). Alternatively, the cartilage extract can be subjected directly to a CsCl density equilibrium centrifugation step under dissociative conditions. The bottom fraction (D_1) of such a gradient contains mostly monomer.

Although proteoglycans are usually extracted under dissociative conditions, they can be extracted under associative conditions from certain tissues (Faltz et al., 1979; Royal et al., 1980). This extraction may be followed by a second extraction under dissociative conditions. A terminology for designating such extraction procedures was introduced by Faltz et al. (1979). According to this method the lowercase letters "a," "d," or "ad" preceding a particular fraction indicates that the extraction was done, respectively, under associative, dissociative, or associative followed by dissociative conditions. It has become a general practice to perform proteoglycan extractions in the presence of protease inhibitors (Oegema et al., 1975).

Proteoglycans are not always first characterized by CsCl centrifugation as described above. In many instances, tissue extracts are characterized directly either by molecular sieve chromatography or by sucrose density gradient centrifugation. As a result of these different methods of characterization, different designations have been introduced to describe the various proteoglycan fractions. From a number of studies it is clear that different tissues synthesize unique types of proteoglycans. The basis for the differences, however, is not evident in any designation based on a single physicochemical criterion. For example, proteoglycans of cartilages with two different developmental histories have been shown to have different sedimentation rates on sucrose gradients (McKeown and Goetinck, 1979), and proteoglycans from a single type of cartilage differ in their sedimentation rates on sucrose gradients depending on the experimental conditions (Kato et al., 1978). In the latter instance the sedimentation rate of a cartilage proteoglycan overlaps with the sedimentation rate of a proteoglycan from noncartilaginous tissues (Kitamura and Yamagata, 1976; Okayama et al., 1976).

The heterogeneity observed in proteoglycans of embryonic tissues may be developmentally significant and may in fact reflect the differential expression of the genetic information in these various tissues. In order to reflect both the developmental and physicochemical basis of proteoglycan heterogeneity, we have adopted a system of identification of proteoglycans that is based on their chromatographic behavior on molecular sieve chromatography on controlled pore glass (CPG) and on the tissue of origin (McKeown and Goetinck, 1979). For example, when a sternal cartilage (SC) extract is chromatographed on CPG 1400, we obtain a profile consisting of an excluded peak [PGS(SC)-I] and an included peak [PGS(SC)-II]. Peak PGS(SC)-I consists of both proteoglycan aggregates and monomers, which can be separated from each other by chromatographing the CPG 1400 V_0 material on CPG 2500. On these columns, aggregates are excluded [PGS(SC)-Ia], and monomers are included [PGS(SC)-Ib]. A general scheme describing our system of identification is presented in Figure 3. The profile on the right in Figure 3 is that of cartilage proteoglycan. For comparison, our identification system, along with that adopted by others, is described in Table I. The fractions obtained after molecular sieve chromatography can be characterized further using functional, chemical, physical, or immunochemical criteria. Functional properties include the ability to interact with

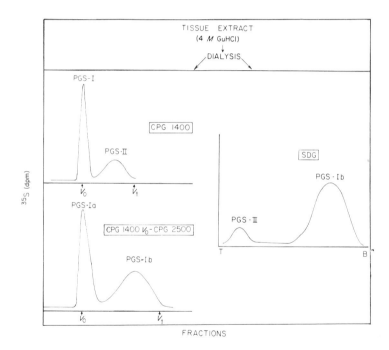

Figure 3 General scheme of designation of proteoglycan fractions obtained by molecular sieve chromatography on controlled pore glass (CPG) (left) and 5–20% dissociative sucrose density gradient (SDG) (right). The CPG 1400 V_0 fraction (PGS-I) contains both monomer and aggregates of mesenchymal and cartilage extracts. When the CPG 1400 V_0 material is chromatographed on CPG 2500, aggregates (PGS-Ia) elute in the void volume and monomer (PGS-Ib) is included. Since the 5–20% sucrose gradients are made up under dissociative conditions (4 M guanidine-HCl), the PGS-I is seen as a monomer (PGS-Ib). PGS-II sediments at the top of the gradient. The sucrose density gradient profile shown is that of a cartilage extract (T, top; B, bottom).

TABLE I
Comparison of Proteoglycan Nomenclature Adopted by Various Investigators

Tissue of origin	Method of separation				
	Controlled pore glass[a]	Agarose (A-50m)[b]	Sucrose[c]	Sucrose[d]	Sepharose 2B[e]
Sternal cartilage	PGS(SC)-I	—	—	—	—
	PGS(SC)-II	—	—	—	—
Meckel's cartilage	PGS(MC)-I	—	—	—	—
	PGS(MCJ)-II	—	—	—	—
Skin fibroblasts	PGS(SF)-I	—	III	—	—
	PGS(SF)-II	—	I	—	—
Limb mesenchyme	PGS(LM)-I	—	III	PCS M	K_{av} 0.41–0.57
	PGS(LM)-II	CSPG-1	I	PCS L	K_{av} 0.92
Limb cartilage	PGS(LC)-I	CSPG-2	IV	PCS H	K_{av} 0.23
	PGS(LC)-II	CSPG-1	I	PCS L	K_{av} 0.92
Somites	—	—	III	PCS M	—
	—	—	I	PCS L	—
Vertebral cartilage	—	—	IV	PCS H	—
	—	—	I	PCS L	—

[a] McKeown and Goetinck (1979).
[b] Levitt and Dorfman (1974).
[c] Okayama et al. (1976).
[d] Kitamura and Yamagata (1976).
[e] DeLuca et al. (1977).

other macromolecules such as hyaluronic acid or link protein. Chemical and physical properties include the composition and size of the core protein and of the polysaccharide and oligosaccharide side chains, and immunochemical characteristics measure the ability of the various proteoglycans to interact with specific antisera. Since more and more criteria will be used to characterize the various proteoglycans, some common denominator will be found which will allow one to categorize these macromolecules in a manner reflecting the expression of the genetic material. Judged from the material that will be presented on proteoglycans during chondrogenesis and on work based on proteoglycans from tissues not in the chondrogenic lineage, this common denominator may turn out to be the core protein of proteoglycan.

VI. PROTEOGLYCAN IMMUNOCHEMISTRY

Early attempts to probe the structure of cartilage proteoglycans by immunochemical techniques were hampered by the fact that the antisera were prepared against poorly characterized antigens. With the development of nondisruptive extractions and purifications procedures (Sajdera and Hascall, 1969; Hascall and Sajdera, 1969), the use of more standardized antigen preparations has helped to clarify the immunological characteristics of proteoglycans. Cartilage proteoglycans from various species have been reported to contain some cross-reactive determinants (Sandson *et al.*, 1970; Keiser *et al.*, 1972; Keiser and Sandson, 1974), which have been shown to be on the protein core in the chondroitin sulfate attachment domain of the molecule (Baxter and Muir, 1975; Muir *et al.*, 1973). Keiser and DeVito (1974) reported another antigenic determinant in the keratan sulfate attachment domain of the proteoglycan. Wieslander and Heinegård (1979) reported a third determinant on the hyaluronic acid-binding domain, which may represent a species-specific determinant (Wieslander and Heinegård, 1977; Sugahara and Dorfman, 1979; Sparks and Goetinck, 1979). Therefore, the bovine cartilage proteoglycans contain at least three antigenic determinants, one in each of the three domains of the macromolecule. Since native chondroitin sulfate and keratan sulfate are not immunogenic and since the enzymatic removal of the chondroitin sulfate chains does not destroy the antigenicity of the proteoglycans, whereas papain digestion does, the antigenicity of proteoglycans is attributed to the protein moiety (Muir *et al.*, 1973; Keiser and Sandson, 1973). The contribution of the N- and O-linked oligosaccharides to the antigenicity of proteoglycans has not been investigated, however.

Radioimmunoassays for proteoglycan fragments, each containing one of these antigenic determinants of proteoglycans, have been developed (Wieslander and Heinegård, 1980). This assay requires the labeling of purified antigens with ^{125}I. An alternative method using proteoglycans labeled endogenously with $[^{35}S]$sulfate has been used in the avian embryonic system (Ho *et al.*, 1977; Sparks *et al.*, 1980; Royal *et al.*, 1980). Although this system cannot be used to measure the three different antigenic determinants, a great deal of specificity is introduced in such an assay since proteoglycans are the only sulfated macromolecules synthesized by embryonic cartilage (Manasek, 1975).

The use of immunological methods in the study of proteoglycans has recently been reviewed by Dorfman *et al.* (1980a), and the preparation of several monoclonal antibodies has been reported for core protein of avian cartilage proteoglycans (Dorfman *et al.*, 1980b).

The antiserum used in the results described here were elicited in rabbits by immunization with highly purified, enzymatically undigested, juvenile chick cartilage fraction designated A_1D_1–1400 V_o. The anti-A_1D_1– 1400 V_o serum contains IgG, which binds $[^{35}S]$sulfate cartilage PGS-I only. The antigen-antibody interaction does not require prior enzymatic removal of the chondroitin sulfate chains and is demonstrated using the precipitation method of Farr (1958). Binding of cartilage proteoglycan takes place in either its monomeric or aggregated state (Fig. 4) (Sparks *et al.*, 1980).

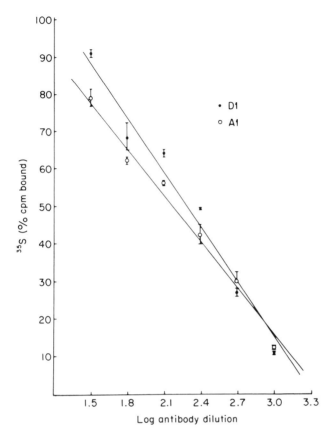

Figure 4 Regression curves of corrected specific binding of antigen plotted against the logarithm of dilutions of anti-A_1 D_1-1400 V_0 serum. The antigens are ($\bullet$) D_1 (monomer) and ($\bigcirc$) A_1 (aggregate) fractions obtained from 14-day embryonic sterna labeled with [^{35}S]sulfate. Antigen–antibody binding was measured by the Farr (1958) test.

When aggregates (A_1) are mixed with anti-A_1D_1-1400 V_0 serum and then layered on a dissociative sucrose density gradient, there is a dramatic shift of the proteoglycan toward the denser region of the gradient (Sparks and Goetinck, 1979) such that more than 30% of the ^{35}S-labeled proteoglycans are recovered in the bottom fraction. Because of the sharpness of the peak in the profile, it is referred to as a "spike." The spiking phenomenon appears to require proteoglycan aggregates since no difference can be seen between anti-A_1D_1-1400 V_0 serum and normal rabbit serum when monomer (D_1) is used as antigen (Fig. 5). These results are interpreted to mean that the anti-A_1D_1-1400 V_0 serum is capable of stabilizing avian aggregate proteoglycans in the presence of 4 M GuHCl.

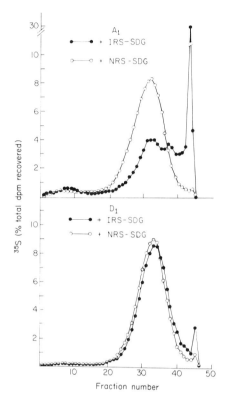

Figure 5 Stabilization of anti-A_1 D_1-1400 V_o serum (IRS) of chick embryonic cartilage proteoglycan aggregates (A_1) in the presence of 4 M guanidine-HCl. The addition of IRS to A_1 preparations results in the shift of the ^{35}S-labeled proteoglycan profile toward the denser region of the gradient. The bottom fraction contains both immunoglobulin and proteoglycan aggregates. This "spiking" phenomenon requires aggregates since it is not seen with a D_1 preparation (●————●, + IRS-SDG; ○————○, NRS-SDG). NRS, normal rabbit serum.

VII. PROTEOGLYCANS IN EMBRYONIC CARTILAGE

Chondrocytes synthesize proteoglycans at a much higher rate than does any other cell type. The high concentration of anionic polymers outside the cell results in the retention of a large amount of solvent, which in turn results in large extracellular spaces. The separation of the chondrocytes by the increasing volume of extracellular matrix is one mechanism by which cartilaginous rudiments grow. The important role played by proteoglycans in growth is demonstrated in studies showing that abnormalities of proteoglycans have a pronounced effect on the morphology of cartilaginous rudiments. This has been demonstrated in hereditary abnormalities that affect proteoglycan structure (Palmoski and

Goetinck, 1972; Orkin *et al.*, 1976) and also in experimental situations in which the proteoglycan structure is specifically disturbed. One such condition involves the use of β-D-xylosides, which compete with core proteins as sites for the initiation of chondroitin sulfate synthesis. Those proteoglycans synthesized in the presence of β-D-xylosides have fewer and shorter chondroitin sulfate side chains (Kato *et al.*, 1978; Lohmander *et al.*, 1979). Because of the absence of a core protein, the β-D-xyloside-bound chondroitin sulfate cannot become part of the cartilage matrix. Although the quantity of core protein synthesis seems to be higher in the presence of β-D-xylosides (Lohmander *et al.*, 1979), the quality of that which is synthesized is indistinguishable from controls (Schwartz, 1977, 1979).

When a β-D-xyloside is injected into fertile eggs, it acts as a potent teratogen producing severe dwarfism as well as edema of the soft tissues (Gibson *et al.*, 1978, 1979). The dwarfism might be readily explained on the basis of the known effect of xylosides on proteoglycan structure. Such abnormal proteoglycans would result in an underhydrated matrix, which would cause the cells in the cartilage of the treated embryo to be closer to one another. This, however, is not the case. There are no observable changes in the extracellular spaces in the cartilage of the embryos treated with the teratogen (Hjelle and Gibson, 1979). However, the xylosides have the predicted effect on the matrix of the limb bud cell cultures, and as a result the cells are in closer apposition to one another (Lohmander *et al.*, 1979).

The undersulfation of the glycosaminoglycan portion of the cartilage proteoglycan of brachymorphic mice has effects that are similar to that of the xyloside on limb bud cell cultures. These mice have a reduced extracellular space in their cartilage and shortened limbs (Orkin *et al.*, 1976). These teratological and genetical abnormalities are restricted to the polysaccharide component of proteoglycans, and, at least in the case of the xyloside treatment, all proteoglycans of cartilage are affected.

When cartilage proteoglycans from a number of avian sources (sternum, tibia, femur, vertebra, and Meckel's cartilage) of genetically normal embryos are examined, about 90% of the proteoglycans are seen as PGS-I and the remaining 10% as PGS-II. The PGS-I can exist as an aggregate by virtue of its ability to interact with hyaluronic acid and link proteins. Femoral and tibial PGS-II have been shown to contain several subpopulations (Kimata *et al.*, 1978). One of these subpopulations has a lower buoyant density than other cartilage proteoglycans, and it contains dermatan sulfate–chondroitin sulfate copolymer chains (Kimata *et al.*, 1978). It is not known whether PGS-II populations found in other cartilage and in noncartilaginous tissues contain similar subpopulations.

Characterization of extracts from sternal cartilage and Meckel's cartilage (MC) on sucrose density gradients under dissociative conditions reveals that the PGS(MC)-I has a faster sedimentation rate than PGS(SC)-I (Fig. 6a and c). The

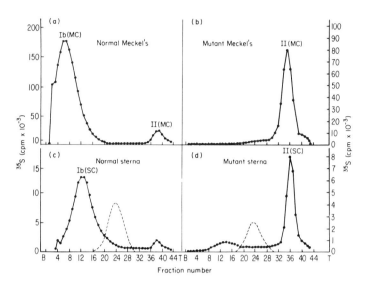

Figure 6 Dissociative sucrose density gradient profile of [^{35}S]sulfate-labeled proteoglycans from extracts of (a) normal Meckel's cartilage, (b) nanomelic Meckel's cartilage, (c) normal sternal cartilage, and (d) nanomelic sternal cartilage. For comparison, the sedimentation profile of PGS-I of skin fibroblasts (SF) is indicated (B, bottom; T, top).

two sedimentation rates cannot be explained on the basis of a difference in size of the sulfated glycosaminoglycans. The glycosaminoglycans of PGS(MC)-I, however, have a slightly higher charge density than those of PGS(SC)-I (McKeown and Goetinck, 1979). Although these two proteoglycans seem to be different from each other on the basis of their sedimentation rate on sucrose gradients, they have several characteristics in common. Both can interact with hyaluronic acid to form aggregates (Lever and Goetinck, 1976; McKeown and Goetinck, 1979), and both bind antibody elicited against proteoglycan monomer (A_1D_1–1400V_o) to the same extent. The same antiserum can also stabilize proteoglycan aggregates from the two types of cartilage in the presence of 4 M GuHCl. Finally, the PGS-I of both cartilages is affected by the nanomelic mutation (Fig. 6b and d).

Nanomelic embryos have cartilaginous structures that are greatly reduced in size (Landauer, 1965) as a result of a reduced extracellular matrix. The molecular basis for the abnormality lies in the greatly reduced levels of PGS-I in all cartilages studied (Mathews, 1967; Fraser and Goetinck, 1971; Palmoski and Goetinck, 1972; Pennypacker and Goetinck, 1976; McKeown and Goetinck, 1979).

Sternal nanomelic chondrocytes, for example, synthesize PGS(SC)-I at only about 2–3% of normal levels. A more detailed analysis of that which is synthesized has revealed that the small quantities of nanomelic PGS(SC)-I are not normal. When PGS(SC)-I from normal and mutant are compared on dissociative

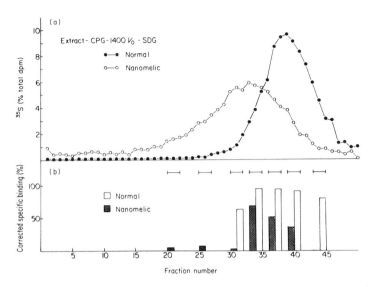

Figure 7 Dissociative sucrose density gradient and immunochemical analysis of PGS(SC)-I of normal and nanomelic embryos. (a) Sucrose density gradient profiles. Sedimentation is from left to right ($\bullet$——$\bullet$, normal; $\bigcirc$——$\bigcirc$, nanomelic). (b) Anti-A_1D_1-1400 V_o serum binding to designated fractions ($\square$, normal; $\blacksquare$, nanomelic).

sucrose gradients, a significantly larger proportion of the nanomelic PGS(SC)-I is found to have a slower sedimentation rate than normal PGS(SC)-I. When the nanomelic proteoglycan fractions from this gradient are tested with anti-A_1D_1–1400 V_o serum, only the material found in the dense region of the distribution shows significant antibody binding (Fig. 7). Similar results are obtained when column fractions of the normal and mutant proteoglycans are challenged with the antibody (Fig. 8*b* and *d*). Only 50% of the PGS(SC)-I of the mutant can be bound by antibody, in contrast to 95% binding of PGS(SC)-I from normal cartilage. When the CPG 1400 V_o material is chromatographed on CPG 2500, reduced binding to antibody is again demonstrated in the mutant, although the molecules that elute as aggregates show a greater specific binding than those that are included. Chromatography of CPG 1400 V_o material from normal and nanomelic extracts on CPG 2500 reveals further differences between the normal and mutant PGS(SC)-I. Whereas 75% of the normal material chromatographs as aggregates, only 25% of the mutant material is in the aggregated state. Since proteoglycan aggregation involves the interaction of hyaluronic acid with a specialized domain of the core protein, the results suggest that the nanomelic proteoglycan, which is included in the CPG 2500, does not have a functional hyaluronic acid-binding domain.

The interpretation of an abnormal core protein in nanomelic PGS(SC)-I is

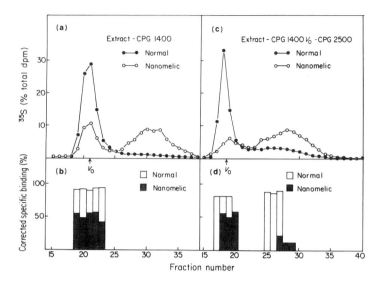

Figure 8 Chromatographic and immunochemical analyses of PGS synthesized by normal and nanomelic sternal cartilage. (a) Molecular sieve chromatography of sternal extracts of CPG 1400 (●——●, normal; ○——○, nanomelic). (b) Anti-A_1D_1-1400 V_0 serum binding of PGS(SC)-I (□, normal; □, nanomelic). (c) Molecular sieve chromatography of PGS(SC)-I (CPG 1400 V_0) on CPG 2500 (●——●, normal; ○——○, nanomelic). (d) Anti-A_1D_1-1400 V_0 serum binding to the various fractions which chromatograph as aggregates and monomers (□, normal; □, nanomelic).

consistent with the antibody-binding results. The antiserum is specific for cartilage proteoglycans and contains no antibody to chondroitin sulfate and hyaluronic acid. In addition, since the antibody can stabilize proteoglycan aggregates in 4 M GuHCl, it is believed that some of the antigenic determinants recognized by the antibodies lie on or near the core protein. The nanomelic PGS(SC)-I, which is not bound by antibody, is interpreted to represent cartilage proteoglycan that is smaller than normal and is therefore missing antigenic determinants. Taken together, these results are interpreted to mean that the small amount of PGS(SC)-I synthesized by the nanomelic chondrocytes is abnormal and that the mutation results in a defective proteoglycan core protein. This interpretation is consistent with previous results which have shown that nanomelic chondrocytes possess all the enzymatic machinery to synthesize chondroitin sulfate and have normal levels of xylosyltransferase (Stearns and Goetinck, 1979; Goetinck et al., 1981).

VIII. PROTEOGLYCAN CHANGES DURING CHONDROGENESIS

Precartilaginous tissue and the cartilage into which it differentiates have been reported by many laboratories to have differences in their proteoglycans (Table I)

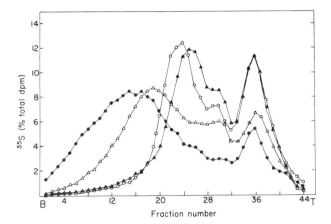

Figure 9 Profiles of [^{35}S]sulfate-labeled proteoglycans sedimented on a 5–20% sucrose gradient under dissociative conditions. The proteoglycans were obtained from intact limb buds at 4 (▲———▲), 5 (○———○), 6 (△———△), and 7 (●———●) days of incubation. Direction of sedimentation is from right to left (B, bottom; T, top). From Royal *et al.* (1980) with permission.

(Levitt and Dorfman, 1973; Goetinck *et al.*, 1974, 1981; Kitamura and Yamagata, 1976; Okayama *et al.*, 1976; DeLuca *et al.*, 1977, 1978; Vasan and Lash, 1977; Lash and Vasan, 1978; Sawyer and Goetinck, 1981). The changes in proteoglycans that take place during chondrogenesis in the avian limb are evident from the sedimentation profiles of limb extracts from embryos ranging from 4 through 7 days of development (Fig. 9). The differences that appear with increasing developmental age are particularly evident between the 4- and 5-day profiles and the 6- and 7-day profiles. The 4- and 5-day limb buds represent precartilaginous stages of the limb, whereas 6- and 7-day embryos contain differentiated cartilage (Royal *et al.*, 1980).

When 4- and 7-day limb extracts are chromatographed on CGP 1400, similar profiles are obtained for both ages (Figs. 10*a* and 11*a*, respectively). When each of the CPG 1400 fractions denoted in Figures 10*a* and 11*a* are sedimented on dissociative sucrose gradients, it is evident that the differences between 4- and 7-day limb extracts are in the CPG 1400 V_0 material (Figs. 10*b* and 11*b*, respectively). The CPG 1400 included fractions of the 4-day limb extracts (Fig. 10*c* and *d*) are indistinguishable from those of the 7-day extract (Fig. 11*c* and *d*) when they are sedimented on dissociative sucrose gradients.

A comparison of the profiles seen in Figures 10*b* and 11*b* indicates that the CPG 1400 V_0 material of the 4-day limb buds [PGS(LM)-I] has a slower sedimentation rate on sucrose gradients than the same material from the 7-day limbs [PGS(LC)-I]. PGS(LM)-I also shows 0% corrected specific binding with anti-A_1D_1–1400 V_0 serum, whereas PGS(LC)-I has 69% corrected specific binding with the same antiserum. The very broad peak observed when PGS(LC)-I is sedimented on sucrose gradients probably reflects the mixture of proteoglycans

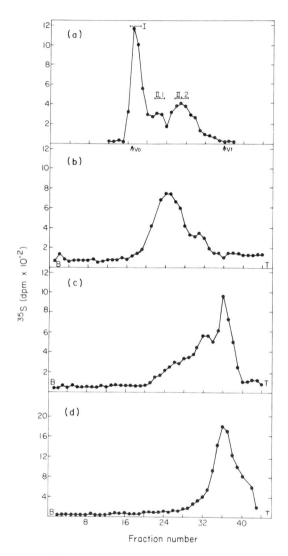

Figure 10 Analyses of [^{35}S]sulfate-labeled proteoglycan of 4-day limb buds extracted in 0.5 *M* guanidine-HCl. (a) Profile of extract chromatographed on CPG 1400. The fractions identified in part (a) were sedimented on a 5–20% sucrose gradient. (b) Fraction I. (c) Fraction II,1. (d) Fraction II,2. B, bottom; T, top. From Royal *et al.* (1980) with permission.

synthesized by the 7-day limb, which contains soft tissues as well as cartilage. The same explanation is advanced for the antibody-binding results in which 69% of PGS(LC)-I is bound, whereas binding to sternal cartilage proteoglycans is more than 90%.

The functional properties of PGS(LM)-I have also been investigated. When

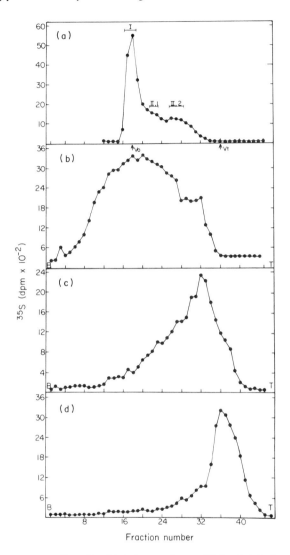

Figure 11 Analyses of [^{35}S]sulfate-labeled proteoglycan of 7-day limb buds extracted in 0.5 M guanidine-HCl. (a) Profile of extract chromatographed on CPG 1400. The fractions identified in part (a) were sedimented on a 5–20% sucrose gradient. (b) Fraction I. (c) Fraction II,1. (d) Fraction II,2. B, bottom; T, top. From Royal *et al.* (1980) with permission.

4-day limb bud extracts are obtained under associative conditions and chromatographed on CPG 1400, 59% of the material chromatographs as PGS(LM)-I (Fig. 12A) and 45% of the CPG 1400 V_o material chromatographs in the void volume of a CPG 2500 column (Fig. 12B). This chromatographic behavior suggests the presence of an aggregated form of PGS(LM)-I. To test this

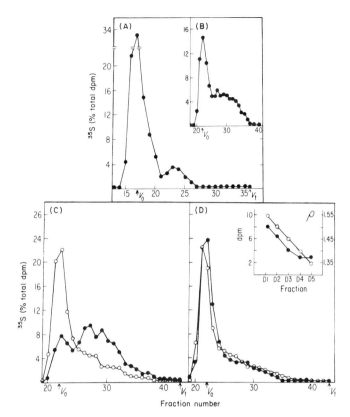

Figure 12 (A) Elution profile of 4-day limb bud [^{35}S]sulfate-labeled proteoglycans on CPG 1400. (B) Elution profile of CPG 1400 V_0 material on CPG 2500. In the inset in D is given the distribution of PGS(LM)-I (CPG 1400 V_0) material on a dissociative CsCl gradient (●——●, ^{35}S-labeled proteoglycans; ○——○, density of fractions. (C) The D_1 (D) D_2 fractions were chromatographed on CPG 2500 after dialysis to associative conditions in the absence (●——●) and presence (○——○) of exogenous hyaluronic acid.

possibility, PGS(LM)-I was subjected to centrifugation in a dissociative CsCl gradient. The distribution of radioactivity and the density of the fractions are shown in Figure 12D (inset).

To test for the ability of the proteoglycans to form aggregates, all the fractions (D_1 through D_5) were dialyzed to associative conditions in the presence or absence of hyaluronic acid. When the D_1 fraction was dialyzed in the presence of hyaluronic acid, 75% of the proteoglycan chromatographs as an aggregate. However, only 25% chromatographs in the void volume in the absence of hyaluronic acid. The D_2 fractions give identical profiles whether or not hyaluronic acid is present during the dialysis to associative conditions. In both cases, the material is clearly excluded on a CPG 2500. All other fractions (D_3, D_4, and D_5) give profiles similar to that of the D_2 fraction. The majority of the radioactivity elutes in the void volume, and no effect of exogenous hyaluronic acid can be observed. These results clearly indicate that PGS(LM)-I found in the D_1 fraction can interact with hyaluronic acid to form aggregates. No aggregate monomer relationship could be demonstrated with the D_2, D_3, D_4, and D_5 fractions of PGS(LM)-I. Each of these fractions behaved as an aggregate whether or not exogenous hyaluronic acid was added. The failure to recover monomer from these fractions may result from the presence of endogenous hyaluronic acid. Hyaluronic acid under these experimental conditions has a density of 1.46 gm/ml (Gregory, 1973).

Although the capacity to form aggregates with hyaluronic acid is a property shared by PGS(LM)-I and PGS-I of cartilage, the two proteoglycans are clearly distinct from one another in several respects. As indicated in Table II, PGS(LM)-I and PGS(LM)-I-D_1 cannot bind to the anti-A_1D_1–1400 V_o serum. This antiserum can bind 69% of PGS(LC)-I and 59% of the PGS(LC)-I-D_1. Similar immunochemical changes in proteoglycans associated with chondrogenesis have been described by Ho et al. (1977) and Vertel and Dorfman (1978). Further changes in proteoglycans associated with chondrogenesis include differences in the size of the polysaccharide side chains (DeLuca et al., 1977), in the composition of the chondroitin sulfate chain (Okayama et al., 1976; DeLuca et al., 1977), and in amino acid composition (Kitamura and Yamagata, 1976). The characterization of the proteoglycans synthesized by stage-24 chick limb mesenchymal cell cultures has revealed that the chondroitin sulfate chains are relatively larger early in the culture period. The proteoglycans synthesized on day 2 have a K_{av} of 0.41–0.57 on Sepharose 2B. This proteoglycan that may correspond to PGS(LM)-I has chondroitin sulfate chains with an average molecular weight of 28,000 daltons. On day 8 of the culture, when cartilage-specific proteoglycans are synthesized, the size of the chondroitin sulfate chains is 17,000 daltons. By day 21 the size has been reduced further, to 14,000

TABLE II
Corrected Specific Binding of Anti-A_1D_1 1400 V_o Serum to
^{35}S-Labeled Proteoglycans

Antigen[a]	Binding (%)
PGS(SC)-I-D_1	91.8
PGS(MC)-I-D_1	95.8
PGC(LC)-I	69.0
PGS(LC)-I-D_1	59.0
PGS(LM)-I	0.0
PGS(LM)-I-D_1	2.7

[a] SC, sternal cartilage; MC, Meckel's cartilage; LC, limb cartilage (whole 7-day embryonic limb); LM, limb mesenchyme.

daltons. The proportion of chondroitin 6-sulfate to chondroitin 4-sulfate also changes over the culture period. On day 2 there is 70% chondroitin 6-sulfate. This level drops off to about 55% at day 10, and it remains constant at that level to day 21. In contrast to the decreasing size of chondroitin sulfate, the molecular weight of keratan sulfate increases with increasing age of the cultures (DeLuca *et al.*, 1977). Taken together, the physicochemical, immunochemical, and chemical evidence clearly indicates that major changes take place in the proteoglycans during chondrogenesis. There is definite evidence of changes in the polysaccharide chains, and the immunochemical data strongly suggest changes in core protein.

In view of the clear difference between PGS(LM)-I and PGS-I of cartilage in normal embryos, it became important to find out if the nanomelic mutation affects the proteoglycans of limb mesenchyme in the same manner as those of cartilage. This study was done with high-cell-density micromass cultures derived from limb buds of individual 4-day normal and mutant embryos (Sawyer and Goetinck, 1981). The analyses of the proteoglycans associated with the cell layer of these cultures are given in Figure 13. Essentially the same results were obtained with the proteoglycans found in the culture medium. The results reveal no

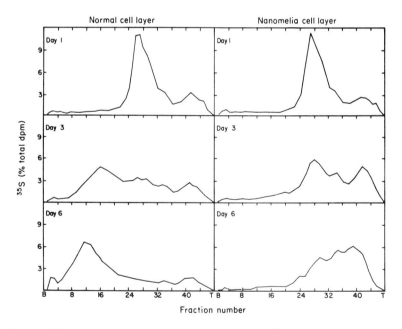

Figure 13 Dissociative sucrose density gradient profiles of [^{35}S]sulfate-labeled PGS extracted from normal and nanomelic cell layers after 1, 3, and 6 days as micromass cultures. Direction of sedimentation is from right to left (B, bottom; T, top).

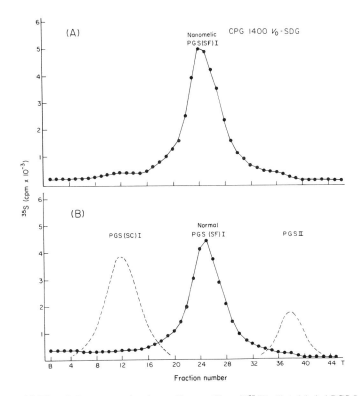

Figure 14 Dissociative sucrose density gradient profiles of [^{35} S]sulfate-labeled PGS-I of nanomelic (A) and normal (B) skin fibroblasts (SF). For comparison PGS-I of sternal cartilage and PGS-□ are shown. Sedimentation is from left to right (B, bottom; T, top).

differences in the sedimentation profile of the proteoglycans obtained from 1-day normal and mutant cultures. The profiles of both genotypes display an identical pattern of proteoglycan heterogeneity at this time. At 3 days a more rapidly sedimenting population of proteoglycans becomes evident in the normal but not in mutant cultures. The appearance of this population of proteoglycans coincides with the differentiation of mesenchymal cells into chondrocytes. The major proteoglycan peak in normal 6-day cultures, which is also absent in mutant 6-day cultures, has a faster sedimentation rate than the major peak in 3-day normal cultures. This observation again demonstrates the changes in proteoglycans during chondrogenesis.

Although the cartilage-specific proteoglycans are not detectable in the mutant cultures, there is a decrease in the synthesis of PGS(LM)-I during chondrogenesis. This suggests that the decrease in PGS(LM)-I is a result of the process of differentiation and not dependent on the appearance of PGS-I of cartilage. The failure of PGS(LM)-I to disappear completely in the micromass cultures may result from the presence in the cultures of a mixture of cell types.

These results indicate that the proteoglycans synthesized by mutant prechondrogenic cells are the same as those of normal prechondrogenic cells. This conclusion is consistent with other results that have shown that there are neither quantitative nor qualitative differences between normal and mutant skin fibroblasts (SF) in their synthesis of proteoglycans (Goetinck and Pennypacker, 1977). The sedimentation profiles on sucrose gradients of PGS(SF)-I of mutant and normal embyros are identical (Fig. 14). For both genotypes the sedimentation rate of PGS(SF)-I is identical to that of PGS(LM)-I.

IX. INTRACELLULAR PROTEINS OF CHONDROCYTES

Evidence has been presented which indicates that there are dramatic differences between the proteoglycans synthesized by normal and nanomelic chondrocytes but not between those synthesized by normal and nanomelic limb mesenchymal cells. The studies comparing normal and nanomelic chondrocytes suggest that the hereditary defect specifically involves the core protein of PGS-I of cartilage.

To test more directly the possibility that the nanomelia mutation affects core protein of cartilage PGS-I, an analysis was made of the intracellular proteins of normal and nanomelic chondrocytes (Argraves, 1981; Argraves *et al.*, 1981). Chondrocytes from 14-day embryonic sterna of both genotypes were labeled for 15 minutes with [^{35}S]cysteine in suspension cultures, and cell extracts were electrophoresed into a 4–15% gradient acrylamide–SDS slab gel (Laemmli, 1970). A fluorograph comparing the extract of normal and nanomelic chondrocytes is shown in Figure 15. The electrophoretic pattern of the proteins synthesized by normal chondrocytes contains a high molecular weight band (lane 1, arrow), which is absent in the extract of the mutant chondrocytes (lane 2). The band has an apparent molecular weight of $(2.46 \pm 0.21) \times 10^5$ daltons.

The 246,000-dalton band could be precipitated from normal chondrocyte extracts with a double antibody method using anti-$A_1 D_1$-1400 V_0 serum and sheep anti-rabbit 1gG (Fc fragment). In Figure 16 are shown the total extract of normal chondrocytes (lane 1), the material precipitated from that extract with the anti-$A_1 D_1$-1400 V_0 serum and the sheep anti-rabbit 1gG (lane 2), and the normal rabbit serum sheep anti-rabbit 1gG control (lane 3).

The 246,000-dalton band, which is immunologically cross-reactive with cartilage PGS-I, is taken to represent intracellular core protein of cartilage PGS-I. The fact that there are no detectable levels of the 246,000-dalton proteins in extracts of nanomelic chondrocytes indicates that the mutation acts at the level of core protein of cartilage PGS-I.

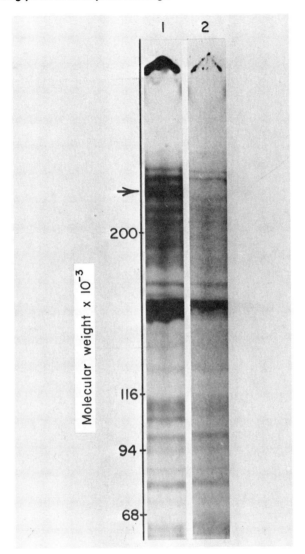

Figure 15 Fluorograph of [^{35}S]cysteine-labeled chondrocyte extracts electrophoresed on a 4–15% gradient acrylamide–SDS slab gel. The normal cell extract (lane 1) contains a radioactive band (arrow) which is missing in the nanomelic chondrocyte extract (lane 2).

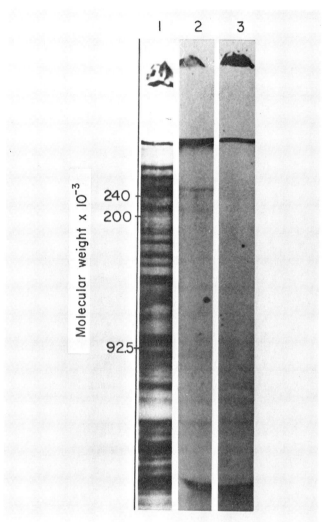

Figure 16 Fluorograph of [^{35}S]cysteine-labeled normal chondrocyte extracts electrophoresed on a 4–15% gradient acrylamide–SDS slab gel. The total extract is shown in lane 1. The material precipitated from the extract with anti-A_1D_1-1400 V_o serum and sheep anti-rabbit IgG is shown in lane 2. The normal rabbit serum sheep anti-rabbit IgG control is shown in lane 3.

X. CONCLUSIONS

The hypotheses associated with the expression of chondrogenesis in terms of glycosaminoglycans or proteoglycans have evolved considerably over the years. Initially, chondrogenesis was monitored by quantifying the increase in chondroitin sulfate synthesis as mesenchyme differentiated into chondrocytes. The observation that there were small amounts of chondroitin sulfate synthesized before chondrogenesis was interpreted to mean that no new macromolecules were being synthesized as differentiation took place. The phenotypic expression of the differentiated state was viewed as a consequence of an increase in the activity of preexisting enzymatic machinery. According to this view, cartilage seemed unique since differentiation in other systems is marked by the synthesis of previously nonexisting macromolecules, which result from the differential activation of genetic material. When chondrogenesis was viewed initially in terms of proteoglycans rather than glycosaminoglycans, a similar augmentation hypothesis was introduced.

The evidence reviewed in this section clearly indicates that the proteoglycans synthesized by precartilaginous tissues are different from those synthesized by chondrocytes. The major proteoglycans synthesized by the two tissues differ from one another in their glycosaminoglycans as well as in their core proteins. Immunological and genetic evidence has been presented which indicates that the core proteins of the two proteoglycans are different. The immunological data are based on the supposition that antibodies against proteoglycans are directed against the core proteins. This is probably a safe assumption when one considers the glycosaminoglycans, since they have been shown not to be antigenic. However, it has not yet been determined if the recently described oligosaccharides contribute to the antigenicity of the proteoglycans.

The genetic evidence is the finding that no core protein for PGS-I of cartilage can be identified in nanomelic chondrocyte extracts. The absence of this core protein could result from a mutation of either a regulatory or a structural gene. However, the presence of small amounts of abnormal PGS-I of cartilage in nanomelic cartilage suggests that the regulatory genes function normally in the mutant cells and that the structural gene for the core protein of PGS-I of cartilage is abnormal. The nanomelic mutation does not affect the quantity and quality of the proteoglycans synthesized by prechondrogenic cells nor by cells that are not in the chondrogenic cell lineage. The core proteins of these proteoglycans are therefore not affected by the mutation. This indicates that the proteoglycans of precartilaginous cells and chondrocytes have different core proteins which are the product of unique genes. In the case of proteoglycans synthesized during chondrogenesis, one gene would be expressed in mesenchymal cells and the other in chondrocytes. The attachment of the polysaccharides and oligosaccharides to the two genetically distinct core proteins may be the result of the activity of the same or different enzyme systems in the two types of tissue.

The activation of previously inactive genes, which results in the initiation of the synthesis of core protein of PGS-I of cartilage, would represent an event of major developmental significance associated with chondrogenesis. Changes in the synthesis of core protein during chondrogenesis would parallel the regulatory events involving the synthesis of the types of collagen associated with cartilage differentiation. In mesenchymal cells, type I collagen is synthesized, and as chondrogenesis takes place the synthesis of this type of collagen ceases and the synthesis of a genetically distinct type II collagen is initiated. These changes in collagen synthesis during chondrogenesis have been reviewed by von der Mark (1980). Furthermore, several tissues have been described as having unique proteoglycans (Yamagishita *et al.*, 1979; Hassell *et al.*, 1979, 1980a,b). The suggestion that these proteoglyglycans may have unique core proteins is consistent with the view that the different core proteins of proteoglycans synthesized by precartilaginous cells and chondrocytes are the product of distinct genes.

ACKNOWLEDGMENTS

The author is grateful to Bridget DeSocio for the illustration of the proteoglycan model shown in Figure 1 and to Pat Timmins for typing the manuscript.

This work was supproted by Grant HD-09174 from the NICHHD and is scientific contribution no. 886 of the Storrs Agricultural Experiment Station, The University of Connecticut, Storrs.

REFERENCES

Abbott, J., and Holtzer, H. (1968). *Proc. Natl. Acad. Sci. U.S.A.* **59,** 1144–1151.
Abbott, J., Mayne, R., and Holtzer, H. (1972). *Dev. Biol.* **28,** 430–442.
Argraves, W. S. (1981). *Fed. Proc., Fed. Am. Soc. Exp. Biol.* **40,** 1760.
Argraves, W. S., McKeown-Longo, P. J., and Goetinck, P. F. (1981). *FEBS Lett.* **131,** 265–268.
Baker, J. R., and Caterson, B. (1977). *Biochem. Biophys. Res. Commun.* **77,** 1–10.
Baxter, E., and Muir, H. (1975). *Biochem. J.* **149,** 657–668.
Cahn, R. D. (1968). In "The Stability of the Differentiated State" (H. Ursprung, ed.), pp. 58–84. Springer-Verlag, Berlin and New York.
Caplan, A. I. (1970). *Exp. Cell Res.* **62,** 341–355.
DeLuca, S., Heinegård, D., Hascall, V. C., Kimura, J. H., and Caplan, A. I. (1977). *J. Biol. Chem.* **252,** 6600–6608.
DeLuca, S., Caplan, A. I., and Hascall, V. C. (1978). *J. Biol. Chem.* **253,** 4713–4720.
DeLuca, S., Lohmander, L. S., Nilsson, B., Hascall, V. C., and Caplan, A. I. (1980). *J. Biol. Chem.* **255,** 6077–6083.
Dorfman, A., Vertel, B. M., and Schwartz, N. B. (1980a). *Curr. Top. Dev. Biol.* **14,** 169–198.
Dorfman, A., Hall, T., Ho, P. L., and Fitch, F. (1980b). *Proc. Natl. Acad. Sci. U.S.A.* **77,** 3970–3973.
Ede, D. A. (1977). *Cell Surf. Rev.* **1,** 495–543.
Faltz, L. L., Reddi, A. H., Hascall, G. K., Martin, D., Pita, J. C., and Hascall, V. C. (1979). *J. Biol. Chem.* **254,** 1375–1380.
Farr, R. S. (1958). *J. Infect. Dis.* **103,** 239–262.
Fell, H. B. (1939). *Proc. R. Soc. London, Ser. B* **229,** 407–463.
Fell, H. B., and Canti, R. B. (1934). *Proc. R. Soc. London, Ser. B* **116,** 316–349.
Fraser, R. A., and Goetinck, P. F. (1971). *Biochem. Biophys. Res. Commun.* **43,** 494–503.
Gibson, K. D., Doller, H. J., and Hoar, R. M. (1978). *Nature (London)* **273,** 151–154.

Gibson, K. D., Seger, B. J., and Doller, H. J. (1979). *Teratology* **19**, 345-356.
Goetinck, P. F., and Pennypacker, J. P. (1977). *In* "Vertebrate Limb and Somite Morphogenesis" (D. A. Ede, J. R. Hinchliffe, and M. Balls, eds.), pp. 139-159. Cambridge Univ. Press, London and New York.
Goetinck, P. F., Pennypacker, J. P., and Royal, P. D. (1974). *Exp. Cell Res.* **87**, 241-248.
Goetinck, P. F., Lever-Fischer, P. L., McKeown-Longo, P. J., Sawyer, L. M., Sparks, K. J., and Argraves, W. S. (1981). *In* "Levels of Genetic Control in Development" (S. Subtelng and U. K. Abbott, eds.), pp. 15-35. Alan R. Liss, Inc., New York.
Gregory, J. D. (1973). *Biochem. J.* **133**, 383-386.
Gross, J. L., and Rifkin, D. B. (1979). *Cell* **18**, 707-718.
Hamburger, V., and Hamilton, H. L. (1951). *J. Morphol.* **28**, 49-92.
Harpaz, N., and Schachter, H. (1980a). *J. Biol. Chem.* **255**, 4885-4893.
Harpaz, N., and Schachter, H. (1980b). *J. Biol. Chem.* **255**, 4894-4902.
Hascall, V. C. (1977). *J. Supramol. Struct.* **7**, 101-120.
Hascall, V. C., and Sajdera, S. W. (1969). *J. Biol. Chem.* **244**, 2384-2396.
Hassell, J. R., Pennypacker, J. P., Yamada, K. M., and Pratt, R. M. (1978). *Ann. N.Y. Acad. Sci.* **312**, 406-409.
Hassell, J. R., Newsome, D. A., and Hascall, V. C. (1979). *J. Biol. Chem.* **254**, 12346-12354.
Hassell, J. R., Newsome, D. A., Krachmer, J. H., and Rodrigues, M. M. (1980a). *Proc. Natl. Acad. Sci. U.S.A.* **77**, 3705-3709.
Hassell, J. R., Gehron Robey, P., Barrack, H. J., Wilczek, J., Rennard, S. J., and Martin, G. R. (1980b). *Proc. Natl. Acad. Sci. U.S.A.* **77**, 4494-4498.
Heinegård, D. (1972). *Biochim. Biophys. Acta* **285**, 181-192.
Hjelle, J. T., and Gibson, K. D. (1979). *Teratology* **53**, 179-202.
Ho, P. L., Levitt, D., and Dorfman, A. (1977). *Dev. Biol.* **55**, 233-243.
Holtzer, H. (1968). *In* "Epithelial-Mesenchymal Interactions" (R. Fleischmajer and R. E. Billingham, eds.), pp. 152-164. Williams & Wilkins, Baltimore, Maryland.
Holtzer, H., and Abbott, J. (1968). *In* "The Stability of the Differentiated State" (H. Ursprung, ed.), pp. 1-16. Springer-Verlag, Berlin and New York.
Holtzer, H., Weintraub, H., Mayne, R., and Mochan, B. (1972). *Curr. Top. Dev. Biol.* **7**, 229-256.
Johnston, M. C. (1966). *Anat. Rec.* **156**, 130-143.
Kato, Y., Kimata, K., Ito, K., Karasawa, K., and Suzuki, S. (1978). *J. Biol. Chem.* **253**, 2784-2789.
Keiser, H., and DeVito, J. (1974). *Connect. Tissue Res.* **2**, 273-282.
Keiser, H., and Sandson, J. I. (1973). *Fed. Proc., Fed. Am. Soc. Exp. Biol.* **32**, 1474-1477.
Keiser, H., and Sandson, J. I. (1974). *Arthritis Rheum.* **17**, 219-228.
Keiser, H., Shulman, H. J., and Sandson, J. I. (1972). *Biochem. J.* **126**, 163-169.
Kimata, J., Oike, Y., Ito, K., Karasawa, K., and Suzuki, S. (1978). *Biochem. Biophys. Res. Commun.* **85**, 1431-1439.
Kitamura, K., and Yamagata, T. (1976). *FEBS Lett.* **71**, 337-340.
Laemmli, U. (1970). *Nature (London)* **227**, 680-685.
Landauer, W. (1965). *J. Hered.* **56**, 121-138.
Lash, J. W. (1968). *In* "Epithelial-Mesenchymal Interactions" (R. Fleischmajer and R. E. Billingham, eds.), pp. 165-172. Williams & Wilkins, Baltimore, Maryland.
Lash, J. W., and Vasan, N. S. (1978). *Dev. Biol.* **66**, 151-171.
Le Lièvre, C. (1974). *J. Embryol. Exp. Morphol.* **31**, 453-477.
Le Lièvre, C., and Le Douarin, N. M. (1975). *J. Embryol. Exp. Morphol.* **34**, 125-154.
Lever, P. L., and Goetinck, P. F. (1976). *Anal. Biochem.* **75**, 67-76.
Levitt, D., and Dorfman, A. (1973). *Proc. Natl. Acad. Sci. U.S.A.* **70**, 2201-2205.
Levitt, D., and Dorfman, A. (1974). *Curr. Top. Dev. Biol.* **8**, 103-149.

Levitt, D., Ho, P. L., and Dorfman, A. (1975). *Dev. Biol.* **43**, 75-90.

Lewis, C. A., Pratt, R. M., Pennypacker, J. P., and Hassell, J. R. (1978). *Dev. Biol.* **64**, 31-47.

Lindahl, U., Bäckström, G., Malmström, A., and Fransson, L. A. (1972). *Biochem. Biophys. Res. Commun.* **46**, 985.

Lohmander, L. S., Hascall, V. C., and Caplan, A. I. (1979). *J. Biol. Chem.* **254**, 10551-10561.

Lohmander, L. S., DeLuca, S., Nilsson, B., Hascall, V. C., Caputo, C. B., Kimura, H., and Heinegård, D. (1980). *J. Biol. Chem.* **255**, 6085-6091.

McKeown, P. J., and Goetinck, P. F. (1979). *Dev. Biol.* **71**, 203-215.

McKeown-Longo, P. J., Sparks, K. J., and Goetinck, P. F. (1981). *Arch. Biochem. Biophys.* **212**, 216-228.

Malmström, A., Fransson, L. A., Höök, L. A., and Lindahl, U. (1975). *J. Biol. Chem.* **250**, 3419. 712-717.

Manasek, F. J. (1975). *Curr. Top. Dev. Biol.* **10**, 35-102.

Mathews, M. B. (1967). *Nature (London)* **213**, 1255-1256.

Mayne, R., Vail, M. S., and Miller, E. J. (1975). *Proc. Natl. Acad. Sci. U.S.A.* **72**, 4511-4515.

Mayne, R., Vail, M. S., and Miller, E. J. (1976). *Dev. Biol.* **54**, 230-240.

Muir, H., Baxter, E., and Brandt, K. D. (1973). *Biochem. Soc. Trans.* **1**, 223-225.

Narasimhan, S., Harpaz, N., Longmore, G., Carver, J. P., Grey, A. A., and Schacter, H. (1980). *J. Biol. Chem.* **255**, 4876-4884.

Nathanson, M. A., and Hay, E. D. (1980a). *Dev. Biol.* **78**, 301-331.

Nathanson, M. A., and Hay, E. D. (1980b). *Dev. Biol.* **78**, 332-351.

Nathanson, M. A., Hilfer, S. R., and Searls, R. L. (1978). *Dev. Biol.* **64**, 99-117.

Nogami, H., and Urist, M. R. (1970). *Exp. Cell Res.* **63**, 404-410.

Nogami, H., and Urist, M. R. (1974). *J. Cell Biol.* **62**, 510-519.

Oegema, T. R., Jr., and Thompson, R. C., Jr. (1981). *J. Biol. Chem.* **256**, 1015-1022.

Oegema, T. R., Jr., Hascall, V. C., and Dziewiatkowski, D. D. (1975). *J. Biol. Chem.* **250**, 6151-6159.

Oegema, T. R., Jr., Brown, M., and Dziewiatkowski, D. D. (1977). *J. Biol. Chem.* **252**, 6470-6477.

Okayama, M., Pacifici, M., and Holtzer, H. (1976). *Proc. Natl. Acad. Sci. U.S.A.* **73**, 3224-3228.

Okayama, M., Yoshimura, M., Muto, M., Chi, J., Roth, S., and Kaji, A. (1977). *Cancer Res.* **37**, 712-717.

Orkin, R. W., Pratt, R. M., and Martin, G. R. (1976). *Dev. Biol.* **50**, 82-94.

Pacifici, M., Boettiger, D., Roby, K., and Holtzer, H. (1977). *Cell* **11**, 891-899.

Palmoski, M. J., and Goetinck, P. F. (1972). *Proc. Natl. Acad. Sci. U.S.A.* **69**, 3385-3388.

Pennypacker, J. P., and Goetinck, P. F. (1976). *Dev. Biol.* **50**, 35-47.

Pennypacker, J. P., and Goetinck, P. F. (1979). *J. Embryol. Exp. Morphol.* **53**, 91-102.

Reddi, A. H., and Anderson, W. A. (1976). *J. Cell Biol.* **69**, 557-572.

Reddi, A. H., and Huggins, C. B. (1972). *Proc. Natl. Acad. Sci. U.S.A.* **69**, 1601-1605.

Reddi, A. H., Hascall, V. C., and Hascall, G. K. (1978). *J. Biol. Chem.* **253**, 2429-2436.

Rodén, L., and Horowitz, M. I. (1978). *In* ''The Glycoconjugates'' (M. I. Horowitz and W. Pigman, eds.), Vol. 2, pp. 3-71. Academic Press, New York.

Royal, P. D., Sparks, K. J., and Goetinck, P. F. (1980). *J. Biol. Chem.* **255**, 9870-9878.

Sajdera, S. W., and Hascall, V. C. (1969). *J. Biol. Chem.* **244**, 77-87.

Sandson, J., Damon, H., and Mathews, M. B. (1970). *In* ''The Chemistry and Molecular Biology of the Intercellular Matrix'' (E. A. Balazs, ed.), Vol. 3, pp. 1563-1567. Academic Press, New York.

Sawyer, L. M., and Goetinck, P. F. (1981). *J. Exp. Zool.* **216**, 121-131.

Schachter, L. P. (1970). *Exp. Cell Res.* **63**, 19-32.

Schwartz, N. B. (1977). *J. Biol. Chem.* **252**, 6316-6321.

Schwartz, N. B. (1979). *J. Biol. Chem.* **254,** 2271-2277.

Schwartz, N. B., Ostrowski, V., Brown, K. S., and Pratt, R. M. (1978). *Biochem. Biophys. Res. Commun.* **82,** 173-177.

Solursh, M., Ahrens, P. B., and Reiter, R. (1978). *In Vitro* **14,** 51-61.

Sparks, K. J., and Goetinck, P. F. (1979). *J. Cell Biol.* **74,** 466a.

Sparks, K. J., Lever, P. L., and Goetinck, P. F. (1980). *Arch. Biochem. Biophys.* **199,** 577-590.

Stearns, K., and Goetinck, P. F. (1979). *J. Cell. Physiol.* **100,** 33-38.

Sugahara, K., and Dorfman, A. (1979). *Biochem. Biophys. Res. Commun.* **89,** 1193-1199.

Thorp, F. K., and Dorfman, A. (1967). *Curr. Top. Dev. Biol.* **2,** 151-190.

Upholt, W. B., Vertel, B. M., and Dorfman, A. (1979). *Proc. Natl. Acad. Sci. U.S.A.* **76,** 4847-4851.

Vasan, N. S., and Lash, J. W. (1977). *Biochem. J.* **164,** 179-183.

Vertel, B. M., and Dorfman, A. (1978). *Dev. Biol.* **62,** 1-12.

von der Mark, K. (1980). *Curr. Top. Dev. Biol.* **14,** 199-225.

Weiss, R. E., and Reddi, A. H. (1981). *J. Cell Biol.* **88,** 630-636.

Wieslander, J., and Heinegård, D. (1977). *Upsala J. Med. Sci.* **82,** 161-169.

Wieslander, J., and Heinegård, D. (1979). *Biochem. J.* **179,** 35-45.

Wieslander, J., and Heinegård, D. (1980). *Biochem. J.* **187,** 687-694.

Yamagishita, M., Rodbard, D., and Hascall, V. C. (1979). *J. Biol. Chem.* **254,** 911-920.

Zimmerman, M., Mumford, R. A., and Steiner, D. F. (1980). *Ann. N.Y. Acad. Sci.* **343,** 1-448.

Zwilling, E. (1968). *Dev. Biol., Suppl.* **2,** 184-207.

SECTION 5

Membrane Glycoconjugates in the Maturation and Activation of T and B Lymphocytes

CARL G. GAHMBERG AND LEIF C. ANDERSSON

I. INTRODUCTION

Characteristic of the immune system of higher organisms is its high reactivity and specificity. These properties devolve from various functionally differentiated cells, which cooperate to develop optimal response against different im-

231

THE GLYCOCONJUGATES, VOL. III

munogenic substances. The communication between immune cells must take place through signals perceived at the plasma membrane, and therefore such cells carry specific surface structures involved in such activities. A large number of "antigens" and "receptors" have been defined on lymphoid cells, but the functional roles and molecular characteristics of these structures are, with a few exceptions, not known.

To understand the mechanisms operating during the immune response it is obviously important to characterize, at the molecular level, the macromolecules located at the surface of the lymphoid cells. Most, if not all, cell-surface proteins of mammalian cells are glycoproteins (Gahmberg, 1976), and the cellular glycolipids are also concentrated at the outer cell surface (Klenk and Choppin, 1970; Renkonen *et al.*, 1970). The hydrophilic oligosaccharides of these glycoconjugates are more or less available to external reagents, and there is a growing amount of data indicating the importance of the carbohydrate portions of such glycoconjugates in lymphocyte functions. Several carbohydrate-specific reagents bind to lymphocyte cell-surface glycoproteins and glycolipids and induce growth stimulation and differentiation.

Nearly 20 years ago Gesner and Ginsburg (1964) showed that when lymphocytes were treated with α-L-fucosidase and injected back into a syngeneic animal, their homing patterns changed as compared to intervenously injected, untreated cells. Except for those indirect data, very little is known about the functional roles of lymphocyte membrane carbohydrate *in vivo*. Several types of carbohydrate-reactive agents induce lymphoid maturation and differentiation *in vitro*. For the carbohydrate chemist, the lymphocyte therefore offers a challenging area of research, and several important results are anticipated in correlating the activity of functionally well defined lymphoid subsets with specific molecules.

The involvements and regulatory roles of T and B lymphocytes and their subsets in the immune system have been extensively reviewed (Katz, 1977; Snell, 1978) and are only briefly summarized here. The emphasis in this section is on the role of carbohydrate in lymphocyte activation and on the various surface glycoproteins and glycolipids that have been characterized in T and B cells and their derivatives.

II. MAJOR LYMPHOCYTE SUBGROUPS

The two major groups of lymphocytes in birds, rodents, and primates are the T (thymus-derived) and B (bursa-derived) cells. On the basis of functional roles in the immune response, several T-cell subsets have been defined: helper T cells, cytotoxic T cells, and suppressor T cells. The T cells regulate the mode and magnitude of the immune response. The B cells are involved mainly in the production of antibodies. Whether clearly definable B-cell subsets exist is still unclear.

Much of the current knowledge on the molecular structure of lymphocyte surface antigens derives from studies in mouse systems. The availability of inbred congeneic mouse strains has made it possible to produce antisera against surface molecules expressing allotype-linked polymorphism. The best known are the antigens coded for by the H-2 complex. Other well-characterized non-H-2-linked alloantigens are Thy-1 (previously called theta antigen) and the antigens of the Lyt series. These are described in more detail below (Section VI,E).

The lymphocyte surface molecules are highly conserved during evolution since corresponding structures are found in mice and men. Therefore, much of our knowledge based on murine systems may be directly applicable to man. It has been difficult to prepare specific antisera against human membrane molecules, but with the development of monoclonal antibody techniques (Köhler and Milstein, 1975) the situation is rapidly changing.

III. SEPARATION OF T AND B LYMPHOCYTES

Several methods have been developed for the purification of lymphocyte subsets. They are based either on different physical parameters such as differences in negative surface charge, density, or size, or on differences in surface structures (Natvig *et al.*, 1976).

Rodent (rat, mouse, guinea pig) T and B lymphocytes have remarkable differences in their net negative surface charge and can easily be fractionated by free-flow cell electrophoresis (Häyry *et al.*, 1975). Density gradient centrifugation is usually applied to the isolation of human blood mononuclear cells (Böyum, 1968), as well as to the purification of discrete subsets of lymphocytes, and monocytes. Velocity sedimentation at unit gravity fractionates according to particle size and enables one to separate activated lymphocytes (large blasts) from resting lymphocytes (small cells) (Andersson and Häyry, 1975).

The presence of various membrane "receptors" is used for the identification and fractionation of defined lymphocyte populations. Human T cells rosette with sheep erythrocytes, whereas B cells bind mouse erythrocytes. The surface membrane immunoglobulin on B cells makes them adhere to anti-immunoglobulin-coated columns or surfaces. Selected lymphocyte populations can be lysed by treatment with specific immune serum plus complement or fractionated by cell sorting (using the Fluorescence Activated Cell Sorter) with fluorochrome-conjugated antibodies.

Lectins specific for discrete carbohydrate structures are useful for agglutination, affinity chromatography, or sorting of lymphocytes (Reisner and Sharon, 1980; Reisner *et al.*, 1980).

IV. LYMPHOCYTIC CELL LINES AND LEUKEMIC CELLS AS SOURCES OF PURE LYMPHOID CELLS

A number of permanent, usually malignant lymphoid cell lines with T-cell, B-cell or non-T-cell, non-B-cell phenotypes, of both murine and human origin, are available. Such cell lines are especially useful for biochemical studies because they are of clonal origin and therefore are uniform, or readily available, and can be grown in quantity [see Nilsson and Pontén (1975) and Nilsson (1977) for reviews].

Another excellent source of large quantities of essentially pure lymphoid cells are patients with lymphoid malignances. Both lymphoid cell lines and leukemic cells are blocked in their differentiation, and from these differentiation antigens can be characterized and purified. The cell surface glycoproteins of cultured human lymphoid cell lines (Andersson *et al.*, 1977; Nilsson *et al.*, 1977; Trowbridge *et al.*, 1977; Gahmberg *et al.*, 1980a) and from cells of patients with various types of leukemia have been extensively studied (Andersson *et al.*, 1979a,b) and are not discussed in detail here. Most importantly, cell surface glycoprotein analysis provides a means whereby the type and stage of differentiation of leukemic cells can be determined.

V. INDUCTION OF LYMPHOID CELL DIVISION AND DIFFERENTIATION *IN VITRO*

During the *in vivo* immune response, lymphocytes with selected receptors are triggered to undergo blast transformation and clonal expansion, leading to the generation of specific effector mechanisms. This event of lymphocyte activation and differentiation can be mimicked in tissue culture.

A. Induction by Mitogenic Lectins

A number of plant and animal lectins bind to lymphocytes, but only some of them are mitogenic. With a few exceptions, mitogenic lectins are polyclonal activators of T lymphocytes. Table I summarizes the properties of some of the best-characterized lectins. Phytohemagglutinin (Nowell, 1960) and concanavalin A are potent mitogens and are most commonly used for the induction of lymphocyte division. Some lectins, such as soybean agglutinin, are mitogenic only after treatment of the cells with neuraminidase to expose penultimate galactosyl residues (Novogrodsky and Katchalsky, 1973b; Dillner-Zetterlind *et al.*, 1980). Pokeweed lectins activate T cells and induce T-cell-dependent B-lymphocyte proliferation. Wheat germ agglutinin strongly binds to T cells, and several receptors for the lectin have been identified (Axelsson *et al.*, 1978). Even so, it is poorly mitogenic, if at all (Boldt *et al.*, 1975; Axelsson *et al.*, 1978). All these

TABLE I
Binding Characteristics and Mitogenicity of Lectins

Lectin	Sugar specificity	Mitogenicity
Concanavalin A	α-D-Mannose, α-D-glucose	+
Leucoagglutinin (phyto-hemagglutinin)	Unclear	+
Soybean agglutinin	N-Acetyl-α,β-D-galactosamine, α,β-D-galactose	+ (after neuraminidase treatment)
Lima bean lectin	N-Acetyl-α-D-galactosamine, α-D-galactose	+
Helix pomatia lectin	N-Acetyl α-D-galactosamine	−
Peanut agglutinin	β-D-Galactose-1→3-D-N-acetyl-galactosamine	−
Wheat germ agglutinin	N-Acetyl-β-D-glucosamine, N-acetylneuraminic acid	±

lectins bind to several surface glycoproteins, and therefore it has not been possible to correlate activation with binding to some specific surface component.

The N-acetyl-α-D-galactosamine-specific lectin from *Helix pomatia* preferentially and strongly binds to a high molecular weight surface glycoprotein of human T cells (Axelsson *et al.*, 1978) after neuraminidase treatment, but it is not mitogenic. Evidently, at least this glycoprotein, which probably contains a large amount of O-glycosidic oligosaccharide (Saito and Osawa, 1980), is not involved in lectin-induced stimulation.

At present we must admit that no clear-cut correlation has been found between lectin binding to individual surface glycoproteins and mitogenicity.

B. Induction by Antibodies Raised against Lymphocyte Surface Molecules

Antilymphocyte antiserum is a potent inducer of blast transformation of lymphocytes (Gräsbeck *et al.*, 1964). The activated cells are T cells (Ochai *et al.*, 1977).

Smith *et al.* (1980) obtained a strong proliferative response using a rabbit anti-mouse brain antiserum that had been absorbed with mouse kidney and red cells. Immunoprecipitation of surface-labeled cells showed that the reactive antigen was Thy-1. For reasons not understood, allospecific anti-Thy-1.2 or monoclonal anti-Thy-1 antibodies did not activate T cells. Van Wauwe *et al.* (1980) described a mitogenic anti-human T-lymhocyte monoclonal antibody. The receptor for this antibody has not yet been identified. Antiimmunoglobulin antibodies induce mitosis of lymphocytes, which may be B cells (Sell *et al.*, 1970). F(ab)$_2$ and, most interestingly, also Fab fragments of anti-rabbit IgG are also stimulatory. Although monovalent antibody fragments against IgG are mitogenic, monovalent antilymphocyte antibodies are not (Sell *et al.*, 1970). This could mean that cross-linking of surface receptors is necessary for stimulation of T cells but not for B cells.

C. Induction of T Cells through Oxidation with Sodium Periodate or Galactose Oxidase

Novogrodsky and Katchalsky (1971, 1973a) first reported that treatment of lymphocytes either with periodate or with neuraminidase plus galactose oxidase induced blast transformation. The periodate effect can be obtained using a 1–2 mM concentration at 0° (Mitchell and Bowers, 1978). Under such conditions, only surface-located sialic acids are oxidized and aldehyde derivatives of sialic acid are formed (Gahmberg and Andersson, 1977). Galactose oxidase oxidizes surface-located galactosyl/N-acetylgalactosaminyl residues (Gahmberg and Hakomori, 1973), and corresponding C-6 aldehydes are formed (Avigad *et al.*, 1962). If the oxidized cells are treated with NaBH$_4$ or other aldehyde-reactive reagents, the mitogenic effect is reduced. This indicates that free aldehyde groups are necessary for blast transformation. Such aldehyde groups could potentially form cross-links through the formation of Schiff bases with neighboring NH$_2$ groups. However, such cross-links would be stabilized after reduction with NaBH$_4$, and if the mitogenic effect were a result of cross-linking one would expect the effect to remain and even become accentuated.

Both oxidation reagents are mitogenic only for T cells. This was shown using athymic (nude) mice, which congenitally lack the T lineages. The lymphocytes of such mice did not respond (Thurman *et al.*, 1973).

Accessory cells are apparently needed for stimulation by periodate or neuraminidase plus galactose oxidase (Novogrodsky *et al.*, 1977). Different growth factors are also involved in the blastogenesis (Monahan *et al.*, 1976). Some of them derive from the adherent cells. Cytotoxic cells are also induced by the oxidizing agents, but this activity may not depend on adherent cells. A growth factor has also been found in the supernatant of neuraminidase and galactose oxidase-treated lymphocytes (Novogrodsky *et al.*, 1980). This factor does not stimulate unprimed lymphocytes, but rather cells that have been previously activated with lectins or allogeneic cells.

The effects of the carbohydrate-specific oxidizing reagents on lymphocyte responses give additional support to the importance of cell-surface carbohydrates in these phenomena. However, as with lectins, we still do not know much about the molecular mechanisms leading to the inducing signals. Is there a specific target glycoconjugate that is necessary for these effects, or does it suffice to oxidize almost any surface sialic acid/galactose residue? By the use of tritiated sodium borohydride for labeling oxidized sialyl/galactosyl residues (Gahmberg and Hakomori, 1973), a number of glycoproteins and glycolipids were shown to be labeled (Andersson and Gahmberg, 1978; Andersson *et al.*, 1978). Mitchell and Bowers (1978) showed that treatment of intact cells with various proteases abolished the activation, whereas stimulation by concanavalin A was not affected. The proteases primarily cleaved the high molecular weight glycoproteins, and these were therefore assumed to be important in the activation by the oxidizing agents (Mitchell and Bowers, 1980). However, it has not been possible to correlate the activation with a single protein, and further insight into the mechanisms involved would seem difficult to obtain directly by these methods.

D. Induction of T Cells by Allogeneic Cells

When lymphocytes from two individuals with different transplantation antigens are cultivated together, cell blast transformation is induced. This mixed lymphocyte reaction has been regarded as an *in vitro* model for the allograft rejection (Häyry *et al.*, 1972). A molecular dichotomy has been demonstrated among the mixed lymphocyte culture activating histocompatibility antigens. The Ia region (HLA-D locus in man) encoded specificities (LD antigens) stimulate helper T cells, which enhance the differentiation of killer T cells. The killer T cells recognize primarily the specificities on the heavy chains of the major transplantation antigens (SD antigens H-2 K and H-2 D in mouse, HLA-A, -B, and -C in man).

Essentially identical responses have been found with virus-infected or hapten-modified stimulator cells in syngeneic cultures. This strong immunity against modified cells provides an evolutionary explanation for the paradoxical alloreactivity (Zinkernagel *et al.*, 1976).

E. Induction of B Cells

There are various compounds that, in the absence of T cells, induce proliferation and/or differentiation of B lymphocytes. Such polyclonal B activators include bacterial lipopolysaccharide (Andersson *et al.*, 1972), some polysaccharides, and purified tuberculin derivative in murine systems. Human blood B lymphocytes respond poorly to most lipopolysaccharide preparations, whereas Epstein–Barr virus, protein A-containing *Staphylococcus aureus* strains, and antibodies to β_2-microglobulin are potent activators (Goodman, 1980).

VI. CHARACTERIZATION AND STRUCTURE OF NORMAL LYMPHOCYTE SURFACE GLYCOPROTEINS

The introduction of sensitive analytical techniques has made it possible to analyze with high accuracy the membrane constituents of small amounts of cells. We and others have been involved in developing radioactive surface-labeling methods, which have proved to be especially useful for studying surface proteins. The most commonly used methods are described below.

A. Methods for Radioactive Surface Labeling of Lymphoid Cells

1. Lactoperoxidase-Catalyzed Cell-Surface Iodination

When intact cells are incubated with lactoperoxidase and ^{125}I in the presence of H_2O_2, exposed tyrosine residues are labeled (Phillips and Morrison, 1971). The H_2O_2 can elegantly be generated by glucose oxidase in the presence of D-glucose (Hubbard and Cohn, 1972). The high molecular weight of the lactoperoxidase inhibits its penetration into cells, and therefore only surface-located polypeptides are labeled. It is naturally important to avoid the presence of broken cells in the preparation. One important advantage of this technique is that the labeled proteins can be cleaved by various treatments and the radioactive peptides further analyzed by peptide mapping techniques.

2. Galactose Oxidase–Tritiated Borohydride Surface-Labeling Technique

Galactose oxidase from *Dactylium dendroides* oxidizes nonreducing terminal D-galactose and N-acetyl-D-galactosamine residues in glycoproteins and glycolipids to the corresponding C-6 aldehydes (Avigad *et al.*, 1962). Because of its macromolecular nature, the enzyme does not penetrate the intact cell membrane, and therefore only surface-exposed carbohydrate is oxidized. Penultimate galactosyl/N-acetylgalactosaminyl residues can be exposed by treatment with neuraminidase, which increases the labeling efficiency. After the enzymes are washed away, the aldehydes are reduced with NaB^3H_4 or KB^3H_4, resulting in

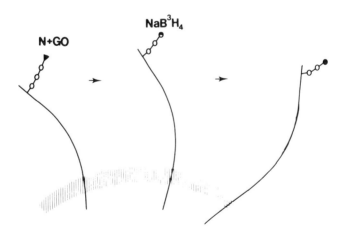

Figure 1 Schematic view of the galactose oxidase–NaB^3H_4 surface-labeling technique: N, neuraminidase; GO, galactose oxidase; ▼, sialic acids; ○, D-galactose; ◓, D-galactose C-6 aldehyde; ●, D-[³H]galactose.

radioactive galactosyl and N-acetylgalactosaminyl residues (Gahmberg and Hakomori, 1973; Steck and Dawson, 1974; Gahmberg, 1976) (Fig. 1). Because most, if not all, surface proteins seem to contain galactosyl/N-acetylgalactosaminyl residues available at the cell surface, the surface proteins become labeled. The reader is referred to recent articles and reviews for more detailed descriptions of the technique (Gahmberg and Hakomori, 1976; Gahmberg et al., 1976a,b; Gahmberg, 1978).

3. Periodate–Tritiated Borohydride Surface Labeling Technique

When intact cells are incubated with 1–2 mM concentrations of sodium metaperiodate at 0° for 10 minutes, only surface-exposed sialic acids are oxidized, and after reduction with NaB^{3}H$_4$ tritiated 5-acetamido-3,5-dideoxy-L-arabino-2-heptulosonic or 5-acetamido-3,5-dideoxy-L-arabino-2-octulosonic acid is formed (Gahmberg and Andersson, 1977). The negative charge of the sialic acids is retained, and the glycoproteins/glycolipids are only slightly modified. The periodate/NaB^{3}H$_4$ labeling technique is schematically shown in Figure 2.

After labeling, the radioactive proteins can be directly analyzed, for example, by polyacrylamide gel electrophoresis in the presence of sodium dodecylsulfate (Laemmli, 1970) or by isoelectric focusing on two-dimensional gels (O'Farrell, 1975). When necessary the radioactive cells may be solubilized in nonionic detergents such as Triton X-100 and the radioactive glycoproteins immuno-precipitated (Kessler, 1975; Gahmberg and Andersson, 1978). The introduction of fluorography has greatly enhanced the sensitivity and substantially improved the resolution (Bonner and Laskey, 1974).

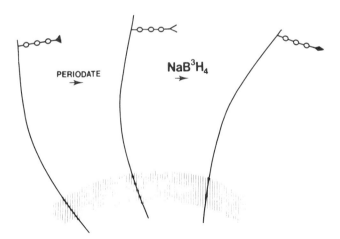

Figure 2 Schematic view of the periodate–NaB^{3}H$_4$ surface-labeling technique: ▼, sialic acid; V, aldehyde derivative of sialic acid; ◆, [^{3}H]sialic acid derivative.

B. Major Transplantation Antigens

Most of the original work on lymphocyte surface glycoproteins was done on surface alloantigens. There are several reasons for this. Antisera obtained by alloimmunization have facilitated the identification and isolation of these molecules. Furthermore, some of these structures constitute the clinically important transplantation antigens, and it is therefore obvious that much attention has been focused on these structures. But even these immunologically well characterized antigens have been very difficult to characterize in molecular detail. Only limited amounts have been available, and their hydrophobic protein nature, resulting in insolubility in ordinary buffers, has created major technical problems. Only during the past few years has substantial development occurred, and now several previously poorly known lymphocyte surface proteins are being characterized (Andersson and Gahmberg, 1979).

A schematic structure of the major transplantation antigens (HLA-A, -B, and -C in man; H-2 K, H-2D in mouse) is shown in Figure 3 (see Owen and Crumpton, 1980, for review). They are glycoproteins comprised of two different subunits, a heavy glycosylated chain and a light chain, which are noncovalently linked. The light chain has been identified as β_2-microglobulin in several species (Nakamuro *et al.*, 1973; Peterson *et al.*, 1974). This nonglycosylated protein was first isolated from urine (Berggård and Bearn, 1968), and at that time its association with the transplantation antigens was not realized. The β_2-

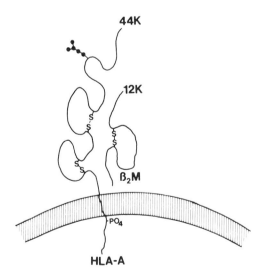

Figure 3 Structure of the HLA-A, -B, -C antigen. The heavy chain (44K) and the light chain (12K) are shown. The locations of the disulfide loops are given. The phosphate is in the heavy chain on the cytoplasmic aspect of the lipid bilayer. β_2M, β_2-microglobulin; ●——●, oligosaccharide.

microglobulin is invariant within a species, whereas the heavy chain contains the allogeneic sites.

Much of the structural work on HLA-A, -B, and -C antigens has been done using material solubilized after papain digestion. Papain cleaves the HLA molecules at one site on the external aspect of the membrane, resulting in a heavy chain fragment with a molecular weight of about 34,000, to which β_2-microglobulin is attached (Mann et al., 1969; Trägårdh et al., 1979b). This has greatly facilitated structural work because it is obviously much easier to work with solubilized molecules than with the membrane-bound antigens (Helenius and Simons, 1975).

The general structure and location of the HLA-A, -B, and -C heavy chain in the membrane actually resembles that of glycophorin A (Tomita and Marchesi, 1975). The amino terminal is externally located, and the single N-glycosidic oligosaccharide is located in this portion of the molecule (Trägårdh et al., 1979b). The complete structure of the oligosaccharide has not been worked out, but it seems to be of a common complex type (Parham et al., 1977). The alloantigenic site(s) is clearly not associated with the carbohydrate portion but with the peptide part, as seen by sequence analysis (Orr et al., 1979b; Trägårdh et al., 1979b).

The intramembrane amino acid sequence is hydrophobic. The carboxy terminal is located in the cytoplasm, and the heavy chain of H-2 may be associated with actin at the inner membrane surface (Koch and Smith, 1978). A fraction of the heavy chains is phosphorylated, and the phosphate is located in the carboxy terminal portion, probably near the inner aspect of the lipid bilayer (Pober et al., 1978).

Recently, several different groups have studied the biosynthesis of H-2 and HLA molecules. Several interesting results have emerged from these studies. Because the heavy chains and β_2-microglobulins are coded for by genes located in different chromosomes, chromosomes 6 and 15 in man, respectively, the gene products must associate after translation. Both the heavy chain and β_2-microglobulin are synthesized as larger precursors containing amino terminal "signal sequences" (Ploegh et al., 1979; Dobberstein et al., 1979; Algranati et al., 1980). After cleavage of the amino terminal extensions, the newly synthesized chains become associated in the endoplasmic reticulum. β_2-Microgloublin is apparently required for the transport of the heavy chain to the cell surface (Dobberstein et al., 1979). In the Daudi cell line established from a Burkitt lymphoma β_2-microglobulin is not synthesized, the heavy chains are synthesized but remain intracellular and are degraded (Ploegh et al., 1979). The carbohydrate of the heavy chain is also needed for proper intracellular migration. When the glycosylation was blocked by tunicamycin, an inhibitor of dolichol phosphate-dependent N-glycosylation (see Chapter 1, this volume), the transplantation antigens remained intracellular (Algranati et al., 1980) and subsequently were degraded.

The HLA heavy chain has been sequenced, and interestingly it shows clear homology with the constant region of immunoglobulins (Peterson *et al.*, 1972; Orr *et al.*, 1979a; Trägårdh *et al.*, 1979a, 1980). The portion of the HLA molecule at the second disulfide loop near the external aspect of the bilayer contains the homology region. Likewise, β_2-microglobulin shows sequence homology with IgG. These findings indicate that these proteins have evolved from a common ancestor protein.

Some functions of the H-2 and HLA-A, -B, and -C antigens are becoming apparent. Virus-infected syngeneic cells induce cytotoxic T cells, which recognize the viral *and* the syngeneic transplantation antigen in a self-restricted manner (Zinkernagel *et al.*, 1976). A physical association between rat major transplantation antigens and an adenovirus-encoded surface protein has in fact been demonstrated (Kvist *et al.*, 1978).

It has been shown that the HLA and H-2 antigens may in some instances function as receptors for Semliki Forest virus (Helenius *et al.*, 1978) and bacteria (Klareskog *et al.*, 1978). Whether these receptor functions are more general remains to be established. On the other hand, it is also clear that the major transplantation antigens do not constitute the only receptors for Semliki Forest virus (Oldstone *et al.*, 1980).

C. Ia Lymphocyte Antigens

The mitogenic response in mixed lymphocyte culture resulting mainly in the induction of T helper cells is due primarily to the disparity of the murine Ia and the corresponding human HLA-D antigens. Antisera against these components block the mixed lymphocyte reaction (see Möller, 1976). These antigens are composed of two [or perhaps three (Charron and McDevitt, 1979)] glycosylated polypeptides with approximate molecular weights of 31,000–34,000 and 24,000–28,000, respectively (Humphreys *et al.*, 1976; Billing *et al.*, 1976; Möller, 1976; Klareskog *et al.*, 1977) (Fig. 4). These polypeptides are noncovalently linked to each other and probably exist as a complex in the membrane. The carboxy terminal portion of the heavy chain is phosphorylated (Kaufman and Strominger, 1979). These antigens are present mainly on B cells and were long considered to be B specific. However, certain T lymphocytes, especially upon activation, also express the antigens (Ko *et al.*, 1979). On the other hand, some B-cell lines lack or contain low levels of the antigens (Trowbridge *et al.*, 1977). These proteins are further enriched on leukemic cells (Janossy *et al.*, 1977; Anderson *et al.*, 1979; Andersson *et al.*, 1979a) and early myeloid cells (Ross *et al.*, 1978). The alloantigenic site(s) showing structural polymorphism seems to be located in the polypeptide portion of the light chains (Silver and Ferrone, 1979; Shackelford and Strominger, 1980), but the carbohydrate may, at least in some instances, also contain alloantigenic sites (Higgins *et al.*, 1980).

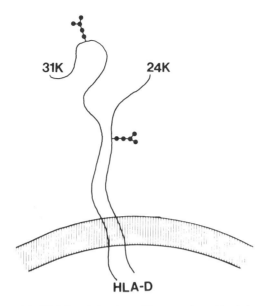

Figure 4 Structure of the HLA-D coded antigens. The two polypeptide chains (31K and 24K) are noncovalently linked to each other, and both are glycosylated. ●——●, oligosaccharide.

Actually very little is known about the carbohydrate structure of these antigens. Most of the carbohydrate is evidently asparagine-linked because tunicamycin blocked its synthesis (Nishikawa *et al.*, 1979; Shackelford and Strominger, 1980), resulting in lower apparent molecular weight of the proteins on gels. However, after tunicamycin treatment, the heavy chain still incorporated some [³H]glucosamine (Nishikawa *et al.*, 1979), indicating the presence of O-linked carbohydrate. The biosynthesis of O-linked oligosaccharides is not inhibited by tunicamycin (Gahmberg *et al.*, 1980b).

D. High Molecular Weight Surface Glycoproteins of Lymphoid Cells

Human cortical and medullary thymocytes, T cells, T lymphoblasts obtained after mitogen stimulation, T lymphoblasts obtained after stimulation with concanavalin A or in mixed lymphocyte culture, B cells, and non-T, non-B lymphocytes all have characteristic surface glycoprotein patterns (Fig. 5) (Andersson and Gahmberg, 1978; Andersson *et al.*, 1978). The corresponding populations of lymphocytes from mice show similar patterns (Trowbridge *et al.*, 1975; Gahmberg *et al.*, 1976; Kimura and Wigzell, 1978). The patterns obtained after oxidation with periodate are similar, but with one major difference. The gp120–gp130 protein now moves as a gp100 or gp110 protein (Fig. 6). This is probably

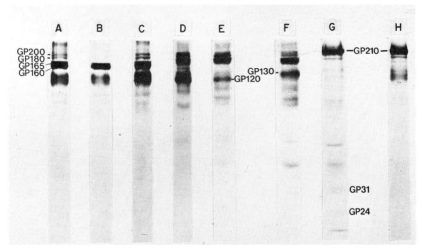

Figure 5 Surface glycoprotein patterns of human thymocytes and lymphocytes. (A) Unfractionated thymocytes; (B) cortical thymocytes; (C) medullary thymocytes; (D) T lymphocytes; (E) concanavalin A-activated T lymphocytes; (F) T lymphocytes activated in mixed lymphocyte culture; (G) spleen B cells; (H) non-T, non-B lymphocytes. GP200 denotes glycoprotein with an apparent molecular weight of 200,000; GP180 denotes glycoprotein with an apparent molecular weight of 180,000; etc. GP31 and GP24 are the polypeptides of the HLA-D coded antigens. The cells were labeled with NaB^3H_4 after treatment with neuraminidase and galactose oxidase.

because of the high content of O-glycosidic oligosaccharides in this protein (Saito and Osawa, 1980). Glycophorin A of red blood cells shows a similar behavior on electrophoresis. After removal of sialic acids, it moves more slowly on gels in the presence of sodium dodecylsulfate, although the molecular weight of the protein is reduced (Gahmberg and Andersson, 1977).

On the basis of the glycoprotein patterns in the high molecular weight region of the gels, the different types of lymphoid cells can be distinguished. The T cells show four closely spaced labeled glycoprotein bands with apparent molecular weights of 160,000–200,000 (Fig. 5D), whereas the B cells contain only one major component with an apparent molecular weight of 210,000 (Fig. 5G). Upon activation with mitogens, gp165 is relatively strongly labeled (Fig. 5E), and gp200 more weakly labeled, whereas gp120 remains essentially unchanged.

In contrast, when T cells are activated by allogeneic cells (Fig. 5F), a strongly labeled gp130 glycoprotein appears, and gp200 and gp180 are weakly labeled (Andersson et al., 1978). Similar glycoprotein changes were seen in murine T lymphoblasts after activation in mixed lymphocyte culture (Gahmberg et al., 1976a; Kimura and Wigzell, 1978; Dunlap et al., 1978).

Van Eijk and co-workers (Van Eijk and Mühlradt, 1977; Van Eijk et al., 1979) labeled the carbohydrate moieties of murine lymphocytes with different radioactive precursors. The polyacrylamide gel electrophoresis patterns obtained resembled those seen after surface labeling.

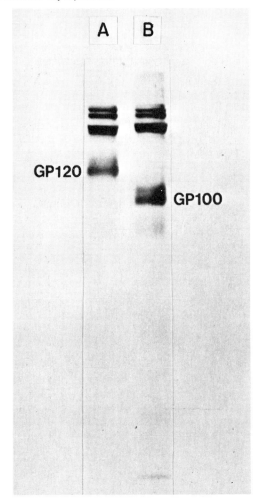

Figure 6 Comparison of the surface glycoprotein patterns of human T cells labeled with ³H after neuraminidase plus galactose oxidase (A) or periodate (B) treatments. Note the shift of GP100 to GP120 after neuraminidase treatment.

The murine glycoprotein T145 corresponding to the human gp130 has been characterized in more detail (Kimura and Wigzell, 1978; Conzelmann *et al.*, 1980). This protein apparently has a unique carbohydrate composition since it selectively binds to the lectin from *Vicia villosa* (Kimura and Wigzell, 1978; Kimura *et al.*, 1979). *N*-Acetyl-D-galactosamine inhibits the binding. The lectin specificity indicates the presence of a blood group A-like carbohydrate structure on the T145 protein. The human counterpart (gp130) also binds to *Vicia villosa* lectin after neuraminidase treatment (K. Karhi, personal communication).

The thymocyte gp 160, the T-cell proteins gp160–gp200, and the B-cell protein gp210 show interesting structural relationships. A monoclonal antibody reacted with all of these proteins from rat lymphoid cells. The proteins were referred to as the common leukocyte antigen (Standring et al., 1978). A similar immunological cross-reactivity has now been found for the corresponding human and mouse proteins (Dalchau et al., 1980; Hoessli and Vassalli, 1980). Also, peptide mapping of the murine common leukocyte antigens from T and B cells showed obvious similarities (Dunlap et al., 1980). Furthermore, the murine common T-leukocyte antigen or the T200 protein shows structural homology with the corresponding protein from human cells (Omary et al., 1980b). It is polymorphic and corresponds to the previously immunologically defined Lyt-5 antigen (Omary et al., 1980a). The rat thymic common leukocyte antigen also shows polymorphism (Carter and Sunderland, 1979).

Although these high molecular weight proteins are similar in structure, they are obviously not identical. This is immediately obvious from their differences in apparent molecular weights. In addition, a T-cell-specific antiserum reacted only with the high molecular weight proteins from thymocytes and T cells but not with the gp210 of B cells (Anderson and Metzgar, 1979; Andersson et al., 1980; Gross and Bron, 1980) (Fig. 7B–D). Dalchau and Fabre (1981) reported a monoclonal antibody that specifically recognizes the human common leukocyte antigen (gp210) of human B cells. This shows that in addition to common antigenic determinants the proteins contain individual specific determinants.

Hauptman and colleagues found a cold-insoluble surface glycoprotein on murine (Hauptman et al., 1979c; Kansu et al., 1979) and human T cells and "null" cells (Hauptman and Kansu, 1978; Hauptman et al., 1979a,b). They estimated that the T-cell-specific, cold-insoluble protein had an apparent molecular weight of 225,000 and that the "null"-cell-specific protein had an apparent molecular weight of 185,000. The T-cell cold-insoluble protein apparently corresponds to the proteins recognized by T-cell-specific antisera.

We do not yet know the extent to which the structures recognized as differentiation antigens reflect differences in the polypeptide portion of the high molecular weight proteins or the carbohydrate portions. Some indication for the relative importance of the carbohydrates was obtained when we compared the total surface-labeled glycopeptides/oligosaccharides of murine T and B cells. The T cells had more O-glycosidic oligosaccharides, whereas B cells contained more of the complex-type N-glycosidic oligosaccharides (Krusius et al., 1979). A substantial portion of these oligosaccharides/glycopeptides apparently derive from the high molecular surface proteins discussed above.

The functions of the high molecular weight glycoproteins are not well known, although the B-cell glycoprotein gp210 has recently been shown to carry the C3 complement receptor (see Section VI,F,2).

The enzyme 5′-nucleotidase is an ectoenzyme, or cell-surface enzyme, of

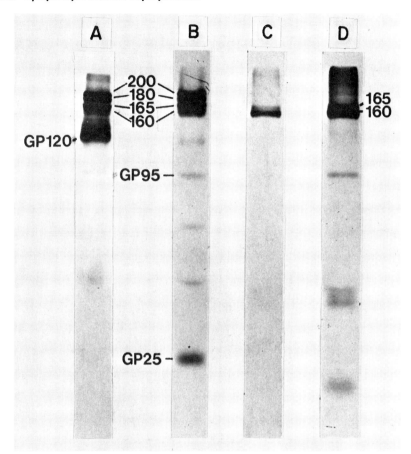

Figure 7 Human T-cell- and thymocyte-specific surface glycoproteins. (A) Surface glycoproteins of normal T cells; (B) surface glycoproteins immunoprecipitated with anti-T-cell-specific antiserum from T lymphocytes; (C) surface glycoproteins immunoprecipitated with anti-T-cell antiserum from thymocytes; (D) same as in C but exposed for a longer time. The cells were labeled with ^{3}H after treatment with neuraminidase and galactose oxidase. The apparent molecular weights of the proteins are indicated.

many tissues (see, for example, DePierre and Karnovsky, 1974). This enzyme displays different levels of activity in various subpopulations of lymphocytes. Low activities were found in thymocytes, with the lowest levels in the cortical thymocytes (Dornand *et al.*, 1980). The activity was higher in T cells and T lymphoblasts, and even higher in B cells (Edwards *et al.*, 1979; Gregory and Kern, 1979). Patients with agammaglobulinemia, who lack B cells, therefore have lower total levels of 5′-nucleotidase activity in their leukocytes (Edwards *et al.*, 1978).

5'-Nucleotidase was purified from pig lymphocytes by means of affinity chromatography on AMP-Sepharose. The purified enzyme had an apparent molecular weight of 130,000 (Dornand *et al.*, 1978a). Concanavalin A strongly inhibited its nucleotidase activity, whereas wheat germ agglutinin, which is not mitogenic, had no effect (Dornand *et al.*, 1978b). This shows that the enzyme is a glycoprotein, and, moreover, because concanavalin A is strongly mitogenic, whereas wheat germ agglutinin is not, it may have some function in mitogen-induced activation.

E. Surface Glycoproteins Specific for Subsets of T Lymphocytes

Alloantigens specific for T cells and their subpopulations were first defined in the murine system. It is now apparent that very similar antigens occur not only in lymphocytes of other rodents, but also in man. We have already discussed the high molecular weight surface glycoproteins, some of which are T specific and show polymorphism, such as the Lyt-5 antigen. Several initially serologically defined T antigens exist, some of which have now been biochemically characterized.

1. Thy-1 Antigen

The Thy-1 antigen exists in two allelic forms in mice, Thy-1.1 and Thy-1.2 (Reif and Allen, 1964). It is found in thymocytes of mice and rat, to a smaller degree on mature T cells, and also in brain (Barclay *et al.*, 1976; Williams *et al.*, 1977). There has been much confusion about the molecular nature of the Thy-1 antigen, but it now seems established that it is a surface glycoprotein with an apparent molecular weight of 25,000 (Barclay *et al.*, 1976; Williams *et al.*, 1977). In addition, gangliosides have been claimed to be Thy-1 active. This discrepancy has not yet been resolved (see Section VII). The rat protein has been purified from thymocytes and brain and contains about 30% carbohydrate. The polypeptide portions of the thymocyte and brain Thy-1 glycoproteins are very similar, if not identical, whereas the carbohydrate portions differ considerably (Williams *et al.*, 1977). For example, the brain molecule contains *N*-acetylgalactosamine, which is not found in the thymocyte protein. Hoessli *et al.* (1980) showed that, during differentiation of thymocytes to T cells, the Thy-1 antigen acquires an increased content of sialic acid. This observation that an individual polypeptide changes its carbohydrate composition during differentiation might be very important for the generation of differentiation antigen specificities and may be more general than commonly believed (compare the common leukocyte antigen).

When membrane lysates from radioactively surface-labeled human T cells were immunoprecipitated with rabbit anti-T-specific antiserum, a glycoprotein with an apparent molecular weight of 25,000 was precipitated (Fig. 7B). This may be the human homolog of Thy-1 (Dalchau and Fabre, 1979; Andersson *et al.*, 1980). Actually, Ades *et al.* (1980) have now purified apparently the same protein from the human T-cell line MOLT-3 and showed its immunological cross-reactivity with the murine Thy-1 antigen.

2. Thymus–Leukemia Antigen

The murine thymus–leukemia (TL) antigen is found on thymocytes and certain leukemic cells (Boyse and Old, 1969; Vitetta *et al.*, 1972). It consists of two polypeptides: one heavy chain, which is glycosylated and has an apparent molecular weight of 45,000, and a light chain, which evidently is β_2-microglobulin (Anundi *et al.*, 1975; Rothenberg and Boyse, 1979). A very similar antigen has been found in human cells. Here, the heavy chain has an apparent molecular weight of 49,000 (Ziegler and Milstein, 1979), and the light chain is similar to β_2-microglobulin, but not identical (Ziegler and Milstein, 1979).

3. Lyt-1 Antigens

The murine Lyt-1, -2, and -3 alloantigens have proved to be very useful for distinguishing maturationally and functionally different T-cell populations. The Lyt-1 is more strongly expressed on helper/inducer T cells, whereas Lyt-2 and Lyt-3 are markers for killer/suppressor T cells. The Lyt-1 molecule has been identified as a surface glycoprotein with an apparent molecular weight of 67,000 using either conventional antibodies (Durda *et al.*, 1978) or monoclonal antibodies (Hogarth *et al.*, 1980). After labeling with galactose oxidase/NaB^{3}H$_4$ a second labeled glycoprotein with an apparent molecular weight of 87,000 was observed (Durda *et al.*, 1978). The relation of this protein to the Lyt-1 antigen remains unclear.

A monoclonal antibody reactive with human helper T cells specifically precipitated a surface glycoprotein (T4) from thymocytes with an apparent molecular weight of 62,000 (Terhorst *et al.*, 1980). This protein may be the human counterpart of the Lyt-1 antigen.

4. Lyt-2 and Lyt-3 Antigens

The Lyt-2 and Lyt-3 antigens were identified from surface-labeled murine thymocytes by means of immunoprecipitation with specific antiserum. In initial experiments a glycoprotein with an apparent molecular weight of 35,000 was specifically precipitated (Durda and Gottlieb, 1978). In later experiments using monoclonal antibodies, two polypeptides with apparent molecular weights of 30,000 and 35,000 were precipitated (Ledbetter and Herzenberg, 1979). Ledbetter *et al.* (1981) have characterized the antigens further and shown that the mouse Lyt-2 and Lyt-3 antigens are associated with three different polypeptide chains with molecular weights of 38,000, 34,000 and 30,000. These form disulfide-bound dimers, tetramers, and hexamers. The Lyt-2 antigen specificity is present on the 38,000 and 34,000 molecular weight components, whereas the Lyt-3 specificity is carried by the 30,000 molecular weight polypeptide.

In human beings a corresponding antigen, T5, which is present on cytotoxic T cells, was immunoprecipitated, also by the use of a monoclonal antibody (Terhorst *et al.*, 1980). Two polypeptides were identified with apparent molecular weights of 30,000 and 32,000. Apparently, the murine and human antigens are closely related.

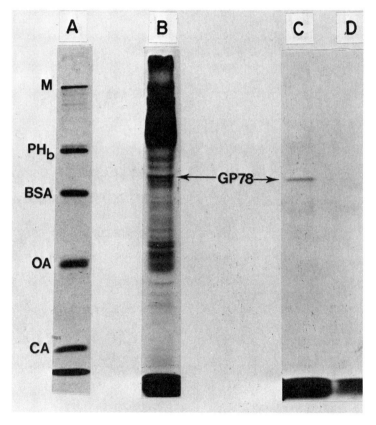

Figure 8 Identification of the Lyt-6 antigen. (A) Standard proteins (M, myosin; PH$_b$, phosphorylase *b;* BSA, bovine serum albumin; OA, ovalbumin; CA, carbonic anhydrase); (B) surface glycoprotein pattern of C57/B6 T lymphoblasts obtained in mixed lymphocyte culture and labeled with ³H after treatment with neuraminidase and galactose oxidase; (C) immunoprecipitate obtained with anti-Lyt-6 antiserum from the cells described in B; (D) control with preimmune serum.

TABLE II
Molecular Characteristics of Human T Cell Surface Proteins

Protein	Molecular weight of subunits	Remarks	Reference
HLA-A, -B, -C	44,000 (heavy chain) 12,000 (light chain = β_2-microglobulin)	Major transplantation antigen	See text
Thy-1	25,000		Andersson et al. (1980), Ades et al. (1980), Dalchau and Fabre (1979)
TL	49,000 (heavy chain) 12,000 (not β_2-microglobulin)		Ziegler and Milstein (1979)
T4	62,000	May correspond to murine Lyt-1	Terhorst et al. (1980)
T5	30,000; 32,000	May correspond to murine Lyt-2 and Lyt-3	Terhorst et al. (1980)
gp52	52,000	Precursor of serum α_1-acid glycoprotein	Gahmberg and Andersson (1978)
gp120 (gp150)	120,000	Moves more slowly on polyacrylamide gels after neuraminidase treatment	Andersson and Gahmberg (1978)
gp130	130,000	Moves more slowly on polyacrylamide gels after neuraminidase treatment; may be specific for cytotoxic T cells	Andersson et al. (1978)
gp160, gp165, gp180, gp200	160,000–200,000	T-Cell specific, T-cell counterparts of the common leukocyte antigen	Andersson et al. (1980)
gp225	225,000	T-Cell cold-insoluble globulin; may be part of gp165–200	Hauptman et al. (1979b)

TABLE III
Molecular Characteristics of Murine T Cell Surface Proteins

Protein	Molecular weight of subunits	Remarks	Reference
H-2 K, D	45,000, 12,000 (β_2-micro-globulin)	Major transplantation antigens	See text
Thy-1	25,000	Previously called theta (θ)	Barclay et al. (1976)
TL	45,000		Rothenberg and Boyse (1979)
Lyt-1	67,000 (possibly other component, 87,000)		Durda et al. (1978); Hogarth et al. (1980)
Lyt-2, Lyt-3	30,000, 32,000, 35,000		Durda and Gottlieb (1978), Ledbetter and Herzenberg (1979), Ledbetter et al. (1981)
Lyt-5	200,000	T-Cell counterpart of the common leukocyte antigen	Omary et al. (1980a,b)
Lyt-6	78,000	Low on thymus cells; more on peripheral T cells	Horton et al. (1979)
T145	145,000	Selectively expressed on killer T cells; binds to Vicia villosa lectin	Kimura et al. (1979)

5. Lyt-6 Antigen

The Lyt-6.2 antigen is present on T cells of appropriate mouse strains (McKenzie and Potter, 1979). It is found at higher levels on lymphoblasts, whereas thymocytes contain smaller amounts (Horton et al., 1979). The antigen was identified by precipitation with antiserum (Horton et al., 1979) from neuraminidase–galactose oxidase/NaB^3H_4-labeled cells (Fig. 8) (Andersson et al., 1981). It is a surface glycoprotein with an apparent molecular weight of 78,000. In BW5147 thymoma cells, two closely spaced bands were precipitated.

A summary of the human and murine T-lymphocyte surface glycoproteins is given in Tables II and III, respectively.

F. Surface Glycoproteins Enriched on B Lymphocytes

In contrast to the several characterized T-lymphocyte- and thymocyte-specific surface molecules, rather few biochemically characterized antigens have been described which clearly are enriched on B cells (Table IV) (Katz, 1977). The immune-associated Ia transplantation antigens are found on most B cells and also on other cells and have been discussed above (Section VI,C). Surface immunoglobulin is a marker of B cells, but T cells may also contain some immunoglobulin or at least immunoglobulin-like molecules (Katz, 1977).

TABLE IV
Molecular Characteristics of Human B Cell Surface Proteins

Protein	Molecular weight of subunits	Remarks	Reference
HLA-D, Ia	31,000, 24,000	Light chain evidently contains the allogeneic sites; molecular weight estimates vary between 31,000–35,000 and 24,000–28,000, respectively	Humphreys et al. (1976), Billing et al. (1976), Klareskog et al. (1977)
gp210	210,000	B-Cell counterpart of the common leukocyte antigen, complement C3 receptor	Andersson and Gahmberg (1978), Dalchau and Fabre (1981), Fearon (1980)
IgG Fc receptor	28,000 or 46,000–70,000	28,000 value after salt extraction; detergent solubilization gives two proteins with molecular weights of 46,000–70,000	Thoenes and Stein (1979), Takacs (1980), Mellman and Unkeless (1980)
gp54	54,000	Single chain	Wang et al. (1979)

Wang et al. (1979) isolated and characterized a B-cell-specific surface glyco-protein with an apparent molecular weight of 54,000. An antiserum made against the protein was mitogenic for purified tonsillar B lymphocytes. Purified Fab fragments were not active. In addition to these proteins some B-cell proteins, which have been the focus of greater interest, are discussed below.

1. Receptors for the Fc Portion of Immunoglobulin G

The literature contains several reports on the isolation of Fc receptors from lymphoid and other cells. Molecular weights between 15,000 and 130,000 have been reported. A major problem seems to be that after solubilization of the lymphocyte membrane in nonionic detergents, several proteins adsorb to the IgG aggregates. Thoenes and Stein (1979) used 80 mM EDTA–50 mM 2-mercaptoethanol to solubilize the human tonsillar Fc receptor. Although measures were taken to avoid proteolysis, the possibility that some proteolysis took place cannot be ruled out. A 28,000 molecular weight surface protein was obtained by affinity chromatography using aggregated IgG–Sepharose. A protein peak with the same molecular weight was obtained from detergent-lysed membranes in addition to other protein peaks. In contrast, Takacs obtained a molecular weight value of 46,000 for the human Fc receptor after solubilization in nonionic detergent (Takacs, 1980). Using a rat monoclonal anti-mouse Fc receptor antibody column Mellman and Unkeless (1980) isolated Fc receptors from different mouse cells. Two glycoproteins were obtained in the presence of nonionic detergent with molecular weights of 47,000 to 70,000. The molecular weights varied according to the cellular source and may be due to differences in glycosylation.

2. Receptors for Complement C3 and Epstein–Barr Virus

The Epstein–Barr virus (EBV) selectively infects primate B cells, whereas the majority of the blast cells seen in the blood of patients with infectious mononucleosis are T blasts. These evidently arise because of an *in vivo* mixed lymphocyte reaction against the infected B cells. The molecular nature of the EBV receptors are still to be characterized, but there are several lines of evidence that they are identical to, or at least closely associated with, B cell complement receptors (Jondal *et al.*, 1976). The expression of EBV and C3 receptors coincided on different cells and cell lines, and co-capping of C3 and EBV receptors

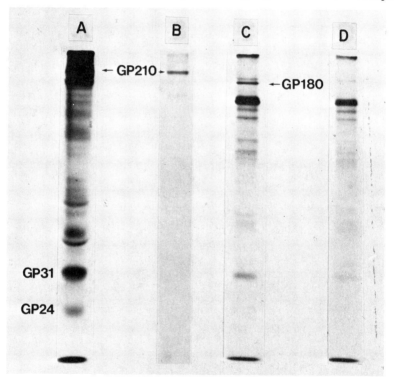

Figure 9 Identification of the B-cell complement C3 receptor. (A) Surface glycoprotein pattern of tonsil lymphocytes; (B) glycoprotein precipitated with anticomplement C3 receptor antiserum from surface-labeled tonsil cells; (C) proteins immunoprecipitated with anticomplement C3 receptor antiserum from [^{35}S]methionine-labeled Raji lymphoma cells; (D) proteins obtained with preimmune serum from [^{35}S]methionine-labeled Raji cells. The samples in A and B were labeled with ^{3}H after treatment with neuraminidase and galactose oxidase.

indicates that they either are present on the same molecule or are closely associated (Yefenof *et al.*, 1976). A close functional proximity of the receptors was further indicated by the finding that EBV absorption was blocked by C3 and immune complexes (Yefenof and Klein, 1977).

Gerdes *et al.* (1980) produced a complement receptor antiserum that was used for immunoprecipitation of surface-labeled human tonsil cells and various B-cell lines. One major surface glycoprotein was precipitated (Fig. 9B) that corresponded in molecular weight to the major B cell surface glycoprotein gp210. Fearon isolated a glycoprotein from erythrocytes with an apparent molecular weight of 205,000. An antiserum against this immunoprecipitated a similar protein from granulocytes, B cells, and monocytes (Fearon, 1980). The purified protein exhibited properties of a complement receptor and is evidently the same as our gp210.

The association of complement receptor activity with gp210 raises some interesting possibilities. The same glycoprotein is evidently the B-cell counterpart of the common leukocyte antigen, whereas the corresponding T-cell antigen is not recognized by the anticomplement receptor antibodies (data not shown). This parallels the monoclonal antibody specificity recognizing only the B counterpart of the common leukocyte antigen discussed in Section VI,D. These findings suggest that the carbohydrate portion of the B cell protein could be responsible for the antibody specificity and the complement binding.

G. gp52: A Membrane Form of α_1-Acid Glycoprotein (Orosomucoid) on Both T and B Cells

α_1-Acid glycoprotein is a serum and urinary glycoprotein. It belongs to the acute-phase proteins that increase during inflammation, during chronic diseases such as tuberculosis, and in cancer patients (Schmid, 1975). It has been extensively studied, and its amino acid sequence has been determined (Schmid *et al.*, 1973). It shows sequence homology with IgG (Schmid *et al.*, 1973). Its real function is unknown, but there are reports that it induces suppression in the mixed lymphocyte reaction, lectin-induced blast transformation, and B-cell differentiation (Chiu *et al.*, 1977; Bennett and Schmid, 1980). It was previously considered to be synthesized exclusively in the liver (Sarcione, 1963), but recently we found that it is made in hematopoietic cells including T and B lymphocytes, granulocytes, and monocytes (Gahmberg and Andersson, 1978). It is found as a cell surface glycoprotein with an apparent molecular weight of 52,000, which apparently is cleaved to give the soluble protein (Fig. 10). The structural homology with IgG (and perhaps HLA-A, -B, and -C) indicates that these proteins belong to the same superfamily of proteins, which are involved in immune reactions.

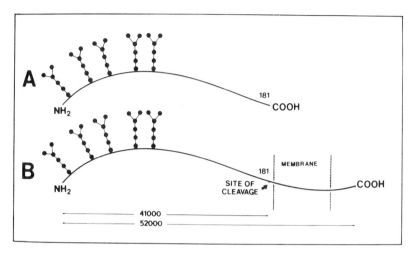

Figure 10 Schematic drawing of α_1-acid glycoprotein (A) and its membrane precursor (B). The site of proteolytic cleavage and the oligosaccharide attachment sites are indicated.

VII. LYMPHOCYTE GLYCOSPHINGOLIPIDS

Glycosphingolipids are enriched in the plasma membrane of mammalian cells and are exclusively located in the outer portion of the lipid bilayer (Gahmberg and Hakomori, 1973; Steck and Dawson, 1974). Detailed studies on the surface exposure of fibroblast and red blood cell glycolipids have shown that only a fraction of the total plasma membrane glycolipids actually is available to external reagents (Gahmberg and Hakomori, 1974; Gahmberg et al., 1976c). Several receptor functions seem to be associated with glycolipids. Some of the more complex glycosphingolipids in red cells carry blood group ABH and Ii antigens (Hakomori and Kobata, 1975), and certain gangliosides, especially GM1, have been implicated as receptors for tetanus and cholera toxins and interferon (Van Heyningen, 1974; Cuatrecasas, 1973; Holmgren et al., 1975; Besançon et al., 1976). For the structures of the neutral glycosphingolipids and gangliosides, see Chapter 6, Volume I.

Rather few studies have been performed on the composition of lymphocyte glycolipids and their changes during cellular differentiation and malignant transformation. Stein and Marcus (1977) analyzed the neutral glycolipids and gangliosides of human thymocytes, peripheral blood lymphocytes, and tonsil cells. The majority of the lymphocytes in peripheral blood were T cells, and those in the tonsils were mainly B cells. The neutral glycolipids of thymocytes and peripheral blood lymphocytes were similar and were composed mainly of ceramide monohexosides, ceramide dihexosides, and complex glycolipids moving on thin-layer chromatography more slowly than globoside (ceramide tetrasaccharide). Tonsil cells contained more ceramide monohexoside, ceramide dihexoside, ceramide trihexoside, and globoside than the T cells and thymocytes. However, they contained lower levels of the more complex glycolipids.

Gangliosides were enriched in peripheral blood lymphocytes. The main component was hematoside (GM3). Tonsil cells contained relatively more complex gangliosides (Stein and Marcus, 1977). Rosenfelder et al. (1978, 1979) labeled murine thymocyte and spleen T and B cells with [^{14}C]galactose and [^{14}C]-glucosamine after mitogen treatments. Again the major labeled neutral glycosphingolipids of thymocytes and T cells were ceramide monohexosides. The B cells seemed to contain more label in ceramide dihexosides. Furthermore, the B cells contained complex neutral glycolipids with unknown structure. Murine thymocytes contained hematoside but also several poorly characterized complex gangliosides (Rosenfelder et al., 1978).

The T and B cells showed characteristic and different ganglioside patterns. The hematoside level was relatively low, whereas both types of cells contained several complex gangliosides. The structures of the "T-specific" and "B-specific" gangliosides remained unclear, however, Importantly, the T and B cells could be distinguished from their ganglioside patterns.

The Thy-1 antigen has been claimed to be associated with complex gangliosides (Wang et al., 1978). Although it is well established that Thy-1 activity is associated with a 25,000 molecular weight glycoprotein of T cells and brain (see Section VI,E), the proposed Thy-1 activity of gangliosides remains controversial. G_{M1} or a minor glycolipid migrating in the G_{M1} position on thin-layer chromatograms has been implicated in the activity. However, anti-G_{M1} antibodies reacted with T cells irrespective of the Thy-1 phenotype (Stein-Douglas et al., 1976). Furthermore, the receptors for anti-Thy-1.2 antibodies and anti-GM1 antibodies did not co-cap (Stein et al., 1978). Asialo-GM1 is present in low levels on thymocytes, but more is present on T cells and the levels increase with age (Schwarting and Summers, 1980). Thus, it seems to be a T-cell differentiation antigen. Interestingly, this glycolipid can be regarded as a marker for natural killer cells. Anti-asialo-GM1 specifically destroyed natural killer cells (Young et al., 1980), and mice low in natural killer cells contained a decreased level of asialo-GM1 (Schwarting and Summers, 1980).

VIII. CONCLUDING REMARKS

Lymphocyte surface carbohydrate is obviously functionally important. Not only do mitogenic lectins activate the cells through initial binding to the carbohydrate portions of cell-surface glycoconjugates, but the clear-cut changes in the surface glycoproteins and glycolipids upon activation indicate functional involvements of these molecules. An interesting example is the common leukocyte antigen. The differences in molecular weights of this antigen in T and B cells could well result from carbohydrate changes occurring during differentiation. These carbohydrate structures may be functionally involved, as seems to be the situation with the B-cell protein GP210, which binds complement and a monoclonal antibody, which the T-cell protein does not.

In this connection it may be relevant to allude to the ABH antigens in red blood cells. These antigens are partially confined to glycoproteins in the red cell membrane (Finne *et al.,* 1978; Järnefelt *et al.,* 1978). Not only one but several glycoproteins carry the antigens (Karhi and Gahmberg, 1980; Finne, 1980). This shows that the same carbohydrate structures, which in this case are of unusual types, may be linked to different protein and lipid backbones (see Gahmberg and Karhi, 1981, for a review). The situation may be similar with the complement receptor and with other types of receptors that have identical specificities but that occur in different types of cells. In this way it would not always be necessary for the cells to synthesize highly specific polypeptides for a certain function, but instead the cells could use the carbohydrate-synthesizing machinery to achieve high specificity.

Several lines of evidence suggest that, through carbohydrate-binding molecules, highly specific intercellular interactions could take place. Kieda *et al.* (1978) found that lymphocytes contain surface glycoproteins with lectin activity. Another, often discussed possibility is that glycosyltransferases are involved in intercellular recognition (Roseman, 1970). In addition, Rauvala and co-workers (Rauvala and Hakomori, 1981), using fibroblasts, have shown that different cell-surface-located glycosidases can mediate cell adhesion. α-Mannosidase seems to be especially important. Therefore, it may very well be that the differentiation-induced carbohydrate changes in cell-surface glycoproteins and glycolipids reflect their involvements in cellular recognition and interactions occurring perhaps through one or more of the above-mentioned mechanisms.

Chemical work on lymphocyte cell-surface proteins is now rapidly advancing because of the availability of (a) cell lines at different stages and directions of differentiation, (b) powerful analytical techniques applicable to small amounts of cells, and (c) monoclonal antibodies for the identification and isolation of the relevant surface structures.

Only when the structure and topology of the different surface glycoproteins and glycolipids have been worked out in molecular detail will it be possible to elucidate the molecular mechanisms involved in lymphocyte differentiation, maturation, and cellular interaction. In the unraveling of this exciting field, the carbohydrate will play a central role.

ACKNOWLEDGMENTS

We thank Anneli Asikainen and Päivi Nieminen for expert technical assistance and Barbara Björnberg for secretarial help. The original work described herein was supported by the Academy of Finland, the Finnish Cancer Society, the Finska Läkaresällskapet, the Association of Finnish Life Insurance Companies, and the National Cancer Institute Grant 1 RO1 CA26294-01A1.

REFERENCES

Ades, E. W., Zwerner, R. K., Acton, R. T., and Balch, C. M. (1980). *J. Exp. Med.* **151,** 400-406.

Algranati, I. D., Milstein, C., and Ziegler, A. (1980). *Eur. J. Biochem.* **103,** 197-207.

Anderson, J. K., and Metzgar, R. S. (1979). *Clin. Exp. Immunol.* **37,** 339-347.

Anderson, J. K., Moore, J. O., and Metzgar, R. S. (1979). *Cancer Res.* **39,** 4810-4815.

Andersson, J., Sjöberg, O., and Möller, G. (1972). *Eur. J. Immunol.* **2,** 349-353.

Andersson, L. C., and Gahmberg, C. G. (1978). *Blood* **52,** 57-67.

Andersson, L. C., and Gahmberg, C. G. (1979). *Mol. Cell. Biochem.* **27,** 117-131.

Andersson, L. C., and Häyry, P. (1975). *Transplant. Rev.* **25,** 121-162.

Andersson, L. C., Gahmberg, C. G., Nilsson, K., and Wigzell, H. (1977). *Int. J. Cancer* **20,** 702-707.

Andersson, L. C., Gahmberg, C. G., Kimura, A. K., and Wigzell, H. (1978). *Proc. Natl. Acad. Sci. U.S.A.* **75,** 3455-3458.

Andersson, L. C., Gahmberg, C. G., Siimes, M. A., Teerenhovi, L., and Vuopio, P. (1979a). *Int. J. Cancer* **23,** 306-311.

Andersson, L. C., Gahmberg, C. G., Teerenhovi, L., and Vuopio, P. (1979b). *Int. J. Cancer* **24,** 717-720.

Andersson, L. C., Karhi, K. K., Gahmberg, C. G., and Rodt, H. (1980). *Eur. J. Immunol.* **10,** 359-362.

Andersson, L. C., Hurme, M., Horton, M. A., Simpson, E., and Gahmberg, C. G. (1981). *Cell Immunol.* **64,** 187-191.

Anundi, H., Rask, L., Östberg, L., and Peterson, P. A. (1975). *Biochemistry* **14,** 5046-5054.

Avigad, G., Amaral, D., Asensio, C., and Horecker, B. L. (1962). *J. Biol. Chem.* **237,** 2736-2743.

Axelsson, B., Kimura, A., Hammarström, S., Wigzell, H., Nilsson, K., and Mellstedt, H. (1978). *Eur. J. Immunol.* **8,** 757-764.

Barclay, A. N., Muirhead-Letarte, M., Williams, A. F., and Faulkes, R. A. (1976). *Nature (London)* **263,** 563-567.

Bennett, M., and Schmid, K. (1980). *Proc. Natl. Acad. Sci. U.S.A.* **77,** 6109-6113.

Berggård, I., and Bearn, A. G. (1968). *J. Biol. Chem.* **243,** 4095-4103.

Besançon, F., Ankel, H., and Basu, S. (1976). *Nature (London)* **259,** 576-578.

Billing, R. J., Safani, M., and Peterson, P. (1976). *J. Immunol.* **117,** 1589-1593.

Boldt, D. H., MacDermott, R. P., and Jorolan, E. (1975). *J. Immunol.* **114,** 1532-1536.

Bonner, W. M., and Laskey, R. A. (1974). *Eur. J. Biochem.* **46,** 83-88.

Boyse, E. A., and Old, L. J. (1969). *Annu. Rev. Genet.* **3,** 269-290.

Böyum, A. (1968). *Scand. J. Clin. Lab. Invest.* **21,** Suppl. 97, 1-107.

Carter, P. B., and Sunderland, C. A. (1979). *Transplant. Proc.* **11,** 1646-1647.

Charron, D. J., and McDevitt, H. O. (1979). *Proc. Natl. Acad. Sci. U.S.A.* **76,** 6567-6571.

Chiu, K. M., Mortensen, R. F., Osmand, A. P., and Gewurz, H. (1977). *Immunology* **32,** 997-1005.

Conzelman, A., Pink, R., Acuto, O., Mach, J.-P., Dolivo, S., and Nabholz, M. (1980). *Eur. J. Immunol.* **10**, 860–868.

Cuatrecasas, P. (1973). *Biochemistry* **12**, 3547–3558.

Dalchau, R., and Fabre, J. W. (1979). *J. Exp. Med.* **149**, 576–591.

Dalchau, R., and Fabre, J. W. (1981). *J. Exp. Med.* **153**, 753–765.

Dalchau, R., Kirkley, J., and Fabre, J. W. (1980). *Eur. J. Immunol.* **10**, 737–744.

DePierre, J. W., and Karnovsky, M. L. (1974). *J. Biol. Chem.* **249**, 7111–7120.

Dillner-Zetterlind, M.-L., Axelsson, B., Hammarström, S., Hellström, S., Hellström, U., and Perlmann, P. (1980). *Eur. J. Immunol.* **10**, 434–442.

Dobberstein, B., Garoff, H., and Warren, G. (1979). *Cell* **17**, 759–769.

Dornand, J., Bonnafous, J.-C., and Mani, J.-C. (1978a). *Eur. J. Biochem.* **87**, 459–465.

Dornand, J., Bonnafous, J.-C., and Mani, J.-C. (1978b). *Biochem. Biophys. Res. Commun.* **82**, 685–692.

Dornand, J., Bonnafous, J.-C., and Mani, J.-C. (1980). *FEBS Lett.* **118**, 225–228.

Dunlap, B., Bach, F. H., and Bach, M. L. (1978). *Nature (London)* **271**, 253–255.

Dunlap, B., Mixter, P. F., Koller, B., Watson, A., Widmer, M. B., and Bach, F. H. (1980). *J. Immunol.* **125**, 1829–1831.

Durda, P. J., and Gottlieb, P. D. (1978). *J. Immunol.* **121**, 983–989.

Durda, P. J., Shapiro, C., and Gottlieb, P. D. (1978). *J. Immunol.* **120**, 53–57.

Edwards, N. L., Magilavy, D. B., Cassidy, J. T., and Fox, I. H. (1978). *Science* **201**, 628–630.

Edwards, N. L., Gelfand, E. W., Burk, L., Dosch, H.-M., and Fox, I. H. (1979). *Proc. Natl. Acad. Sci. U.S.A.* **76**, 3474–3476.

Fearon, D. T. (1980). *J. Exp. Med.* **152**, 20–30.

Finne, J. (1980). *Eur. J. Biochem.* **104**, 181–189.

Finne, J., Krusius, T., Rauvala, H., Kekomäki, R., and Myllylä, G. (1978). *FEBS Lett.* **89**, 111–115.

Gahmberg, C. G. (1976). *J. Biol. Chem.* **251**, 510–515.

Gahmberg, C. G. (1978). *In* "Methods in Enzymology" (V. Ginsburg, ed.), Vol. 50, Part C, pp. 204–206. Academic Press, New York.

Gahmberg, C. G., and Andersson, L. C. (1977). *J. Biol. Chem.* **252**, 5888–5894.

Gahmberg, C. G., and Andersson, L. C. (1978). *J. Exp. Med.* **148**, 507–521.

Gahmberg, C. G., and Hakomori, S.-I. (1973). *J. Biol. Chem.* **248**, 4311–4317.

Gahmberg, C. G., and Hakomori, S.-I. (1974). *Biochem. Biophys. Res. Commun.* **54**, 283–291.

Gahmberg, C. G., and Hakomori, S.-I. (1976). *Biomembranes* **8**, 131–165.

Gahmberg, C. G., and Karhi, K. K. (1981). *In* "Receptors in Biological Systems" (R. M. Gorczynski and G. B. Price, eds.). Dekker, New York.

Gahmberg, C. G., Häyry, P., and Andersson, L. C. (1976a). *J. Cell Biol.* **68**, 642–653.

Gahmberg, C. G., Itaya, K., and Hakomori, S. (1976b). *Methods Membr. Res.* **7**, 175–206.

Gahmberg, C. G., Myllylä, G., Leikola, J., Pirkola, A., and Nordling, S. (1976c). *J. Biol. Chem.* **251**, 6108–6116.

Gahmberg, C. G., Andersson, L. C., and Nilsson, K. (1980a). *Leuk. Res.* **4**, 279–286.

Gahmberg, C. G., Jokinen, M., Karhi, K. K., and Andersson, L. C. (1980b). *J. Biol. Chem.* **255**, 2169–2175.

Gerdes, J., Klatt, U., and Stein, H. (1980). *Immunology* **39**, 75–84.

Gesner, B. M., and Ginsburg, V. (1964). *Proc. Natl. Acad. Sci. U.S.A.* **52**, 750–755.

Goodman, M. G. (1980). *Immunol. Today* **1**, 92–96.

Gräsbeck, R., Nordman, C. G., and de la Chapelle, A. (1964). *Acta Med. Scand., Suppl.* **412**, 39–47.

Gregory, S. H., and Kern, M. (1979). *J. Immunol.* **123**, 1078–1982.

Gross, N., and Bron, C. (1980). *Eur. J. Immunol.* **10**, 417–422.

Hakomori, S., and Kobata, A. (1975). *In* "The Antigens" (M. Sela, ed.), Vol. 2, pp. 80–140. Academic Press, New York.

Hauptman, S. P., and Kansu, E. (1978). *Nature (London)* **276**, 393–394.

Hauptman, S. P., Kansu, E., Sobczak, G., and Serno, M. (1979a). *J. Immunol.* **122**, 1035–1040.

Hauptman, S. P., Kansu, E., Serno, M., and Godfrey, S. (1979b). *J. Exp. Med.* **149**, 158–171.

Hauptman, S. P., Kansu, E., and Godfrey, S. (1979c). *J. Immunol.* **123**, 1007–1013.

Häyry, P., Andersson, L. C., Nordling, S., and Virolainen, M. (1972). *Transplant. Rev.* **12**, 91–140.

Häyry, P., Andersson, L. C., Gahmberg, C. G., Roberts, P., Ranki, A., and Nordling, S. (1975). *Isr. J. Med. Sci.* **11**, 1299–1318.

Helenius, A., and Simons, K. (1975). *Biochim. Biophys. Acta* **415**, 29–79.

Helenius, A., Morein, B., Fries, E., Simons, K., Robinson, P., Schirrmacher, V., Terhorst, C., and Strominger, J. L. (1978). *Proc. Natl. Acad. Sci. U.S.A.* **75**, 3846–3850.

Higgins, T. J., Parish, C. R., Hogarth, P. M., McKenzie, I. F. C., and Hämmerling, G. J. (1980). *Immunogenetics* **11**, 467–482.

Hoessli, D., Bron, C., and Pink, J. R. L. (1980). *Nature (London)* **283**, 576–578.

Hoessli, D. C., and Vassalli, P. (1980). *J. Immunol.* **125**, 1758–1763.

Hogarth, P. M., Potter, T. A., Cornell, F. N., McLachlan, R., and McKenzie, I. F. C. (1980). *J. Immunol.* **125**, 1618–1624.

Holmgren, J., Lönnroth, L., Månsson, J.-E., and Svennerholm, L. (1975). *Proc. Natl. Acad. Sci. U.S.A.* **72**, 2520–2524.

Horton, M. A., Beverley, P. C. L., and Simpson, E. (1979). *Eur. J. Immunol.* **9**, 345–352.

Hubbard, A. L., and Cohn, Z. A. (1972). *J. Cell Biol.* **55**, 390–405.

Humphreys, R. E., McCune, J. M., Chess, L., Herrman, H. C., Malenka, D. J., Mann, D. L., Parham, P., Schlossman, S. F., and Strominger, J. L. (1976). *J. Exp. Med.* **144**, 98–112.

Janossy, G., Goldstone, A. H., Capellaro, D., Greaves, M. F., Kulenkampff, J., Pippard, M., and Welsh, K. (1977). *Br. J. Haematol.* **37**, 391–402.

Järnefelt, J., Rush, J., Li, Y.-T., and Laine, R. A. (1978). *J. Biol. Chem.* **253**, 8006–8009.

Jondal, M., Klein, G., Oldstone, M. B. A., Bokish, V., and Yefenof, E. (1976). *Scand. J. Immunol.* **5**, 401–410.

Kansu, E., Sobczak, G., and Hauptman, S. P. (1979). *J. Immunol.* **122**, 1041–1044.

Karhi, K. K., and Gahmberg, C. G. (1980). *Biochim. Biophys. Acta* **622**, 344–354.

Katz, D. H. (1977). "Lymphocyte Differentiation, Recognition and Regulation." Academic Press, New York.

Kaufman, J. F., and Strominger, J. L. (1979). *Proc. Natl. Acad. Sci. U.S.A.* **76**, 6304–6308.

Kessler, S. W. (1975). *J. Immunol.* **115**, 1617–1624.

Kieda, C. M. T., Bowles, D. J., Ravid, A., and Sharon, N. (1978). *FEBS Lett.* **94**, 391–396.

Kimura, A. K., and Wigzell, H. (1978). *J. Exp. Med.* **147**, 1418–1434.

Kimura, A. K., Wigzell, H., Holmquist, G., Ersson, B., and Carlsson, P. (1979). *J. Exp. Med.* **149**, 473–484.

Klareskog, L., Sandberg-Trädgårdh, L., Rask, L., Lindblom, J. B., Curman, B., and Peterson, P. A. (1977). *Nature (London)* **265**, 248–251.

Klareskog, L., Banck, G., Forsgren, A., and Peterson, P. A. (1978). *Proc. Natl. Acad. Sci. U.S.A.* **75**, 6197–6201.

Klenk, H. D., and Choppin, P. W. (1970). *Proc. Natl. Acad. Sci. U.S.A.* **66**, 57–64.

Ko, H.-S., Fu, S. M., Winchester, R. J., Yu, D. T., and Kunkel, H. G. (1979). *J. Exp. Med.* **150**, 246–255.

Koch, G. L. E., and Smith, M. J. (1978). *Nature (London)* **273**, 274–281.

Köhler, G., and Milstein, C. (1975). *Nature (London)* **256**, 495–497.

Krusius, T., Finne, J., Andersson, L. C., and Gahmberg, C. G. (1979). *Biochem. J.* **181**, 451–456.

Kvist, S., Östberg, L., Persson, H., Philipson, L., and Peterson, P. A. (1978). *Proc. Natl. Acad. Sci. U.S.A.* **75,** 5674–5678.

Laemmli, U. K. (1970). *Nature (London)* **227,** 680–685.

Ledbetter, J. A., and Herzenberg, L. A. (1979). *Immunol. Rev.* **47,** 63–90.

Ledbetter, J. A., Seaman, W. E., Tsu, T. T., and Herzenberg, L. A. (1981). *J. Exp. Med.* **153,** 1503–1516.

McKenzie, I. F. C., and Potter, T. (1979). *Adv. Immunol.* **27,** 179–338.

Mann, D. L., Rogentine, G. N., Fahey, J. L., and Nathenson, S. G. (1969). *J. Immunol.* **103,** 282–292.

Mellman, I. S., and Unkeless, J. C. (1980). *J. Exp. Med.* **152,** 1048–1069.

Mitchell, R. N., and Bowers, W. E. (1978). *J. Immunol.* **121,** 2181–2192.

Mitchell, R. N., and Bowers, W. E. (1980). *J. Immunol.* **124,** 2632–2640.

Möller, G., ed. (1976). *Transplant. Rev.* **30,** 1–322.

Monahan, T. M., Frost, A. F., and Abell, C. W. (1976). *Biochem. Biophys. Res. Commun.* **73,** 1115–1121.

Nakamuro, K., Tanigaki, N., and Pressman, D. (1973). *Proc. Natl. Acad. Sci. U.S.A.* **70,** 2863–2865.

Natvig, J. B., Perlmann, P., and Wigzell, H., eds. (1976). *Scand. J. Immunol.* **5,** Suppl. 5.

Nilsson, K. (1977). *Haematol. Blood Transfus.* **20,** 253–264.

Nilsson, K., and Pontén, J. (1975). *Int. J. Cancer* **15,** 321–341.

Nilsson, K., Andersson, L. C., Gahmberg, C. G., and Wigzell, H. (1977). *Int. J. Cancer* **20,** 708–716.

Nishikawa, Y., Yamamoto, K., Onodera, G., Tamura, G., and Mitsui, H. (1979). *Biochem. Biophys. Res. Commun.* **87,** 1235–1242.

Novogrodsky, A., and Katchalski, E. (1971). *FEBS Lett.* **12,** 297–300.

Novogrodsky, A., and Katchalski, E. (1973a). *Proc. Natl. Acad. Sci. U.S.A.* **70,** 1824–1827.

Novogrodsky, A., and Katchalski, E. (1973b). *Proc. Natl. Acad. Sci. U.S.A.* **70,** 2515–2518.

Novogrodsky, A., Stenzel, K. H., and Rubin, A. L. (1977). *J. Immunol.* **118,** 852–857.

Novogrodsky, A., Suthanthiran, M., Saltz, B., Newman, D., Rubin, A. L., and Stenzel, K. H. (1980). *J. Exp. Med.* **151,** 755–760.

Nowell, P. C. (1960). *Cancer Res.* **20,** 462–466.

Ochai, T., Ahmed, A., Grebbe, S. C., and Sell, K. W. (1977). *Transplant. Proc.* **9,** 1049–1054.

O'Farrell, P. H. (1975). *J. Biol. Chem.* **250,** 4007–4021.

Oldstone, M. B. A., Tishon, A., Dutko, F. J., Kennedy, I. T., Holland, J. J., and Lempert, P. W. (1980). *J. Virol.* **34,** 256–265.

Omary, M. B., Trowbridge, I. S., and Scheid, M. P. (1980a). *J. Exp. Med.* **151,** 1311–1316.

Omary, M. B., Trowbridge, I. S., and Battifora, H. A. (1980b). *J. Exp. Med.* **152,** 842–852.

Orr, H. T., Lancet, D., Robb, R. J., Lopez de Castro, J. A., and Strominger, J. L. (1979a). *Nature (London)* **282,** 266–270.

Orr, H. T., Lopez de Castro, J. A., Parham, P., Ploegh, H. L., and Strominger, J. L. (1979b). *Proc. Natl. Acad. Sci. U.S.A.* **76,** 4395–4399.

Owen, M. J., and Crumpton, M. J. (1980). *Immunol. Today* **1,** 117–122.

Parham, P., Alpert, B. N., Orr, H. T., and Strominger, J. L. (1977). *J. Biol. Chem.* **252,** 7555–7567.

Peterson, P. A., Cunningham, B. A., Berggård, I., and Edelman, G. M. (1972). *Proc. Natl. Acad. Sci. U.S.A.* **69,** 1697–1701.

Peterson, P. A., Rask, L., and Lindblom, J. B. (1974). *Proc. Natl. Acad. Sci. U.S.A.* **71,** 35–39.

Phillips, D. R., and Morrison, M. (1971). *Biochemistry* **10,** 1766–1771.

Ploegh, H. L., Cannon, L. E., and Strominger, J. L. (1979). *Proc. Natl. Acad. Sci. U.S.A.* **76,** 2273–2277.

Pober, J. S., Guild, B. C., and Strominger, J. L. (1978). *Proc. Natl. Acad. Sci. U.S.A.* **75**, 6002–6006.

Rauvala, H., and Hakomori, S. (1981). *J. Cell Biol.* **88**, 149–159.

Rauvala, H., Carter, W. G., and Hakomori, S. (1981). *J. Cell Biol.* **88**, 127–137.

Reif, A. E., and Allen, J. M. V. (1964). *J. Exp. Med.* **120**, 413–433.

Reisner, Y., and Sharon, N. (1980). *Trends Biochem. Sci.* **5**, 29–31.

Reisner, Y., Pahwa, S., Chiao, J. W., Sharon, N., Evans, R. L., and Good, R. A. (1980). *Proc. Natl. Acad. Sci. U.S.A.* **77**, 6778–6782.

Renkonen, O., Gahmberg, C. G., Simons, K., and Kääriäinen, L. (1970). *Acta Chem. Scand.* **24**, 733–735.

Roseman, S. (1970). *Chem. Phys. Lipids* **5**, 270–297.

Rosenfelder, G., Van Eijk, R. V. W., Monner, D. A., and Mühlradt, P. F. (1978). *Eur. J. Biochem.* **83**, 571–580.

Rosenfelder, G., Van Eijk, R. V. W., and Mühlradt, P. F. (1979). *Eur. J. Biochem.* **97**, 229–237.

Ross, G. D., Jarowski, C. I., Rabellino, E. M., and Winchester, R. J. (1978). *J. Exp. Med.* **147**, 730–744.

Rothenberg, E., and Boyse, E. A. (1979). *J. Exp. Med.* **150**, 777–791.

Saito, M., and Osawa, T. (1980). *Carbohydr. Res.* **78**, 341–348.

Sarcione, E. J. (1963). *Arch. Biochem. Biophys.* **100**, 516–519.

Schmid, K. (1975). *In* "The Plasma Proteins: Structure, Function and Genetic Control" (F. W. Putnam, ed.), 2nd ed. Vol. 1, pp. 183–222. Academic Press, New York.

Schmid, K., Kaufmann, H., Isemura, S., Bauer, F., Emura, J., Motoyama, T., Ishiguro, M., and Nanno, S. (1973). *Biochemistry* **12**, 2711–2724.

Schwarting, G. A., and Summers, A. (1980). *J. Immunol.* **124**, 1691–1694.

Sell, S., Mascari, R. A., and Hughes, S. J. (1970). *J. Immunol.* **105**, 1400–1405.

Shackelford, D. A., and Strominger, J. L. (1980). *J. Exp. Med.* **151**, 144–165.

Silver, J., and Ferrone, S. (1979). *Nature (London)* **279**, 436–437.

Smith, R. T., Norcross, M., Mainio, V., and Konaka, Y. (1980). *Immunol. Rev.* **51**, 193–214.

Snell, G. D. (1978). *Immunol. Rev.* **38**, 3–69.

Standring, R., McMaster, W. R., Sunderland, C. A., and Williams, A. F. (1978). *Eur. J. Immunol.* **8**, 832–839.

Steck, T. L., and Dawson, G. (1974). *J. Biol. Chem.* **249**, 2135–2142.

Stein, K. E., and Marcus, D. M. (1977). *Biochemistry* **16**, 5285–5291.

Stein, K. E., Schwarting, G. A., and Marcus, D. M. (1978). *J. Immunol.* **120**, 676–679.

Stein-Douglas, K. E., Schwarting, G. A., Naiki, M., and Marcus, D. M. (1976). *J. Exp. Med.* **143**, 822–832.

Takacs, B. J. (1980). *Mol. Immunol.* **17**, 1293–1314.

Terhorst, C., Van Agthoven, A., Reinherz, E., and Schlossman, S. (1980). *Science* **209**, 520–521.

Thoenes, J., and Stein, H. (1979). *J. Exp. Med.* **150**, 1049–1966.

Thurman, G. B., Giovanella, B., and Goldstein, A. L. (1973). *J. Immunol.* **113**, 810–812.

Tomita, M., and Marchesi, V. T. (1975). *Proc. Natl. Acad. Sci. U.S.A.* **72**, 2964–2968.

Trägårdh, L., Rask, L., Wiman, K., Fohlman, J., and Peterson, P. A. (1979a). *Proc. Natl. Acad. Sci. U.S.A.* **76**, 5839–5842.

Trägårdh, L., Wiman, K., Rask, L., and Peterson, P. A. (1979b). *Biochemistry* **18**, 1322–1328.

Trägårdh, L., Rask, L., Wiman, K., Fohlman, J., and Peterson, P. A. (1980). *Proc. Natl. Acad. Sci. U.S.A.* **77**, 1129–1133.

Trowbridge, I. S., Ralph, P., and Bevan, M. L. (1975). *Proc. Natl. Acad. Sci. U.S.A.* **72**, 157–161.

Trowbridge, I. S., Hyman, R., and Klein, G. (1977). *Eur. J. Immunol.* **7**, 640–645.

Van Eijk, R. V. W., and Mühlradt, P. F. (1977). *Eur. J. Biochem.* **78**, 41–45.

Van Eijk, R. V. W., Rosenfelder, G., and Mühlradt, P. F. (1979). *Eur. J. Biochem.* **101**, 185–193.

Van Heyningen, W. E. (1974). *Nature (London)* **249**, 415–417.

Van Wauwe, J. P., De Mey, J. R., and Goossens, J. G. (1980). *J. Immunol.* **124**, 2708–2713.

Vitetta, E., Uhr, J. W., and Boyse, E. A. (1972). *Cell. Immunol.* **4**, 187–191.

Wang, C. Y., Fu, S. M., and Kunkel, H. G. (1979). *J. Exp. Med.* **149**, 1424–1437.

Wang, T. J., Freimuth, W. W., Miller, H. C., and Esselman, W. J. (1978). *J. Immunol.* **121**, 1361–1365.

Williams, A. F., Barclay, A. N., Muirhead-Letarte, M., and Morris, R. J. (1977). *Cold Spring Harbor Symp. Quant. Biol.* **41**, 51–61.

Yefenof, E., and Klein, G. (1977). *Int. J. Cancer* **20**, 347–352.

Yefenof, E., Klein, G., Jondal, M., and Oldstone, M. B. A. (1976). *Int. J. Cancer* **17**, 693–700.

Young, W. W., Hakomori, S.-I., Durdik, J. M., and Henney, C. S. (1980). *J. Immunol.* **124**, 199–201.

Ziegler, A., and Milstein, C. (1979). *Nature (London)* **279**, 243–244.

Zinkernagel, R. M., Dunlop, M. B. C., Blanden, R. V., Doherty, P. C., and Shreffler, D. C. (1976). *J. Exp. Med.* **144**, 519–532.

SECTION 6

Expression of Glycosphingolipid Glycosyltransferases in Development and Transformation

SUBHASH BASU AND MANJU BASU

I. INTRODUCTION

Glycosphingolipids (GSL's) are widely distributed in animal tissues, particularly in synaptic membranes and cell surfaces. Since the discovery of galactocerebroside by Thudicum (1884), who isolated it from white matter of human brain, many laboratories have studied the chemical structures of GSL's, their distribution in animal tissues, and their biosynthesis *in vivo* (for review see Sweeley and Siddiqui, 1977). This section is limited to a consideration of the biosynthesis *in vitro* of GSL's in normal developing tissues and in some cultured transformed cells.

The commonly occurring GSL's can be classified into two different families (Fig. 1): (a) the acidic, or neuraminic acid-containing, GSL's (gangliosides) and (b) the neutral GSL's, containing mono-, di-, or oligoglycosyl chains attached to ceramide. The oligoglycosylceramides are further classified into two specific groups: (a) a core structure of GlcNAcβ1-3Galβ1-4Glcβ1-1Cer (LcOse$_3$Cer) and (b) a core structure of Galα1-4Galβ1-4Glcβ1-1Cer (GbOse$_3$Cer) (IUPAC-IUB, 1977). Short-chain GSL's of all these families appear to be ubiquitous among eukaryotic cells. However, long-chain GSL's of blood group or globoside

265

THE GLYCOCONJUGATES, VOL. III

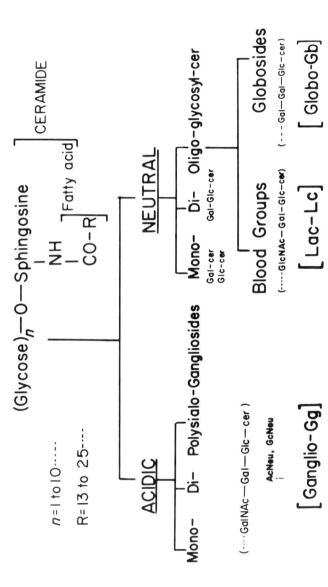

Figure 1 Classification of glycosphingolipids.

families are important constituents of the plasma membranes of normal and transformed cells. Very little is known about the function of these GSL's on cell surfaces (Roseman, 1970; Svennerholm, 1980). Recent studies have indicated that inhibition of antiviral activity by GM2 ganglioside is due to the specific interaction of interferon with the carbohydrate moiety of the ganglioside molecule (Besançon *et al.*, 1976; Ankel *et al.*, 1980). Choleragen or its B subunit binds with GM1 ganglioside (Van Heyningen *et al.*, 1971; Holmgren *et al.*, 1973; Cuatrecasas, 1973; M. Basu *et al.*, 1976), and the A subunit stimulates adenylate cyclase (Moss and Vaughan, 1977, 1979; Gill and Meren, 1978). The inhibition of synaptic junction formation at higher concentrations and a facilitatory effect at low concentrations of GM1 have been observed in 10-day-old chicken embryonic myotubes (Obata and Handa, 1979). The possibility that neutral GSL's, $GgOse_3Cer(GalNAc\beta1\text{-}4Gal\beta1\text{-}4Glc\text{-}Cer)$ and $nLcOse_4Cer\text{-}(Gal\beta1\text{-}4GlcNAc\beta1\text{-}3Gal\beta1\text{-}4Glc\text{-}Cer)$, are tumor-associated surface antigens in Kirsten-virus-transformed tumor cells and polyoma-transformed NIL cells (NILpy), respectively, has been suggested by Hakomori and co-workers (Urdal and Hakomori, 1980; Sundsmo and Hakomori, 1976). Irrespective of blood type, $nLcOse_4Cer$ has also been identified in normal human erythrocytes (Ando *et al.*, 1973) and in elevated quantities in the stroma of patients with congenital dyserythropoietic anemia type II (Joseph and Gockerman, 1975). Are these changes in the content of short-chain GSL's on transformed cells the result of blocked synthesis of longer-chain GSL's (Basu *et al.*, 1979) or elevated activity of any specific glycosyltransferase? Until recently, limitations on the availability of pure acceptor GSL's and donor sugar nucleotides have restricted extensive investigations into this question. Several extensive reviews on the structure and biosynthesis of *N*-acetylgalactosamine-containing gangliosides in normal tissues (Svennerholm, 1964; Kaufman *et al.*, 1966; Roseman, 1970; Fishman and Brady, 1976; Basu *et al.*, 1976b; Fishman and Brady, 1976; Dawson, 1978; S. Basu *et al.*, 1980) and transformed cells (Brady and Fishman, 1974; M. Basu *et al.*, 1980c) have already appeared. The objective of this section is to consider the biosynthesis *in vitro* of GSL's in developing tissues such as animal brains and in chemically transformed cells.

II. BIOSYNTHESIS OF GANGLIOSIDES IN DEVELOPING ANIMAL BRAINS

Acidic glycosphingolipids containing sialic acids (*N*-acetyl or *N*-glycolyl) are usually referred to by the generic name ganglioside. A ganglioside was first isolated from Tay–Sachs diseased brain by Klenk (1939), who named it after its origin, ganglionic cells. The structure of this ganglioside appears to be $GalNAc\beta1\text{-}4$ $(NeuAc\alpha2\text{-}3)Gal\beta1\text{-}4Glc\text{-}Cer$, or GM2 (Ledeen, 1977; Svennerholm, 1962; Kuhn and Wiegandt, 1963). Since the isolation of the first ganglioside

(GM2) from human brain 42 years ago (Klenk, 1939), and the first hematoside (GM3) from horse erythrocytes (Yamakawa and Suzuki, 1953), the structures of at least 20 different gangliosides from neural (Svennerholm *et al.*, 1973; Iwamori and Nagai, 1979; Fredman *et al.*, 1980) and nonneural (Puro, 1969; Wiegandt and Bucking, 1970; Wiegandt, 1973) tissues have been elucidated. A novel fucose-containing GM1 ganglioside, Fucα1-2Galβ1-3GalNAcβ1-4-(NeuAcα2-3)Galβ1-4G1c-Cer, has been isolated from bovine (Ghidoni *et al.*, 1976) and porcine (Sonnio *et al.*, 1978) brains and rat hepatoma H35 cells (Baumann *et al.*, 1979). In addition to the hexosamine-free and *N*-acetylgalactosamine-containing gangliosides, the structures of at least four different gangliosides containing *n*LcOse₄Cer have been reported by several laboratories (Wiegandt and Schulze, 1969; Svennerholm *et al.*, 1972; Li *et al.*, 1973; Ohashi and Yamakawa, 1977; Rauvala *et al.*, 1978; Chien *et al.*, 1978).

A. Studies on *N*-Acetylgalactosamine-Containing Gangliosides

On the basis of studies with an embryonic chicken brain (ECB) membrane system (Roseman, 1970; S. Basu, 1966; Steigerwald *et al.*, 1975; S. Basu *et al.*, 1980), the stepwise biosynthesis of GD1a ganglioside, starting from ceramide,

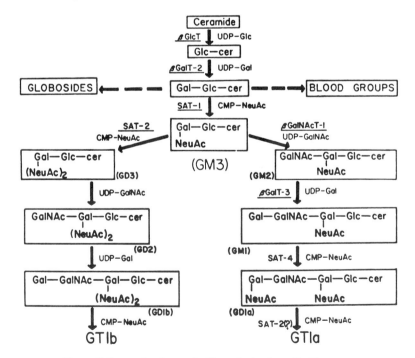

Figure 2 Proposed pathways for biosynthesis of gangliosides.

has been proposed (Fig. 2). Biosynthesis *in vitro* of GD1b starting from GD3 in membrane fragments isolated from 15-day-old rat brain has also been reported (Cumar *et al.*, 1972). Several of these glycolipid glycosyltransferases (Table I; Fig. 2) from different tissues, such as rat brain (Arce *et al.*, 1966; Yip and Dain, 1970), mouse brain (Yu and Lee, 1976), Golgi apparatus of rat liver (Keenan *et al.*, 1974), and bovine thyroid (Pacuszka *et al*, 1978), which are involved in the synthesis of GD1a and GT1a, have been reported by several laboratories. In addition to ECB, embryonic frog brains (Yiamouyiannis, 1967), and newborn rat brains, the chick retina system has been used for studies *in vitro* on ganglioside

TABLE I
Glycosphingolipid Glycosyltransferases

Enzyme name	Abbreviation	Expected linkage	References
UDP-Glc : Cer β-glucosyltransferase	βGlcT	β1-1	S. Basu *et al.* (1973), Vunnam and Radin (1980)
UDP-GlcNAc : LacCer β-GlcNAc-transferase	βGlcNAcT	β1-3	S. Basu *et al.* (1970)
UDP-Gal : 2-OH-Cer β-galactosyltransferase	βGalT-1	β1-1	Morell and Radin (1969), S. Basu *et al.* (1969, 1971)
UDP-Gal : Glc-Cer β-galactosyltransferase	βGalT-2	β1-4	Hildebrand and Houser (1969), S. Basu *et al.* (1968)
UDP-Gal : GM2 β-galactosyltransferase	βGalT-3	β1-3	S. Basu *et al.* (1965)
UDP-Gal : LcOse$_3$Cer β-galactosyltransferase	βGalT-4	β1-4	Basu and Basu (1972)
UDP-Gal : nLcOse$_4$Cer α-galactosyltransferase	αGalT-5	α1-3	Basu and Basu (1973)
UDP-Gal : LacCer α-galactosyltransferase	αGalT-6	α1-4	Stoffyn *et al.* (1974)
UDP-GalNAc : GM3 β-GalNAc-transferase	βGalNAcT-1	β1-4	Steigerwald *et al.* (1975)
UDP-GalNAc : LacCer β-GalNAc-transferase	βGalNAcT-1	β1-4	M. Basu *et al.* (1974)
UDP-GalNAc : GbOse$_3$Cer β-GalNAc-transferase	βGalNAcT-2	β1-3	Chien *et al.* (1973) Ishibashi *et al.* (1974)
UDP-GalNAc : GbOse$_4$Cer α-GalNAc-transferase	αGalNAcT-3	α1-3	Kijimoto *et al.* (1974) Yeung *et al.* (1974) Yokosawa *et al.* (1978)
GDP-Fuc : nLcOse$_4$Cer α-fucosyltransferase	αFucT-2	α1-2	Basu *et al.* (1975)
GDP-Fuc : nLcOse$_5$Cer α-fucosyltransferase	αFucT-2	α1-2	Presper *et al.* (1982)
GDP-Fuc : LcOse$_3$Cer α-fucosyltransferase	αFucT-3	α1-3	Presper *et al.* (1978)
CMP-NeuAc : Lac-Cer α-sialyltransferase	SAT-1	α2-3	S. Basu (1966), Kaufman *et al.* (1966), Arce *et al.* (1966)
CMP-NeuAc : GM3 α-sialyltransferase	SAT-2	α2-8	Kaufman *et al.* (1968)
CMP-NeuAc : nLcOse$_4$Cer α-sialyltransferase	SAT-3	α2-3	S. Basu *et al.* (1980)
CMP-NeuAc : NeuGc-nLcOse$_4$ Cer α-sialyltransferase	SAT-2	α2-8	Higashi *et al.* (1981), Basu *et al.* (1981b)
CMP-NeuAc : GgOse$_4$Cer α-sialyltransferase	SAT-3	α2-3	S. Basu (1966), Stoffyn and Stoffyn (1980)
CMP-NeuAc : GM1 α-sialyltransferase	SAT-4	α2-3	S. Basu (1966) Kaufman *et al.* (1966), Ng and Dain (1977a)

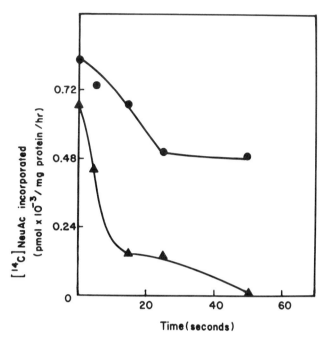

Figure 3 Heat inactivation of sialyltransferase activities. Heating was carried out at 55°. Closed circles, SAT-1 activity; closed triangles, SAT-4 activity.

biosynthesis. Very little is known about the biosynthesis of gangliosides *de novo* and the substrate selectivity of the glycosyltransferases during the development of brain tissues and differentiating neuronal cells. The specific activities of all four GSL sialyltransferases (SAT-1, SAT-2, SAT-3, and SAT-4, Table I) are maximal in 7- to 11-day-old embryonic chicken brains (Basu and Kaufman, 1965; Basu, 1966; Kaufman *et al.*, 1966; Kaufman and Basu, 1966). Heat inactivation at 55° (Fig. 3) and substrate competition studies indicate that these four reactions are probably catalyzed by different proteins. The exact physiological role of SAT-3 was not clear until recently, when it was found that SAT-3, which catalyzes the transfer of NeuAc to GgOse$_4$Cer (S. Basu, 1966; Kaufman and Basu, 1966), also catalyzes the transfer of sialic acid to the terminal galactose unit of glycoconjugates containing terminal lactosamine (Galβ1-4GlcNAc) residues (S. Basu *et al.*, 1980, 1981b). Whether the same sialyltransferase catalyzes the transfer of sialic acid to the GM1 ganglioside in ECB (SAT-4, Table I) has not yet been established. However, unlike the other three sialyltransferase activities, the activity of SAT-4 was maximal in 11-day-old embryonic chicken brain (Fig. 4). When membrane preparations isolated from 11-day-old ECB were used, the apparent K_m values were 0.15 mM for GM1

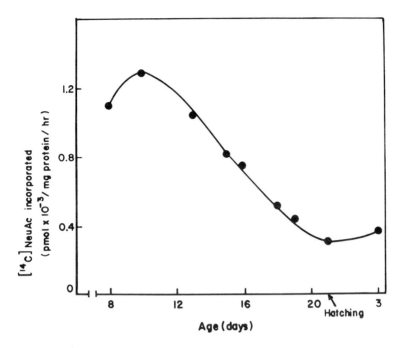

Figure 4 Effect of chicken embryonic age on SAT-4 activity.

ganglioside and 0.05 mM for CMP-NeuAc during the formation of GD1a ganglioside. Substrate competition studies with lactosylceramide and GM1 ganglioside in a rat brain membrane system (Ng and Dain, 1977a) have also suggested that these SAT-1 and SAT-4 reactions are probably catalyzed by two different proteins. Neither exogenously added asialofetuin nor α_1-acid glycoprotein (containing Galβ1-4GlcNAc-R) competes with lactosylceramide or GM1 as substrate. These results indicate the presence of a separate SAT-4 activity in the rat brain membrane system. No metal ion requirement could be demonstrated for any of the four sialyltransferase reactions. By the use of membrane preparations isolated from 9-day-old ECB, the biosynthesis *in vitro* of GT1a was achieved (Yohe and Yu, 1980). Whether the transfer of sialic acid from CMP-NeuAc to the terminal sialic acid of GD1a is catalyzed by the previously studied (Kaufman *et al.*, 1968) ECB SAT-2 (which catalyzes the formation of NeuAcα2-8NeuAc-Gal-Glc-Cer from GM3) is not known. Recently, biosynthesis *in vitro* of disialoneolactotetraosylceramide (NeuAcα2-8NeuGcα2-3nLcOse$_4$ Cer) has been achieved (Higashi *et al.*, 1981; Basu *et al.*, 1981b) using solubilized SAT-2 activity isolated from 9-day-old embryonic chicken brain.

Panzetta *et al.* (1980) studied the labeling of individual gangliosides in embryonic chicken retina incubated with [³H]GlcNH$_2$. During retinal development the maximal incorporation of radioactivity into the shorter-chain ganglioside GM3 appears at 8 days, whereas the maximal incorporation of radioactivity into the longer-chain ganglioside GD1a occurs at a much later stage of development. These data, along with the results obtained from studies *in vitro* (Fig. 4), suggest that SAT-4 may remain active at much later stages of development to complete biosynthesis *de novo* of the longer-chain ganglioside GD1a from GM1. It is also possible that the substrate GM1 is available only at a later age, although SAT-4 activity is present in the early stages of embryonic brain development.

The subcellular localization of these GSL glycosyltransferases in developing brain cells has not been unambiguously established. It has been suggested that the synthesis of glycoconjugates at synaptic membranes is controlled by surface glycosyltransferases (Den *et al.*, 1975; Roseman, 1970). The combined Golgi apparatus and endoplasmic reticulum fractions isolated from rat liver accounted for more than 80% of the GSL glycosyltransferase activities present in the total homogenate (Keenan *et al.*, 1974). However, SAT-1 activity has been recognized as a synaptosomal membrane-bound enzyme in calf brain cortex (Preti *et al.*, 1980). The presence of other GSL glycosyltransferase activities that catalyze the synthesis of N-acetylgalactosamine-containing gangliosides in synaptosomal membrane is still controversial (Ng and Dain, 1977b).

The structures of three new N-acetylgalactosamine-containing gangliosides from the fat body of the frog have been reported (Ohashi, 1980). Whether the biosynthesis of these gangliosides (Galα1-3GM1; Galβ1-3Galα1-3GM1; Galα1-3Galβ1-3Galα1-3GM1) could be achieved from GM1 by the action of already discovered GSL β1-3 galactosyltransferase (βGalT-3, Table I) is not yet known. The existence of a terminal galactose unit linked β1-3 to the penultimate galactose residue in a glycosphingolipid is novel.

B. Studies on Short-Chain Glycosphingolipids

Glycosyltransferases involved in the synthesis of galactosylceramide (βGalT-1) and glucosylceramide (βGlcT) have been characterized in ECB's (S. Basu *et al.*, 1968, 1969, 1971, 1973) and developing rat brains (Morell and Radin, 1969; Shah, 1971, 1973). The βGalT-1 activity (Table I) appears between 17 and 21 days of embryonic age (Basu *et al.*, 1971). The activity is barely detectable in young embryos (7–15 days old) or in microsomal membranes isolated from newborn rat brains up to day 15 (Brenkert and Radin, 1972; Cestelli *et al.*, 1980). However, glucosyltransferase (βGlcT, Table I), which transfers glucose from UDP-glucose to ceramide containing normal fatty acids (Basu *et al.*, 1973), appears to be present quite early in the chick embryos (11–15 days) and newborn rat brains. A threefold increase in the specific activity of this enzyme has been observed between the seventh and thirteenth days of embryonic age (Basu *et al.*, 1973), whereas activity is reduced to 20% during hatching. It has been suggested

(Roseman, 1970; Basu *et al.*, 1968) that glucosylceramide is a precursor of the next step in the *de novo* synthesis of gangliosides (Fig. 2), whereas galactosylceramide is a precursor of sulfatides (Sarlieve *et al.*, 1976) and G7 ganglioside (Yu and Lee, 1976). The activity profiles of βGlcT (Basu *et al.*, 1973) and βGalT-1 (Basu *et al.*, 1971) coincide with the onset of ganglioside deposition and myelination in embryonic chicken brains, the latter process starting in embryonic development, between 15 and 17 days. The βGalT-1 activity increases during myelination, peaking at age 16–17 days in rat (Vunnam and Radin, 1979, 1980) and mouse brain (Mandel *et al.*, 1972). It has also been shown that the activity of this enzyme is four times higher in the microsomal preparation isolated from white matter than in that from gray matter (Shah, 1971). A similar relationship has been found when activity is compared in the peripheral nerves of 15-day-old normal and quaking mouse brains (Cestelli *et al.*, 1980). During development of rat brains, the specific activity of βGalT-1 in the microsomal fraction reached its maximum at 20 days, whereas the activity in heavy myelin reached its maximum at 16 days and then declined to a constant level between 85 and 90 days (Koul *et al.*, 1980). Additional studies are necessary before it can be concluded unambiguously that βGalT-1 is a predominantly glial enzyme and βGlcT is only of neuronal origin.

Biosynthesis *in vitro* of lactosylceramide has been achieved in ECB (Basu *et al.*, 1968) and rat spleen (Hildebrand and Hauser, 1969). The galactosyltransferase activity (βGalT-2, Table I) also appears in the early days of embryonic development (12 days). This activity has been distinguished from βGalT-3 activity (UDP-galactose : GM2 β1-3 galactosyltransferase) based on pH optimum and metal requirements (Table II). It is expected that βGalT-2 catalyzes the transfer of galactose to position C-4 of glucosylceramide, but the exact chemical structure of the enzymatic product has not been reported.

The activity of βGalT-4 (UDP-galactose : LcOse$_3$Cer β1-4 galactosyltransferase) has been characterized in ECB (S. Basu *et al.*, 1980) and rabbit bone marrow (Basu and Basu, 1972, 1973). It has been shown that βGalT-4, which catalyzes the transfer of galactose to glycoconjugates containing terminal N-acetylglucosamine, also appears at an early stage of embryonic development (Table II; J. W. Kyle and M. Basu, unpublished). On the basis of heat inactivation

TABLE II
Properties of Embryonic Chicken Brain GSL : Galactosyltransferases

Enzyme	Optimal activity	Metal requirement	pH optimum
βGalT-1	19-day	$Mg^{2+} > Mn^{2+} > Ca^{2+}$	7.5
βGalT-2	15-day	$Mn^{2+} > Mg^{2+}$	6.8
βGalT-3	13-day	$Mn^{2+} > Co^{2+}$	7.3
βGalT-4	11-day	Mn^{2+}	6.4

TABLE III
Embryonic Chicken Brain Galactosyltranferace Activities—
Heat Inactivation Studies[a]

		[^{14}C]Galactose incorporated[b] (after heating at 55° for indicated seconds)				
Substrate	Linkage	0	5	10	20	40
GM2	β1-3	100	94	78	42	21
(monosialoganglioside)						
α_1-Acid glycoprotein	β1-4	100	106	53	4	0.5
(asialo-, agalacto-GP)						

[a] Complete incubation mixtures contained the following components (in micromoles) in final volumes of 0.05 ml: GM2, 0.035, or α_1-acid GP (asialo-, agalacto-), 0.2; mixture of Triton CF-54 and Tween 80 (2 : 1), 0.15 mg; cacodylate-HCl buffer, pH 7.3, 5; MnCl$_2$, 0.5; UDP-[^{14}C]galactose, 0.03 (0.8 × 10^6 cpm per μmole); enzyme fractions (heated at 55° for the times indicated and cooled to 4°), 0.2 mg of protein. ^{14}C-Labeled product formation was quantitatively determined by an assay method published previously (Basu *et al*. 1965). Under these reaction conditions product formation was proportional to enzyme concentration up to 5 mg/ml. The rate of reaction remained constant for 2 hours of incubation at 37° and was independent of substrate concentration; 100% means 2.9–3.0 nmoles/mg protein per hour.
[b] Expressed as percent activity.

studies (Table III) and substrate competition results (Table IV), it is suggested that ECB βGalT-4 activity, which transfers galactose to the terminal *N*-acetylglucosamine residue of α_1-glycoprotein, is different from βGalT-3 activity, which catalyzes the synthesis of GM1 from GM2 ganglioside (S. Basu *et al.*, 1965, 1980). On the other hand, a distinction between βGalT-2 and βGalT-4 has

TABLE IV
Embryonic Chicken Brain Galactosyltransferase Activities—Substrate
Competition Experiment[a]

			[^{14}C]Galactose incorporated (pmoles/mg protein/hr)	
Substrate	Concentration (mM)	Linkage	Observed	Theoretical value for two enzymes
GM2	0.7	β1-3	5,700	
(monosialoganglioside)				
α_1-Acid glycoprotein	4.0	β1-4	5,830	
(asialo-, agalacto-GP)				
GM$_2$ + α_1-acid GP	0.7 + 4.0		11,030	11,530

[a] Conditions were the same as those described in Table III.

not been established unambiguously because of the membrane-bound nature of these enzymes. It appears that embryonic brains (chicken, pig, etc.) are rich sources of these enzyme activities, and further purification and separation of the solubilized enzymes are necessary before their exact physiological roles can be assigned. A higher βGalT-4 activity and little or almost no βGalT-3 activity have been detected in an established culture of brain cells (TSD) derived from the cerebrum of a fetus with Tay-Sachs disease (Basu *et al.*, 1979). Furthermore, βGalT-4 and βGalT-3 activities isolated from ECB are not inhibited in the presence of TSD extract. It is speculated that in TSD-transformed cells there might be a genetic block in the production of βGalT-3, whereas the synthesis of βGalT-4 remains uninhibited. A greater availability of such mutant glial or neuronal cells will also help in distinguishing the properties of the various GSL galactosyltransferases.

III. BIOSYNTHESIS OF BLOOD-GROUP-RELATED GLYCOSPHINGOLIPIDS IN CHEMICALLY DIFFERENTIATED CULTURED CELLS

Neutral, fucose-free blood-group-active glycosphingolipids containing the common core structure of the tetraglycosylceramide nLcOse$_4$Cer(Galβ1-4GlcNAcβ1-3Galβ1-4Glc-Cer) are found in erythrocytes (Eto *et al.*, 1968), brain (Li *et al.*, 1973), and tumor cells (Sundsmo and Hakomori, 1976). The stepwise biosynthesis *in vitro* of nLcOse$_4$Cer and its conversion to B-active penta-glycosylceramide starting from lactosylceramide (Table V) were first reported in rabbit bone marrow cells (Basu *et al.*, 1970; M. Basu, 1974a; M. Basu and S. Basu, 1972, 1973) and more recently have been described in mouse (Moskal, 1977) and human (Presper, 1979) neuroblastoma cells before and after chemically induced differentiation (Moskal *et al.*, 1974; S. Basu *et al.*, 1976b), in Vero cells (M. Basu *et al.*, 1975), in chemically transformed guinea pig embryonic cells (M. Basu *et al.*, 1980b), and in mouse B and T lymphomas (M. Basu *et al.*, 1980c). The conversion of nLcOse$_4$Cer to blood group H_I-active glycosphingolipid (Fucα1-2nLcOse$_4$ Cer, Fig. 5) by the action of αFucT-2 (GDP- fucose : nLcOse$_4$Cer α1- 2 fucosyltransferase, Table I) (S. Basu *et al.*, 1975; Pacuszka and Koscielak, 1976) has been established in human IMR-32 neuroblastoma cells (Presper *et al.*, 1977, 1978). In addition to αFucT-2 activity, these cells contain αFucT-3 activity (GDP-fucose : LcOse$_3$Cer α1-3 fucosyl-transferase). αFucT-3 is believed to take part in the synthesis of a novel Le^a-type glycosphingolipid (Fig. 5) previously isolated from human adenocarcinomas by Hakomori and co-workers (Yang and Hakomori, 1971). Recently it has been observed that rat prostate adenocarcinoma cell cultures have five times greater levels of βGalT-4 activity (Fig. 5), which catalyzes the biosynthesis of

TABLE V
Structures of Fucose-Containing Glycosphingolipids

Type	Source[a]	Structure
H_I ($n=1$)	Human RBC	(Galβ1-4GlcNAcβ1-3)$_n$ Galβ1-4Glc—Cer
H_{II} ($n=2$)	Human RBC	| (α1-2)
		Fuc
		Gal
B_I ($n=1$)	Human RBC	| (α1-3)
		(Galβ1-4GlcNAcβ1-3)$_n$ Galβ1-4Glc—Cer
B_{II} ($n=2$)	Human RBC	| (α1-2)
		Fuc
		GalNAc
A^a ($n=1$)	Human RBC	| (α1-3)
		(Galβ1-4GlcNAcβ1-3)$_n$ Galβ1-4Glc—Cer
A^b ($n=2$)	Human RBC	| (α1-2) $^{1\ or\ 2}$
		Fuc
Novel Le^a	Adenocarcinoma (human)	Galβ1-4GlcNAcβ1-3Galβ1-4Glc—Cer (α1-3)
Megaloglycolipids	Human RBC (A,O) ($n=10$–27)	(Fuc)$_3$—(Gal)$_n$ Fuc(GlcNAc)$_{n-2}$—Glc—Cer

[a] RBC, red blood cell.

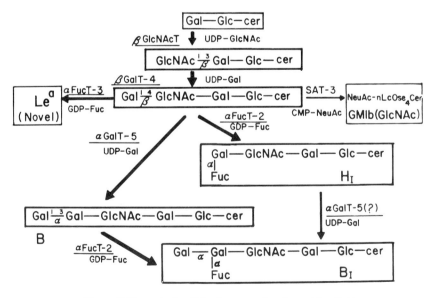

Figure 5 Biosynthesis of blood-group-active glycosphingolipids.

nLcOse$_4$Cer, than the tumor cell infiltrated lymph nodes from rats bearing the prostate adenocarcinoma (Jenis *et al.,* 1981). In IMR-32 cells as well as in bovine spleen (Presper *et al.,* 1982), αFucT-2 appears to catalyze the transfer of fucose from GDP-β-L-fucose to nLcOse$_4$Cer and nLcOse$_5$Cer to form H$_1$- and B$_1$-active glycosphingolipids, respectively. However, differentiation induced by 6-mercaptoguanosine does not alter the specific activity of αFucT-2 when tested with each substrate separately. In view of the recent isolation of (α1-3)-linked fucose-containing glycoprotein from human neuroblastoma (Santer and Glick, 1980), it can be speculated that the αFucT-3 level might be altered during tumorigenesis or differentiation in these cells. It is interesting that the activities of both αFucT-2 and αFucT-3 are very low in the mouse neuroblastoma clones N-18 (Presper *et al.,* 1978), N1E-115, and NS-20 (M. Basu, unpublished), whereas moderate activity of αFucT-2 has been detected in the mouse B lymphomas ABLS-140 and TEPC-15 produced by induction with Abelson virus or pristane (M. Basu *et al.,* 1980c). Biosynthesis *in vitro* of blood group A-related GSL's has not been studied in differentiated neuronal cells; studies with chemically transformed cells are discussed in Section IV.

IV. BIOSYNTHESIS *IN VITRO* OF GLYCOSPHINGOLIPIDS OF THE FORSSMAN FAMILY IN CHEMICALLY TRANSFORMED CULTURED CELLS

Neoplastic transformation of cultured eukaryotic cells can be achieved in the laboratory by radiation, tumor virus infection, or exposure to specific chemical carcinogens. The colonies of nontransformed cells usually grow *in vitro* as monolayers, whereas transformed cells grow in a crowded fashion and exhibit cell-to-cell association in a random fashion. Reports from different laboratories have established that the virally transformed cultured eukaryotic cells 3T3py are agglutinated more rapidly in the presence of specific lectins than their non-transformed counterparts (Inbar and Sachs, 1969; Burger and Noonan, 1970). It appears that transformation is expressed on the cell surface by the accumulation of specific glycoconjugates, perhaps through a change in the biosynthesis of these macromolecules (Den *et al.,* 1974). The changes in GSL glycosyl-transferase activities involved in the biosynthesis of gangliosides in virally transformed cells have been reviewed (Fishman and Brady, 1976; Sweeley *et al.,* 1978). This section focuses specifically on the properties and changes in the levels of GSL glycosyltransferases involved in the synthesis of Forssman-active GSL's (Table VI) in chemically transformed cells.

The biosynthesis of Forssman GSL's also proceeds by the stepwise addition of monosaccharide units starting from a common intermediate, lactosylceramide (Fig. 6), in embryonic chicken brain (Chien *et al.,* 1973; Chien, 1975), normal

TABLE VI
Structures of Neutral Glycosphingolipids

Name	Structure	Source[a]
GbOse$_3$Cer	$Gal\frac{1\ 4}{\alpha}Gal$—Glc—Cer	Kidney—Human(Fabry) Serum—Human (Fabry) RBC—Porcine Heart—Porcine
Globoside (GbOse$_4$Cer)	$GalNAc\frac{1\ 3}{\beta}Gal\frac{1\ 4}{\alpha}Gal$—Glc—Cer	RBC—Human, porcine Serum—Human, porcine Heart—Porcine
Cytolipin-R	$GalNAc\frac{1\ 3}{\beta}Gal\frac{1\ 3}{\alpha}Gal$—Glc—Cer	Lymphosarcoma—Rat Kidney—Rat
Forssman (GbOse$_5$Cer)	$GalNAc\frac{1\ 3}{\alpha}GalNAc\frac{1\ 3}{\beta}Gal\frac{1\ 4}{\alpha}Gal$—Glc—Cer	Spleen—Equine RBC—Caprine Adenocarcinoma
Blood group A type	$GalNAc\frac{1\ 3}{\alpha}\underset{Fuc}{Gal}\frac{1\ 4}{\beta}GlcNAc$—Gal—Glc—Cer	RBC—Human

[a] RBC, red blood cell.

guinea pig tissues (Ishibashi *et al.*, 1974; Kijimoto *et al.*, 1974), adrenal Y1-K cells (Yeung *et al.*, 1974), chemically transformed embryonic guinea pig cells (M. Basu *et al.*, 1980b) and N1L-2K cells (Kijimoto *et al.*, 1980). Thymus tissue from lymphoma-bearing hamsters also contains (Ishibashi *et al.*, 1975) higher activities of the GSL glycosyltransferases that lead to the synthesis of Forssman GSL's. Three distinct GSL *N*-acetylgalactosaminyltransferase activities (Table I) have been characterized in normal tissues (Chien *et al.*, 1973; Ishibashi *et al.*, 1974) and cultured tumor cells (M. Basu *et al.*, 1976, 1980b). The βGalNAcT activity (Table I) isolated from 11-day-old ECB catalyzes the transfer of *N*-acetylgalactosamine to GbOse$_3$ Cer (Galα1-4Galβ1-4Glc-Cer) to form a globoside (Chien *et al.*, 1973). From competition studies it has been concluded that the same protein may also catalyze the transfer of *N*-acetylgalactosamine to position C-4 of the terminal galactose unit of GM3 ganglioside (Steigerwald *et al.*, 1975). Whether the same enzyme transfers *N*-acetylgalactosamine to position C-4 of the terminal galactose of all three GSL's (LacCer, GM3, and GbOse$_3$Cer) has yet to be proved with purified βGalNAcT-1 activity. A membrane preparation isolated from guinea pig bone marrow appears to catalyze the transfer of a βGalNAc unit to the terminal galactose of LacCer and GM3 more efficiently than that of GbOse$_3$Cer (M. Basu *et al.*, 1974). This suggests that perhaps guinea pig bone marrow contains only βGalNAcT-1 activity, which catalyzes the formation of (β1-4)-linked GalNAc-

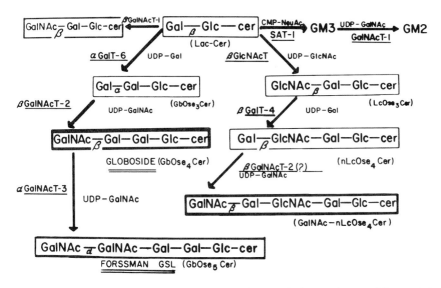

Figure 6 Proposed pathways for biosynthesis of *N*-acetylgalactosamine-containing neutral glycosphingolipids.

containing GSL's, whereas ECB might contain both βGalNAcT-1 and βGalNAcT-2 activities. It is important that guinea pig erythrocytes contain GgOse$_3$ Cer(GalNAcβ1-4Galβ1-4Glc-Cer) as the surface GSL (Seyama and Yamakawa, 1974) (Fig. 6), and it is the βGalNAcT-1 activity present in the bone marrow of this animal that most likely catalyzes its biosynthesis from lactosylceramide (M. Basu *et al.*, 1974b). However, lactosylceramide is a poor substrate for a βN-acetylgalactosaminyltransferase (βGalNAcT-2) isolated (M. Basu *et al.*, 1976, 1980b, 1980c) from embryonic guinea pig 103 cells and their chemically transformed clones, 104C1, 106B, and 107C3 (Evans and DiPaolo, 1975). Purified membrane preparations are isolated from these clonal lines and under single-assay conditions the V_{max} values with various probable substrates relative to GbOse$_3$Cer are as follows: GbOse$_3$Cer, 100%; GbOse$_4$Cer, 43%; nLcOse$_4$Cer, 48%; LacCer, 1%; and GM3, 1%. The enzymatic products obtained from GbOse$_3$Cer and GbOse$_4$Cer have been characterized as globoside and Forssman GSL, respectively. This conclusion is based primarily on their cleavage pattern with purified β- and α-hexosaminidases. The presence of at least two specific GSL *N*-acetylgalactosaminyltransferases (βGalNAcT-2 and αGalNAcT-3, Table I) in these cell lines has been postulated. Using purified membrane fractions from guinea pig bone marrow (M. Basu *et al.*, 1974b) and 104C1 (benzo[*a*]pyrene-transformed) cells (M. Basu *et al.*, 1976, 1980c), we achieved the biosynthesis of a pentaglycosylceramide (GalNAc-nLcOse$_4$Cer) from

nLcOse$_4$Cer(Galβ1-4GlcNAcβ1-3Galβ1-4Glc-Cer) and UDP-[^{14}C]GalNAc. Unlike blood group A-active GSL, this enzymatic product contains a β-linked terminal N-acetylgalactosamine, as evidenced by its cleavage by purified jack bean β-hexosaminidase (Li and Li, 1970). A novel ganglioside containing both GalNAc and GlcNAc (NeuAcα2-3GalNAcβ1-3Galβ1-4GlcNAcβ1-3Galβ1-4Glc-Cer) has been isolated from human erythrocytes (Watanabe and Hakomori, 1979). It is probable that GalNAcβ-nLcOse$_4$Cer is an intermediate in the biosynthesis of this newly discovered ganglioside.

Possible relationships between tumorigenicity and expression of GalNAcT-2 and GalNAcT-3 activities involved in Forssman GSL biosynthesis would be an important area of research in the future. Nontumorigenic cells (103 and 106B) appear to contain more Forssman GSL (GalNAcα1-3GalNAcβ1-3Galα1-4Galβ1-4Clc-Cer) on their surfaces than do their tumorigenic variants (104C1 and 107C3), as determined by binding studies with anti-Forssman antibody, *Bandeiraea simplicifolia* A$_4$ isolectin, and *Dolichos biflorus* lectin (Basu *et al.*, 1981a). Using [^{14}C]galactose as labeling compound, we observed that synchronized 104C1 and 106B cells synthesize maximum amounts of ^{14}C-labeled Forssman GSL 8 hours after release from arrest at the G$_1$ growth phase (Basu *et al.*, 1981a).

V. CONCLUSIONS

The biosynthesis *in vitro* of at least 20 different glycosphingolipids has been achieved using membrane-bound enzymes isolated from animal brain, bone marrow, liver, and spleen. Membrane preparations isolated from cultured cells before and after transformation also contain various levels of activities of these GSL glycosyltransferases. The structures of the enzymatic products have been characterized on the basis of migration on thin-layer chromatograms and chemical and enzymatic analyses of their components. The linkages have been determined by the use of specific glycosidases. In view of the difficulties involved in obtaining the enzymatic products in even low quantities (nanomoles), the permethylation technique (Stoffyn and Stoffyn, 1979, 1980) appears to be a very useful method for identifying their terminal linkages. This method is limited to the substrates containing terminal galactose only.

On the basis of species specificities, kinetic studies, substrate competition, and heat inactivation studies, it appears that the activities listed in Table I are probably catalyzed by different proteins. Solubilization of these activities and further determination of their kinetic parameters would confirm this statement. Some of these GSL glycosyltransferase activities [αFucT-2 (Presper *et al.*, 1982)] have already been solubilized and separated from other activities, but many more remain to be purified. It is possible that some of these GSL glycosyltransferases catalyze the transfer of a specific sugar to both glycosphingolipid and glycoprotein conjugates at equal (M. Basu and S. Basu, unpublished; Presper, 1979) or different rates (Sadler *et al.*, 1979; Rearick *et al.*, 1979). However, the ultimate and penultimate sugars of a glycoconjugate chain are the important determinants of the activity of any specific glycosyltransferase. Most of the kinetic studies *in vitro* have been done in the presence of specific detergents, the exact roles of which are not yet known. It is obvious that *de novo* biosynthesis of these glycosphingolipids occurs in tissues or cultured cells in the absence of detergents. Perhaps in the presence of liposomes (Cestelli *et al.*, 1979) or some specific GSL-binding proteins (Metz and Radin, 1980), the GSL substrates would be more accessible to the active sites of these enzymes. A search for such binding proteins would be an important part of additional studies on the control of GSL glycosyltransferase activities.

The structures of long-chain GSL's and glycoproteins containing 8–15 oligosaccharides, with specific blood group activities, have been established in human erythrocytes (Watanabe *et al.*, 1979a,b), hog gastric mucosa (Slomiany and Horowitz, 1973; A. Slomiany and Slomiany, 1977; A. Slomiany and Slomiany, 1980), and Ehrlich ascites tumor cells (Eckhardt and Goldstein, 1979). Biosynthesis *de novo* of these new long-chain GSL's may be catalyzed by GSL glycosyltransferases already characterized or by glycosyltransferases that have yet to be discovered.

REFERENCES

Ando, S., Kon, K., Isobe, M., and Yamakawa, T. (1973). *J. Biochem. (Tokyo)* **73**, 893–895.

Ankel, H., Krishnamurti, C., Besançon, F., Steffanos, S., and Falcoff, E. (1980). *Proc. Natl. Acad. Sci. U.S.A.* **77**, 2528–2532.

Arce, A., Maccioni, H. F., and Caputto, R. (1966). *Arch. Biochem. Biophys.* **116**, 52–58.

Basu, M. (1974a). D.Sc. Thesis University of Calcutta, Calcutta.

Basu, M., and Basu, S. (1972). *J. Biol. Chem.* **247**, 1489–1495.

Basu, M., and Basu, S. (1973). *J. Biol. Chem.* **248**, 1700–1706.

Basu, M., Chien, J.-L., and Basu, S. (1974b). *Biochem. Biophys. Res. Commun.* **60**, 1097–1104.

Basu, M., Moskal, J. R., Gardner, D. A., and Basu, S. (1975). *Biochem. Biophys. Res. Commun.* **66**, 1380–1388.

Basu, M., Basu, S., Schanabruch, W. G., Moskal, J. R., and Evans, C. H. (1976). *Biochem. Biophys. Res. Commun.* **71**, 385–392.

Basu, M., Presper, K. A., Basu, S., Hoffman, L. M., and Brooks, S. E. (1979). *Proc. Natl. Acad. Sci. U.S.A.* **76**, 4270–4274.

Basu, M., Higashi, H., Basu, S., Stoffyn, A., and Stoffyn, P. (1980a). *Proc. Soc. Complex Carbohydrate, 1980* p. 17.

Basu, M., Higashi, H., Basu, S., and Evans, C. H. (1980b). *Fed. Proc., Fed. Am. Soc. Exp. Biol.* **39**, 2184.

Basu, M., Basu, S., and Potter, M. (1980c). *ACS Symp. Ser.* **128**, 187–212.

Basu, S. (1966). Ph.D. Thesis, University of Michigan, Ann Arbor.

Basu, S., and Kaufman, B. (1965). *Fed. Proc., Fed. Am. Soc. Exp. Biol.* **24**, 479.

Basu, S., Kaufman, B., and Roseman, S. (1965). *J. Biol. Chem.* **240**, 4115-4117.

Basu, S., Kaufman, B., and Roseman, S. (1968). *J. Biol. Chem.* **243**, 5802-5804.

Basu, S., Schultz, A., and Basu, M. (1969). *Fed. Proc., Fed. Am. Soc. Exp. Biol.* **28**, 540.

Basu, S., Basu, M., Den, H., and Roseman, S. (1970). *Fed. Proc., Fed. Am. Soc. Exp. Biol.* **29**, 410.

Basu, S., Schultz, A., Basu, M., and Roseman, S. (1971). *J. Biol. Chem.* **246**, 4272-4279.

Basu, S., Kaufman, B., and Roseman, S. (1973). *J. Biol. Chem.* **248**, 1388-1394.

Basu, S., Basu, M., and Chien, J.-L. (1975). *J. Biol. Chem.* **250**, 2956-2962.

Basu, S., Basu, M., Moskal, J. R., and Chien, J.-L. (1976a) *In* "Glycolipid Methodology" (L. A. Witting, ed.), pp. 123-138. Am. Oil Chem. Soc. Press, Champaign, Illinois.

Basu, S., Moskal, J. R., and Gardner, D. A. (1976b). *In* "Ganglioside Function" (G. Porcellati, B. Ceccarelli, and G. Tettamanti, eds.), pp. 45-63. Plenum, New York.

Basu, S., Basu, M., Chien, J.-L., and Presper, K. A. (1980). *In* "Structure and Function of Gangliosides" (L. Svennerholm, H. Dreyfus, P. Urban, and P. Mandel, eds.), pp. 213-226. Plenum, New York.

Basu, S., Basu, M., and Higashi, H. (1981a). *In* "Glycolipids in Biology" (A. Makita and S. Handa, eds.), Plenum, New York (in press).

Basu, S., Basu, M., and Higashi, H. (1981b). *Proc. 6th Internatl. Glycoconjugates, 1981*, p. 41.

Baumann, H., Nudelman, E., Watanabe, K., and Hakomori, S. (1979). *Cancer Res.* **39**, 2637-2643.

Besançon, F., Ankel, H., and Basu, S. (1976). *Nature (London)* **259**, 576-578.

Brady, R. O., and Fishman, P. H. (1974). *Biochim. Biophys. Acta* **355**, 121-148.

Brenkert, A., and Radin, N. S. (1972). *Brain Res.* **36**, 183-193.

Burger, M. M., and Noonan, K. D. (1970). *Nature (London)* **228**, 512-515.

Cestelli, A., White, F. V., and Costantino-Ceccarini, E. (1970). *Biochim. Biophys. Acta* **572**, 283-292.

Cestelli, A., Suzuki, K., Siegel, D., Suzuki, K., and Costantino-Ceccarini, E. (1980). *Brain Res.* **186**, 185-194.

Chien, J.-L. (1975). Ph.D. Thesis, University of Notre Dame, Notre Dame, Indiana.

Chien, J.-L., Williams, T. J., and Basu, S. (1973). *J. Biol. Chem.* **248**, 1778-1785.

Chien, J.-L., Li, S.-C., Laine, R. A., and Li, Y.-T. (1978). *J. Biol. Chem.* **253**, 4031-4035.

Cuatrecasas, P. (1973). *Biochemistry* **12**, 3558-3566.

Cumar, F. A., Tallman, J. F., and Brady, R. O. (1972). *J. Biol. Chem.* **247**, 2322-2327.

Dawson, G. (1978). *In* "The Glycoconjugates" (M. I. Horowitz and W. Pigman, eds.), Vol. 2, pp. 255-284. Academic Press, New York.

Den, H., Sela, B.-A., Roseman, S., Sachs, L. (1974). J. Biol. Chem. 249, 659-661.

Den, H., Kaufman, B., McGuire, E. J., and Roseman, S. (1975). *J. Biol. Chem.* **250**, 739-746.

Eckhardt, A. E., and Goldstein, I. J. (1979). *In* "Glycoconjugate Research" (J. D. Gregory and R. W. Jeanloz, eds.), Vol. 2, pp. 1043-1045. Academic Press, New York.

Eto, T., Ichikawa, Y., Nishimura, K., Ando, S., and Yamakawa, T. (1968). *J. Biochem. (Tokyo)* **64**, 205-213.

Evans, C. H., and DiPaolo, J. A. (1975). *Cancer Res.* **35**, 1035-1044.

Fishman, P. H., and Brady, R. O. (1976). *Science* **194**, 906-915.

Fredman, P., Mansson, J. E., Svennerholm, L., Karlsson, K. A., Pascher, I., and Samuelsson, B. E. (1980). *FEBS Lett.* **110**, 80-84.

Ghidoni, R., Sonnio, S., Tettamanti, G., Wiegandt, H., and Zambotti, V. (1976). *J. Neurochem.* **27**, 511-515.

Gill, D. M., and Meren, P. (1978). *Proc. Natl. Acad. Sci. U.S.A.* **75**, 3050-3054.

Higashi, H., Basu, M., and Basu, S. (1981). *Fed. Proc., Fed. Am. Soc. Exp. Biol. Proc.* **40**, 1743.

Hildebrand, J., and Hauser, G. (1969). *J. Biol. Chem.* **244,** 5170–5180.

Holmgren, J., Lonnorth, I., and Svennerholm, L. (1973). *Infect. Immunol.* **8,** 208–214.

Inbar, M., and Sachs, L. (1969). *Proc. Natl. Acad. Sci. U.S.A.* **63,** 1418–1425.

Ishibashi, T., Kijimoto, S., and Makita, A. (1974). *Biochim. Biophys. Acta* **337,** 92–106.

Ishibashi, T., Atsuta, T., and Makita, A. (1975). *J. Natl. Cancer Inst.* **55,** 1433–1436.

IUPAC-IUB (1977). *Lipids* **12,** 455–468.

Iwamori, M., and Nagai, Y. (1979). *J. Neurochem.* **32,** 767–777.

Jenis, D., Basu, S., and Pollard, M. (1981). *Cancer Biochem. Biophys.* **6** (in press).

Joseph, K. C., and Gockerman, J. P. (1975). *Biochem. Biophys. Res. Commun.* **65,** 146–152.

Kaufman, B., and Basu, S. (1966). *In* "Methods in Enzymology" (E. F. Neufeld and V. Ginsburg, eds.), Vol. 8, pp. 365–368. Academic Press, New York.

Kaufman, B., Basu, S., and Roseman, S. (1966). *In* "Inborn Disorders of Sphingolipid Metabolism" (B. Volk and A. Aaronson, eds.), pp. 193–213. Pergamon, Oxford.

Kaufman, B., Basu, S., and Roseman, S. (1968). *J. Biol. Chem.* **243,** 5804–5806.

Keenan, T. W., Morré, D. J., and Basu, S. (1974). *J. Biol. Chem.* **249,** 310–315.

Kijimoto, S., Ishibashi, T., and Makita, A. (1974). *Biochem. Biophys. Res. Commun.* **56,** 177–184.

Kijimoto, S., Yokosawa, N., and Makita, A. (1980). *J. Biol. Chem.* **225,** 9037–9040.

Klenk, E. (1939). *Hoppe-Seyler's Z. Physiol. Chem.* **263,** 128–143.

Koul, O., Chou, K.-H., and Jungalwala, F. B. (1980). *Biochem. J.* **186,** 959–969.

Kuhn, R., and Wiegandt, H. (1963). *Chem. Ber.* **96,** 866–880.

Ledeen, R. W. (1977). *J. Supramol. Struct.* **8,** 1–17.

Li, S.-C., and Li, Y.-T. (1970). *J. Biol. Chem.* **245,** 5153–5160.

Li, Y.-T., Mansson, J. E., Vanier, M. T., and Svennerholm, L. (1973). *J. Biol. Chem.* **248,** 2634–2636.

Mandel, P., Nussbaum, J. L., Nesković, N. M., Sarlieve, L. L., and Kurihara, T. (1972). *Adv. Enzyme Regul.* **10,** 101–118.

Metz, R. J., and Radin, N. S. (1980). *J. Biol. Chem.* **255,** 4463–4467.

Morell, P., and Radin, N. S. (1969). *Biochemistry* **8,** 506–512.

Moskal, J. R. (1977). Ph.D. Thesis, University of Notre Dame, Notre Dame, Indiana.

Moskal, J. R., Gardner, D. A., and Basu, S. (1974). *Biochem. Biophys. Res. Commun.* **61,** 751–758.

Moss, J., and Vaughan, M. (1977). *J. Biol. Chem.* **252,** 2455–2457.

Moss, J., and Vaughan, M. (1979). *Annu. Rev. Biochem.* **48,** 581–600.

Ng, S. S., and Dain, J. A. (1977a). *J. Neurochem.* **29,** 1075–1083.

Ng, S. S., and Dain, J. A. (1977b). *J. Neurochem.* **29,** 1085–1093.

Obata, K., and Handa, S. (1979). *In* "Integrative Control Functions of the Brain" (M. Ito, ed.), Vol. 2, pp. 5–14. Elsevier, Amsterdam.

Ohashi, M. (1980). *J. Biochem. (Tokyo)* **88,** 583–589.

Ohashi, M., and Yamakawa, T. (1977). *J. Biochem. (Tokyo)* **81,** 1675–1690.

Pacuszka, T., and Koscielak, J. (1976). *Eur. J. Biochem.* **64,** 499–506.

Pacuszka, T., Duffard, R. O., Nishimura, R. N., Brady, R. O., and Fishman, P. H. (1978). *J. Biol. Chem.* **253,** 5839–5846.

Panzetta, P., Maccioni, H. J. F., and Caputto, R. (1980). *J. Neurochem.* **35,** 100–108.

Presper, K. A. (1979). Ph.D. Thesis, University of Notre Dame, Notre Dame, Indiana.

Presper, K. A., Basu, M., and Basu, S. (1976). *Fed. Proc., Fed. Am. Soc. Exp. Biol.* **35,** 1441.

Presper, K. A., Basu, M., and Basu, S. (1977). *Fed. Proc., Fed. Am. Soc. Exp. Biol.* **36,** 731.

Presper, K. A., Basu, M., and Basu, S. (1978). *Proc. Natl. Acad. Sci. U.S.A.* **75,** 289–293.

Presper, K. A., Basu, M., and Basu, S. (1982). *J. Biol. Chem.* **257,** 169–173.

Preti, A., Fiorilli, A., Lombardo, A., Caimi, L., and Tettamanti, G. (1980). *J. Neurochem.* **35,** 281–296.

Puro, K. (1969). *Biochim. Biophys. Acta* **187**, 401–413.

Rauvala, H., Krusius, T., and Finne, J. (1978). *Biochim. Biophys. Acta* **531**, 266–274.

Rearick, J. I., Sadler, J. E., Paulson, J. C., and Hill, R. L. (1979). *J. Biol. Chem.* **254**, 4444–4451.

Roseman, S. (1970). *Chem. Phys. Lipids* **5**, 270–297.

Sadler, J. E., Rearick, J. I., Paulson, J. C., and Hill, R. L. (1979). *J. Biol. Chem.* **254**, 4434–4443.

Santer, U. V., and Glick, M. C. (1980). *Biochem. Biophys. Res. Commun.* **96**, 219–226.

Sarlieve, L. L., Nesković, N. M., Freysz, L., Mandel, P., and Rebel, G. (1976). *Life Sci.* **18**, 251–260.

Seyama, Y., and Yamakawa, T. (1974). *J. Biochem. (Tokyo)* **75**, 837–842.

Shah, S. N. (1971). *J. Neurochem.* **18**, 395–402.

Shah, S. N. (1973). *Arch. Biochem. Biophys.* **159**, 143–150.

Slomiany, A., and Horowitz, M. I. (1973). *J. Biol. Chem.* **248**, 6232–6238.

Slomiany, A., and Slomiany, B. L. (1980). *Biochem. Biophys. Res. Commun.* **93**, 770–775.

Slomiany, B. L., and Slomiany, A. (1977). *FEBS Lett.* **73**, 175–180.

Sonnio, S., Ghidoni, R., Galli, G., and Tettamanti, G. (1978). *J. Neurochem.* **31**, 947–956.

Steigerwald, J. C., Basu, S., Kaufman, B., and Roseman, S. (1975). *J. Biol. Chem.* **258**, 6727–6734.

Stoffyn, A., and Stoffyn, P. (1979). *Carbohydr. Res.* **74**, 279–286.

Stoffyn, A., and Stoffyn, P. (1980). *Carbohydr. Res.* **78** 327–340.

Stoffyn, A., Stoffyn, P., and Hauser, G. (1974). *Biochim. Biophys. Acta* **360**, 174–17 .

Sundsmo, J. S., and Hakomori, S. (1976). *Biochim. Biophys. Res. Commun.* **68**, 799–806.

Svennerholm, L. (1962). *Biochem. Biophys. Res. Commun.* **9**, 436–441.

Svennerholm, L. (1964). *J. Lipid Res.* **5**, 145–155.

Svennerholm, L. (1980). *In* "Cholera and Related Diarrheas" (O. Ouchterloney and J. Holmgren, eds.), pp. 80–87. S. Karger, Basel.

Svennerholm, L., Bruce, A., Rynmark, B. M., and Vanier, M. T. (1972). *Biochim. Biophys. Acta* **280**, 626–636.

Svennerholm, L., Mansson, J.-E., and Li, Y.-T. (1973). *J. Biol. Chem.* **248**, 740–742.

Sweeley, C. C. and Siddiqui, B. (1977). *In* "The Glycoconjugates" (M. I. Horowitz, W. Pigman, eds.), Vol. 1, pp. 459–540. Academic Press, New York.

Sweeley, C. C., Fund, Y. K., Macher, B. A., Moskal, J. R., and Nuñez, G. (1978). *ACS Symp. Ser.* **80**, 47–89.

Thudicum, J. L. W. (1884). "A Treatise on the Chemical Constitution of Brain." Bailliere, London (new reprint: Archon Books, Hamden, Connecticut, 1962).

Urdal, D. L., and Hakomori, S. (1980). *J. Biol. Chem.* **255**, 10509–10516.

Van Heyningen, W. E., Carpenter, C. J., Pierce, N. F., and Greenough, W. B. (1971). *J. Infect. Dis.* **124**, 415–418.

Vunnam, R. R., and Radin, N. S. (1979). *Biochim. Biophys. Acta* **573**, 73–82.

Vunnam, R. R., and Radin, N. S. (1980). *Chem. Phys. Lipids* **26**, 265–278.

Watanabe, K., and Hakomori, S. (1979). *Biochemistry* **24**, 5502–5504.

Watanabe, K., Hakomori, S., Childs, R. A., and Feizi, T. (1979a). *J. Biol. Chem.* **254**, 3221–3228.

Watanabe, K., Powell, M. E., and Hakomori, S. (1979b). *J. Biol. Chem.* **254**, 8223–8229.

Wiegandt, H. (1973). *Hoppe-Seyler's Z. Physiol. Chem.* **354**, 1049–1056.

Wiegandt, H., and Bucking, H. W. (1970). *Eur. J. Biochem.* **15**, 287–292.

Wiegandt, H., and Schulz, B. (1969). *Z. Naturforsch., B: Anorg. Chem., Org. Chem., Biochem., Biophys., Biol.* **24B**, 945–946.

Yamakawa, T., and Suzuki, S. (1953). *J. Biochem. (Tokyo)* **38**, 199–212.

Yang, H., and Hakomori, S. (1971). *J. Biol. Chem.* **246**, 1192–1200.

Yeung, K. K., Moskal, J. R., Chien, J. L., Gardner, D. A., and Basu, S. (1974). *Biochem. Biophys. Res. Commun.* **59**, 252–260.

Yiamouyiannis, J. A. (1967). Ph.D. Thesis, University of Rhode Island, Kingston.

Yip, G. B., and Dain, J. A. (1970). *Biochim. Biophys. Acta* **206,** 252–260.

Yohe, H. C., and Yu, R. K. (1980). *J. Biol. Chem.* **255,** 608–613.

Yokosawa, N., Kijimoto-Ochiai, S., and Makita, A. (1978). *Biochem. Biophys. Acta* **528,** 138–147.

Yu, R. K., and Lee, S. H. (1976). *J. Biol. Chem.* **251,** 198–203.

3

Glycoconjugates in Cellular Adhesion and Aggregation

SECTION 1

Fibronectin–Proteoglycan Binding as the Molecular Basis for Fibroblast Adhesion to Extracellular Matrices

BARRETT J. ROLLINS, MARTHA K. CATHCART,
AND LLOYD A. CULP

I. CELL ADHESION TO EXTRACELLULAR MATRICES AND ITS SIGNIFICANCE

Many normal cell types must adhere to a suitable tissue culture surface for survival and growth *in vitro*. This stringent anchorage dependence is thought to replace the *in vivo* requirements for cell–extracellular matrix or cell–cell interactions in tissues and therefore to allow normal cell behavior *in vitro*. Interest in

THE GLYCOCONJUGATES, VOL. III

this anchorage requirement has stemmed from the desire to understand the basic behavioral properties of cells, their movements, and their interactions with the environment, as well as from the observation of apparent differences in the adherence of tumor cells (Gail and Boone, 1972; Culp and Black, 1972a,b; Weber *et al.*, 1977; Shields and Pollock, 1974; DiPasquale and Bell, 1974; Poste and Fidler, 1980). In this section we focus on the biochemical mechanism of adhesion of several cell types to one type of extracellular matrix—the serum-coated tissue culture dish [for other reviews, see Culp (1978) and Grinnell (1978)]. The cellular and biochemical origins of adhesion sites are reviewed, with emphasis on the interaction of glycoconjugates in these sites.

The *in vitro* tissue culture surface, often glass or plastic, must fulfill certain criteria to serve as a satisfactory substratum for adhesion that promotes normal growth and behavior of cells. One requirement for physiologically compatible adhesion to such surfaces is the binding and enrichment of adhesion-promoting factors found in the serum of animals (see Section III,A). Without such mediators, the adhesion of the cell to the substratum does not allow the cell to exhibit natural activity and behavior. Adhesions that are too weak allow the cell to detach and die, and adhesions that are too strong interfere with basic cellular processes such as movement and mitosis (Takeichi, 1971; Grinnell, 1974; Revel and Wolken, 1973; Culp and Buniel, 1976; Folkman and Moscona, 1978). The type of adhesive interaction in which a cell participates is therefore of fundamental importance for movement, recognition, and survival both at the cellular level and collectively in tissues. Thus, any alteration in the basic adhesive properties of a cell significantly alters its social behavior. The intrinsic anchorage dependence of certain cell types can be altered and even eliminated upon transformation from a normal to a tumorigenic phenotype. Undoubtedly, these basic alterations of adhesion are closely related to the altered behavior of tumorigenic and particularly metastatic cells (Poste and Fidler, 1980).

II. BIOLOGY OF FIBROBLAST ADHESION TO THE TISSUE CULTURE SUBSTRATUM

A. Cell Attachment and Spreading

Cell attachment is an active process inhibited by low temperature, metabolic inhibitors, and microfilament-disrupting agents (Juliano and Gagalang, 1977; Nath and Srere, 1977; McClay *et al.*, 1977). It is essentially a two-step process initially requiring specific binding of cell surface molecules to the extracellular matrix, followed by cytoskeletal reorganization at the adhesion site, which "strengthens" the adherence of the cell (Culp, 1978). Upon exposure to an appropriate tissue culture substratum, a rounded fibroblast attaches by means of small filopodial extensions from the cell surface (Albrecht-Buehler, 1976; Rosen and Culp, 1977). More permanent adhesions are then developed by extension and branching of these filopodial processes across the substratum surface. The membrane fills in between the branching fingers to form footpad adhesion sites several microns in diameter (Fig. 1a). The membranous extensions between the cell body and the adhesion site thicken, with resultant cell body spreading over the tissue culture surface creating broader and more extensive areas of footpad adhesions (Rosen and Culp, 1977). The cell adhesion sites remain focal and tightly apposed to the substratum (Izzard and Lochner, 1976, 1980); they do not consist of the entire undersurface exposed to the substratum but only of discrete focal areas both underneath and at the periphery of the cell (Revel *et al.*, 1974; Rosen and Culp, 1977; Vogel and Kelley, 1977; Vogel, 1978; Britch and Allen, 1980).

B. Cell Movement

During natural cell movement over the substratum, it appears that adhesions are continually being made and broken (Culp, 1978). The leading lamella of the cell tests the nearby substratum by extending microspikes, which "feel" around for a favorable place to adhere (Albrecht-Buehler, 1976). Concomitant with the formation of new adhesion sites, the sites at the posterior of the cell are abandoned. That is, the cells leave behind adherent footpads as "footprints" during the process of normal cell movement (Culp, 1978; Rosen and Culp, 1977; Chen, 1977). These resemble morphologically and biochemically those that are cleaved from the cell body during cell detachment by EGTA* treatment (Culp, 1976a,b; Culp *et al.*, 1979).

C. Cell Detachment—Natural and Induced

The cell retains a basically flat, thinly spread morphology until mitosis or perturbation by various agents. During mitosis the cell takes on a rounded morphology and pulls away from the adherent footpad areas while retaining contact with them through thin retraction fibers (Revel *et al.*, 1974). In this conformation the cell is particularly susceptible to shear forces and is easily detached from the substratum. These rapidly reversible effects during mitosis are probably controlled within the cell by cytoskeletal rearrangements (Revel *et al.*,

*Abbreviations: BSA, bovine serum albumin; CIg, cold-insoluble globulin (the plasma-contained and slightly different form of cellular fibronectin); EGTA, ethylenebis(oxyethylenenitrilo)tetraacetic acid; GAG, glycosaminoglycan; GAP, glycosaminoglycan-associated protein; MSV, murine sarcoma virus; PAGE, polyacrylamide gel electrophoresis; PBS, phosphate-buffered saline; PMSF, phenylmethylsulfonyl fluoride; SAM, substratum-attached material; SDS, sodium dodecyl sulfate; SV40, Simian virus 40.

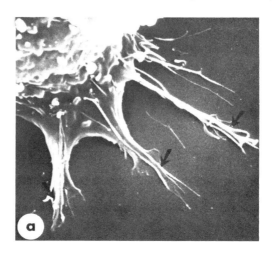

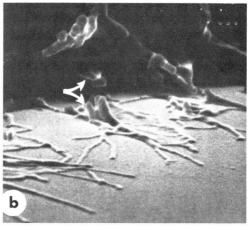

1974; Vogel, 1978) rather than by molecular events on the outer surface of the cell (Culp, 1978).

Flattened cells also assume a rounded morphology after exposure to proteases and chelating agents (Culp and Black, 1972a; Revel *et al.*, 1974; Rosen and Culp, 1977; Vogel, 1978; Culp *et al.*, 1978a). These agents are used for subculturing and detaching cells and exploit the particular susceptibility of the rounded cell to detachment by shear forces caused by rocking the container or pipetting the medium over the surface. The thin retraction fibers that connect the rounded cell body to the adherent footpads are broken and the cell body is released, leaving the footpad areas of the undersurface firmly adherent to the substratum (Fig. 1b) (Rosen and Culp, 1977). The footpad material that is left adherent following EGTA-mediated cell release has been termed substratum-attached material (SAM) and has been studied extensively by Culp and co-workers (Culp, 1978) (Fig. 1c). This material consists of isolated cell–extracellular matrix adhesion sites enriched approximately 80- to 100-fold. The biochemical composition of SAM generated after gentle trypsin-mediated detachment of cells is similar to that of SAM generated by EGTA-mediated detachment, indicating that cytoskeletal perturbation resulting from cell surface disorganization at the periphery of adhesion sites is the principle cause of detachment (Vogel, 1978; Culp *et al.*, 1978a; Britch and Allen, 1980).

In experiments investigating the effect of SAM-coated substrata (prepared by detaching confluent fibroblasts with chelating agents) upon the subsequent attachment and growth of a fresh population of cells, several investigators have observed that this adhesive material promotes attachment and cytoplasmic spreading (Culp, 1974; Weiss *et al.*, 1975; Collins, 1980; Harper and Juliano, 1980). In addition, it appears that it can compensate somewhat for the decreased or altered adhesive properties of oncogenic virus-transformed cells (Culp, 1974; Mapstone and Culp, 1976). The growth and movement behavior of SV40-transformed 3T3 murine fibroblasts were uniquely affected on surfaces coated with 3T3 SAM, allowing these transformed cells to exhibit a behavior somewhat

---

Figure 1 Cellular origin of substratum-attached material. Swiss 3T3 cells were detached by EGTA treatment after growth to confluence and then allowed to reattach to fresh serum-coated glass coverslips in medium containing 10% serum. Specimens were processed for scanning electron microscopy as described by Rosen and Culp (1977) by fixation in buffered glutaraldehyde, dehydration with graded alcohols and Freon, and critical point drying. Samples were then sputter-coated with gold–palladium. (a) A cell 2 hours after attachment, showing well-developed footpad adhesion sites at the arrows (magnification 6500×; 20° tilt); (b) a cell detaching after 10 minutes of treatment in 0.5 mM EGTA in PBS with gentle shaking showing shearing of the labile retraction fiber (at the arrows) that connects the cell body to the footpad (magnification 13,500×; 80° tilt); (c) morphology of a single adhesion site as SAM after detachment of cells that had been grown for 24 hours, showing the remnant of the retraction fiber at the arrow (magnification 13,500×; 80° tilt). Taken from Culp *et al.* (1978b) or Rosen and Culp (1977) with permission.

similar to that of normal 3T3 cells. However, SAM could not induce growth control in the transformed cells (Culp, 1974).

III. BIOCHEMISTRY OF FIBROBLAST ADHESION SITES

A. Serum Components Involved in Adhesion

Several serum components bind to tissue culture surfaces, including specific mediators of the cell–substratum adhesive interaction. These mediators are preadsorbed to the glass or plastic surface and are necessary for appropriate cell–substratum adhesion (Culp and Buniel, 1976; Grinnell, 1976a). The adhesive preference of a cell for one type of artificial matrix over another may be due solely to the types of serum components that bind to the matrix and subsequently mediate adhesion. Analysis of serum components that adhere to glass and plastic tissue culture surfaces (Culp and Buniel, 1976; Haas and Culp, 1979) revealed several Coomassie blue-stainable bands separable upon SDS–PAGE, including albumin. The SDS–PAGE profiles of thiol-reduced adsorbed serum components on glass microbeads, bacteriological plastic dishes, or tissue culture plastic dishes were similar, whereas unreduced preparations generated very different profiles (Haas and Culp, 1979). This indicates that the chemical nature of the substratum modulates either the ability of bound serum components to form disulfide-linked complexes or the binding of differing disulfide-cross-linked complexes out of serum.

Recent studies have drastically reduced the number of serum proteins suspected of actively mediating the adhesion of several cell types including fibroblasts. Grinnell (1976a,b) fractionated whole serum and generated a complex of components that actively mediated cell attachment and cytoplasmic spreading. This "spreading factor" was shown to be greatly enriched for cold-insoluble globulin (CIg) (Culp, 1978). The most important mediator appears to be the 440,000 MW dimeric glycoprotein CIg (Höök et al., 1977; Grinnell and Hays, 1978; Kleinman et al., 1979; Murray et al., 1980). Its importance is indicated by the ability of antibody to CIg to block the spreading of cells on serum-coated dishes (Grinnell and Hays, 1978). Cold-insoluble globulin permits normal adhesion and spreading of cells to take place on artificial tissue culture substrata or collagen (Klebe, 1974; Pearlstein, 1976; Kleinman et al., 1978, 1979). It binds very tightly to the tissue culture surface [much of it is resistant to SDS extraction (Haas and Culp, 1979)] but is readily detectable by immunochemical methods (Mautner and Hynes, 1977; Murray et al., 1980; Murray and Culp, 1981).

The function of CIg in the serum layer appears to be modulated by its association with other molecules in this layer, as suggested by several lines of investigation. Very different disulfide-cross-linked complexes are extracted from chemi-

cally different matrices (Haas and Culp, 1979), and the amounts of immunoac-
cessible CIg in the serum layer absorbed to glass or plastic, as assayed by
^{125}I-labeled staphylococcal protein A binding to antigen–antibody complexes, are
quite different (Murray et al., 1980; Murray and Culp, 1981). Interference
reflection microscopic analyses of rat fibroblast adhesion sites formed on CIg
-coated substrata reveal an inability to form the tightly apposed focal contacts
observed on serum-coated substrata (Thom et al., 1979). Serum macromolecules
other than CIg have been identified in a preliminary fashion as critical factors in
the physiologically compatible adhesion of fibroblasts (Knox and Griffiths,
1979; Thom et al., 1979). There also appears to be a non-CIg-dependent
mechanism for Balb/c 3T3 cells to adhere to a collagenous matrix (Linsenmayer
et al., 1978). Adhesion sites generated on CIg-adsorbed substrata contain much
higher quantities of hyaluronate and chondroitin proteoglycan than do sites on
serum-adsorbed substrata (Murray et al., 1980), whereas the concentrations of
cell-surface fibronectin and the various cytoskeletal proteins in the two classes of
sites are virtually identical. The latter result also suggests that a high concentration of CIg on the substratum cannot functionally substitute for cell-surface
fibronectin in the adhesion process.

B. Cell Surface Components in Adhesion Sites

Culp and co-workers have extensively studied the biochemical composition of
the adhesion sites from the murine cell lines Balb/c 3T3, SV40-transformed 3T3
(SVT2), and concanavalin A-selected revertant variants of the SVT2 cell line
after EGTA-mediated detachment of cells. The transformed cells are less adhesive than their normal counterparts and deposit less SAM than the more adhesive
3T3 and revertant cells (Culp and Black, 1972a). SAM has been shown to be
considerably enriched in several cell-surface-associated components. This evidence, as well as the molecular basis for these selective process, is reviewed
below.

1. Glycoprotein and Protein Components

Adhesion site proteins and glycoproteins are strongly associated with their in
vitro substratum. They require treatment with an anionic detergent such as SDS
for complete solubilization (Cathcart and Culp, 1979b). Figure 2 illustrates a
representative two-dimensional autoradiogram of leucine-radiolabeled SAM that
was solubilized with 0.2% SDS. Similar analyses of surface membrane preparations demonstrate that SAM is enriched in only some of the components found
over the entire cell surface or in its membrane-associated cytoskeleton (Culp,
1976a,b; Vessey and Culp, 1978).

The principal glycoprotein component enriched in SAM is fibronectin (C_0 in
Fig. 2). Many biochemical and immunochemical criteria have been used to

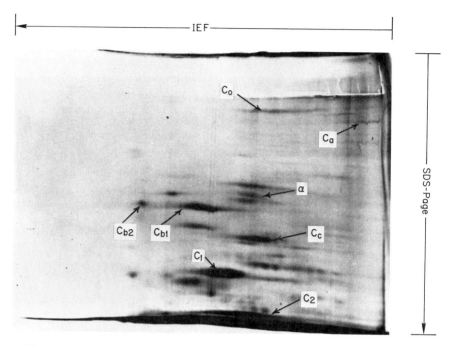

Figure 2 Two-dimensional electrophoresis of the larger proteins in SAM. SVT2 SAM from cells growing in [^{14}C]leucine-containing medium for 72 hours was solubilized with 0.2% SDS, concentrated, reduced with mercaptoethanol, and analyzed as described by Culp *et al.* (1979) by autoradiography. Isoelectric focusing (IEF) as the first dimension was performed in 2 mm diameter (12 cm length) cylindrical gels using a mixture of 1.6% pH 5–7 ampholines and 0.4% pH 3–10 ampholines. The second dimension of slab SDS–PAGE was performed after pouring a 12% ORTEC separating gel (Culp, 1976a), a 2 cm length 5% ORTEC stacking gel, and finally a 0.5 cm length 2% agarose gel in which the disc gel was embedded. After electrophoresis, the slab gel was dried and autoradiographed for 4 months. Basic proteins are distributed on the right side and acidic proteins on the left side of the gel. The following proteins have been biochemically and immunochemically identified in these adhesion site preparations: C_o (MW 220,000) as cell-surface fibronectin; C_a (MW 200,000) as myosin; α (MW 92,000) as α-actinin; C_{b1} (MW 85,000), C_{b2}, and C_c (MW 67,000) as presumptive cytoskeletal proteins greatly enriched in SAM; C_1 (MW 55,000) as the 10 nm diameter filament subunit protein; and C_2 (MW 42,000) as actin (both the β and γ forms). Taken from Culp *et al.* (1979) with permission.

identify this component as authentic cell-derived fibronectin (Culp, 1976a,b; Mautner and Hynes, 1977; Badley *et al.*, 1978; Cathcart and Culp, 1979b; Murray *et al.*, 1980; Murray and Culp, 1981). This cell surface component appears to be very important for the initial adhesive interaction between the cell and the extracellular matrix in that it is the most prominent iodinated cell surface component present in newly formed or long-term-matured adhesion sites (Fig.

3). Several different pools of fibronectin have now been differentiated in these adhesion sites (Culp, 1976b; Culp *et al.*, 1978a; Cathcart and Culp, 1979b; Murray and Culp, 1981): (a) the pulse-radiolabeled and metabolically labile pool, which is readily solubilized by treatment of SAM with trypsin or testicular hyaluronidase; (b) the long-term-radiolabeled and metabolically stable pool, which resists enzymatic solubilization or guanidine hydrochloride extraction coordinately with the heparan proteoglycans and cytoskeletal proteins in SAM (Culp *et al.*, 1978a; Garner and Culp, 1981, in press); (c) a small pool, which is accessible to antibody but which cannot be solubilized with hyaluronidase; and (d) a large pool, which is inaccessible to antibody or lactoperoxidase, even after various extraction or fixation protocols are followed. These different subsets of fibronectin in SAM undoubtedly reflect differing molecular interactions with other components in the adhesion site, particularly with the large amounts of glycosaminoglycan or proteoglycan in these sites (see Sections IV and V).

Substratum-attached material is also enriched in some, but not all, cytoskeletal proteins found in fibroblasts, including the microfilament proteins actin, myosin, and an extremely small amount of α-actinin (Fig. 2) (Culp, 1976a,b; Culp *et al.*, 1979; Badley *et al.*, 1978). In the cell, these components are present in microfilaments (Goldman *et al.*, 1975) found in association with the cell surface membrane, especially at points of cell–cell or cell–extracellular matrix contact (McNutt *et al.*, 1971, 1973; Perdue, 1973; Dermer *et al.*, 1974; Goldman *et al.*, 1975; Pollack *et al.*, 1975; Willingham *et al.*, 1977; Singer, 1979). The organization of actin within the adhesion sites appears to be very different from that observed in whole cells and very similar to that observed in membrane-associated actin (Gruenstein *et al.*, 1975; Cathcart and Culp, 1979b). It is believed that the contractile apparatus must have an anchor to a membrane to allow for cell movement. Consistent with this is the identification of two subsets of adhesion site actin, one of which is extractable from adhesion sites with nonionic detergent or ATP–KCl solutions and a second of which is highly resistant to any extraction condition (Cathcart and Culp, 1979b).

Protein C_1 in Figure 2 appears to be the subunit protein of the 10 nm diameter filaments of these cells (Culp, 1978; Culp *et al.*, 1979). It is extremely resistant to extraction from SAM (Cathcart and Culp, 1979b) and is the most prominent cytoskeletal protein observed in newly formed adhesion sites (Culp, 1976a). Thus, both classes of cellular ''Microfilaments'' are represented in the adhesion sites, perhaps suggesting a close structural and binding interrelationship between them and with the cell membrane.

The relative proportions of glycoprotein and protein components in SAM are different from those observed in whole cells or surface membranes, suggesting a coordinate enrichment for a specialized group of components involved in adhesion on select regions of the cell surface (see Section VI,A). The amounts of

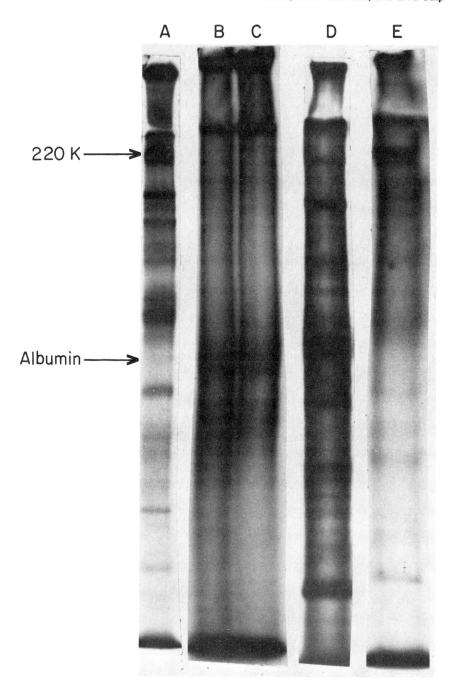

these components are similar in all three cell lines studied, regardless of normal or transformed phenotype. This suggests that quantitative rather than qualitiative alterations in these components accompany transformation in this system.

2. Phospholipid Components

The membranous nature of SAM as depicted by scanning electron microscopy (Rosen and Culp, 1977) has been confirmed by biochemical studies (Cathcart and Culp, 1979a). Several classes of phospholipids have been identified in SAM from the three murine fibroblast cell lines (Table I). The adhesion sites are noticeably enriched in phosphatidylethanolamine and phosphatidylserine and depleted of phosphatidylcholine as compared to whole cell or surface membrane preparations. Again, this demonstrates the specialized composition of the adhesion sites. Unlike the protein and glycoprotein components in SAM, which are similar in the three cell lines, the phospholipid components of SAM are markedly different in the transformed (SVT2) cell line as compared to the normal 3T3 and revertant cell lines. This altered composition may to some extent reflect the altered adhesive properties of transformed cells.

IV. GLYCOSAMINOGLYCANS IN ADHESION

Since the first ultrastructural evidence was obtained for GAG's on the surface of cultured cells (Gasić and Baydak, 1962; Defendi and Gasić, 1963; Dorfman, 1963), this class of macromolecules has held a fascination for investigators in the fields of cellular adhesion and oncogenesis. Numerous studies have attempted to correlate the presence or absence of GAG's with the normal or transformed (i.e., more or less adhesive) phenotype (although decreased adhesiveness and tumorigenicity do not necessarily go hand in hand). For example, increased amounts of GAG's have been seen histochemically after transformation of fibroblasts by polyoma virus (Defendi and Gasić, 1963), by Rous sarcoma virus (Ericksen *et al.*, 1961; Temin, 1965), and by SV40 virus (Makita and Shimojo,

---

Figure 3 Iodinated cell surface proteins enriched in SAM. Cell, SAM, and serum components adsorbed to the plastic tissue culture dishes were radiolabeled with ^{131}I by the lactoperoxidase technique described by Cathcart and Culp (1979b). Samples were solubilized in SDS, electrophoresed on a 12% slab ORTEC SDS–PAGE gel, and autoradiographed. Wells A and D: cell and SAM, respectively, from cultures iodinated *prior to* cell detachment with EGTA. Well B: adsorbed serum proteins iodinated on the dish without cells. Well C: SAM iodinated *after* EGTA-mediated detachment of cells. Well E: SAM fraction after the iodinated cells in A (detached previously with EGTA treatment) were permitted to reattach to fresh serum-coated dishes for 2 hours (followed by their detachment with EGTA). The arrow at 220 K indicates the cell-surface fibronectin component, and the lower arrow indicates the migration location of bovine albumin. Taken from Cathcart and Culp (1979b) with permission.

TABLE I
Enrichment of Specific Phospholipids in Adhesion Sites of
Balb/c 3T3 Cells[a]

Phospholipid component[b]	% of total phospholipid		
	Cell	Surface membrane	SAM
LPC	1.2	1.0	0.6
SM	3.7	5.8	2.4
PC	61.2	56.7	37.4
LPE	0.3	0.4	2.4
PI	8.0	8.5	3.0
PS	3.3	4.1	17.9
PA	0.5	0.1	0.5
PE	15.9	20.2	28.6
PG	0.6	0.2	0.8
CL	2.8	2.6	5.4

[a] Cells were radiolabeled for 48 hours in medium containing [^{32}P]orthophosphate (100 μCi/ml) and detached with EGTA treatment, followed by two rinses with PBS for analysis as the cell fraction. Surface membranes were isolated from the EGTA-detached cells by the two-phase polymer method described by Cathcart and Culp (1979a). The SAM fraction was harvested with SDS. Data are presented as percentage of total phospholipid in each fraction and are taken from Cathcart and Culp (1979a) with permission.

[b] Abbreviations: LPC, lysophosphatidylcholine; SM, sphingomyelin; PC, phosphatidylcholine; LPE, lysophosphatidylethanolamine; PI, phosphatidylinositol; PS, phosphatidylserine; PA, phosphatidic acid; PE, phosphatidylethanolamine; PG, phosphatidylglycerol; CL, cardiolipin.

1973). Others, however, have reported a relative reduction in total GAG content after such transformation (Cohn *et al.*, 1976; Saito and Uzman, 1971).

Studies on the synthesis of specific GAG's after transformation are also in disagreement. Initial reports of reduced hyaluronic acid production by various cell lines after SV40 transformation (Hamerman *et al.*, 1965) were contradicted by later reports of increased production after transformation by SV40 (Makita and Shimojo, 1973; Satoh *et al.*, 1973) or avian sarcoma virus (Ishimato *et al.*, 1966; Bader, 1972). Increased hyaluronic acid production has also been correlated with increased tumorigenicity (Winterbourne and Mora, 1977). Similarly, transformation has been reported to increase (Makita and Shimojo, 1973; Dietrich *et al.*, 1977), decrease (Saito and Uzman, 1971; Goggins *et al.*, 1972; Chiarugi *et al.*, 1974; Roblin *et al.*, 1975), and not affect (Satoh *et al.*, 1973; Cohn *et al.*, 1976) the synthesis and steady-state levels of sulfated GAG's. Contradictory data have also been presented for heparan sulfate production upon oncogenic virus transformation of various cell types (Makita and Shimojo, 1973; Satoh *et al.*, 1973; Saito and Uzman, 1971; Chiarugi *et al.*, 1974). There appears to be no generalized correlation with increased or decreased synthesis of any GAG class and the transformed or tumorigenic capacity of the cell. It has been observed that the heparan sulfate from two different tumorigenic cell lines is more highly sulfated than that from their normal counterparts (Underhill and Keller, 1975; Winterbourne and Mora, 1977, 1978; Johnston *et al.*, 1979).

Another approach has been to investigate the role of these macromolecules in adhesion per se, regardless of the relation to the malignant phenotype. This has been done both in three-dimensional cell aggregation and in two-dimensional tissue culture systems. In cell aggregation systems, the isolation of factors that enhance intercellular adhesion has provided evidence for GAG's playing a direct role in adhesion. The adhesion factor from *Microciona parthena* is a proteoglycan complex containing uronic acid residues (Cauldwell *et al.*, 1973; Henkart *et al.*, 1973; Humphreys *et al.*, 1977). A similar factor from *M. prolifera* also has GAG-like characteristics (Turner and Burger, 1973; Weinbaum and Burger, 1973). In mammalian systems, some aggregation factors have been found to be hyaluronidase sensitive (Pessac and Defendi, 1972; Underhill and Dorfman, 1978). An increased production of hyaluronic acid (Koyoma *et al.*, 1975; Tomida *et al.*, 1975) and sulfated GAG's (Goggins *et al.*, 1972) is observed concomitant with the increased adhesiveness to the tissue culture substratum upon the addition of dibutyryl-AMP to some cells. This effect may be fortuitous, however, since the effect of cAMP on adhesion may be mediated by its effect on subcortical cytoskeletal organization, leading to increased adhesion (Willingham *et al.*, 1977; Badley *et al.*, 1980).

A. Glycosaminoglycans in Adhesion Sites

Early biochemical analysis had revealed the presence of hyaluronic acid in SAM (Terry and Culp, 1974), and metabolic radiolabeling with $^{35}SO_4^{2-}$ revealed sulfated GAG as well (Terry and Culp, 1974; Roblin *et al.*, 1975). When glucosamine- *or* sulfate-radiolabeled SAM was analyzed by SDS–PAGE, most of the radioactivity was found in three high molecular weight GAG-associated bands (Culp, 1976a). This GAG could be efficiently extracted from SAM only by the use of ionic detergents (Cathcart and Culp, 1979b), suggesting that the GAG's form an integral part of the adhesion site and are tightly "anchored" in SAM, probably to the cellular cytoskeletal elements and the substratum-bound serum-derived CIg, both of which are also resistant to nonionic detergent extraction (Brown *et al.*, 1976; Cathcart and Culp, 1979b; Haas and Culp, 1979).

By comparing the GAG's in SAM to those in the rest of the cell in a chemically rigorous fashion, it was hoped that a significant redistribution of certain of these carbohydrates in SAM might reveal their presumptive functional roles in the adhesion process. Methods were devised for isolating and purifying the GAG's free of the SDS used to extract SAM quantitatively (Rollins and Culp,

1979a) so that the highly specific bacterial chondroitinases could be used to digest the galactosaminoglycans and hyaluronic acid (Saito *et al.*, 1968; Yamagata *et al.*, 1968). Appreciable amounts of hyaluronate, chondroitin 4-sulfate, and unsulfated chondroitin, along with small amounts of chondroitin 6-sulfate and dermatan sulfate, were identified in this way. The effect of depolymerization of heparan sulfate by nitrous acid degradation (Lindahl *et al.*, 1973) was monitored by molecular sieve chromatography, and in this way two pools of heparan sulfate (a highly *N*-sulfated pool and an undersulfated, highly *N*-acetylated pool) were identified (Rollins and Culp, 1979a). The quantity of glycoprotein-derived glycopeptide in SAM that had been generated by exhaustive digestion of SAM with Pronase could also be monitored by this chromatographic technique.

Glycosaminoglycans in SAM, in the EGTA-detached cells, and in EGTA-

TABLE II

Distribution of Polysaccharides in Balb/c 3T3 Adhesion Sites under Various Physiological Conditions[a]

	Radioactivity (%)[c]					
	Long-term		Pulse		Reattaching	
Polysaccharide[b]	Cell	SAM	Cell	SAM	Cell	SAM
Glycopeptide	73.6	27.9	56.3	8.3	59.7	66.6
GAG	26.4	72.1	43.7	91.2	40.3	33.4
	100.0	100.0	100.0	100.0	100.0	100.0
GAG class						
HS	48.8	26.3	46.5	15.0	56.8	80.2
C6S	0.8	2.3	0.9	2.5	2.5	1.6
C4S	5.7	22.4	3.7	21.7	4.3	8.0
DS	24.0	2.7	16.4	0.0	17.7	0.9
COS	2.1	23.6	8.8	34.5	2.5	4.8
HA	17.6	22.7	23.7	26.3	16.2	4.5
	100.0	100.0	100.0	100.0	100.0	100.0

[a] Cells were grown in medium containing [^{3}H]glucosamine for 72 hours (long-term) or 2 hours (pulse), after which cell fractions were collected. The cell fraction was obtained after EGTA-mediated detachment of cells and rinsing twice with PBS. A portion of the cells was permitted to reattach to fresh dishes in serum-containing medium for 1 hour (reattaching), followed by EGTA-mediated detachment to obtain the reattaching cell fraction. Substratum-attached material (SAM) was quantitatively solubilized with 0.2% SDS. Analysis for specific GAG's or glycoprotein-derived glycopeptide is described by Rollins and Culp (1979a). Data taken from Rollins and Culp (1979a) with permission.

[b] Abbreviations: HS, heparan sulfate; C6S, chondroitin 6-sulfate; C4S, chondroitin 4-sulfate; DS, dermatan sulfate; COS, unsulfated chondroitin; HA, hyaluronic acid.

[c] Glycopeptide and GAG are shown as the percentage of total polysaccharide radioactivity, whereas the individual GAG's are shown as the percentage of total GAG radioactivity.

solubilized material were examined under three different radiolabeling conditions (Table II). (a) In cells radiolabeled *in situ* for 72 hours, the SAM would represent footpad plus footprint material; (b) in well-spread cells pulse-radiolabeled for 2 hours, the radiolabeled elements of SAM would represent molecules being incorporated into already functional footpads (that is, pulse radiolabeling would give an indication of the components of the "maturing" or "aging" footpad); and (c) in preradiolabeled cells attaching to dishes for only 1 hour, SAM would represent newly formed adhesion sites. The results for the cell-associated and SAM-associated GAG's in Balb/c 3T3 cells are summarized in Table II. These data reveal that there is a strikingly different distribution of GAG's in SAM as opposed to the rest of the cell. This lends support to the concept, presented earlier, of the adhesion site as a biochemically differentiated area of the cell surface. In analyzing the specific GAG's, the most significant finding is that when adhesion sites are newly formed (i.e., in reattaching cell SAM), they are highly enriched in heparan sulfate and depleted of hyaluronic acid and the chondroitins. During "maturation" of adhesion sites (as assessed by the 2-hour radiolabel pulse) and their abandonment as footprints (included in the long-term radiolabeling experiments), there is a progressive accumulation of the chondroitins and hyaluronic acid in a nearly constant 2:1 ratio. Dermatan sulfate is present in minimal amounts in SAM [EGTA-soluble material is highly enriched in dermatan sulfate and hyaluronate and depleted of heparan sulfate (Rollins and Culp, 1979a)]. It is important to note that the chondroitins and hyaluronate are accumulating *specifically* in SAM during adhesion site maturation since the GAG distribution in the cell-associated material remains essentially identical under all three radiolabeling protocols. These results were essentially reproduced in SVT2 cells and in concanavalin A-selected revertant cells of SVT2 (Rollins and Culp, 1979a), which have regained their pretransformation adhesive and growth control properties (Culp and Black, 1972b).

B. Heparan Sulfate in Adhesion Sites

The presence of large amounts of heparan sulfate in adhesion sites was not an unexpected finding (Kraemer, 1971a,b), considering the cell surface origins of SAM (Terry and Culp, 1974; Culp, 1976a,b; Rosen and Culp, 1977). The intrinsic nature of the presence of heparan sulfate in the adhesion site is suggested by its resistance to solubilization from SAM upon treatment with trypsin (Culp *et al.*, 1978a), testicular hyaluronidase (Culp *et al.*, 1978a), or guanidine hydrochloride (Rollins and Culp, 1979b). Hyaluronic acid and the various chondroitin species, in contrast, are coordinately removed from SAM by gentle *in situ* trypsinization or hyaluronidase treatment under conditions in which 85–90% of the cellular fibronectin, heparan sulfate, and cytoskeletal elements resist solubilization (Culp *et al.*, 1978a). This suggests that hyaluronate and the chondroitins are not stringently required elements of adhesive bond formation. Similarly, digestion of cell surface hyaluronate or chondroitins with enzymes does not adversely affect the ability of fibroblasts to adhere and spread (Bruns and Gross, 1980).

The fact that heparan sulfate makes up over 80–90% of the GAG in the newly formed adhesion site of reattaching cells, however, gives an indication of the possible importance of this polysaccharide in the initial adhesive interaction between the cell surface and the serum-coated substratum. The interactions between heparins (usually of hog mucosal or beef lung origin) and various plasma proteins, especially antithrombin III (Rosenberg and Damus, 1973) and CIg (Stathakis and Mosesson, 1977; Jilek and Hörmann, 1979; Laterra *et al.*, 1980), are well documented (Rosenberg, 1978; Lindahl and Höök, 1978). Although the material in SAM and other nonhematological systems has been consistently referred to as heparan sulfate, it is not unreasonable to assume major areas of structural homology between these molecules (the heparan sulfates) and the classically defined heparins (Linker and Hovingh, 1973). The heparins are extremely heterogeneous, even intramolecularly, consisting of almost randomly distributed regions of high or low concentrations of iduronate, glucuronate, and N-sulfated or N-acetylated glucosamine (Linker and Hovingh, 1973, 1975; Lindahl *et al.*, 1977; Lindhal and Höök, 1978). Actually, because of the affinity for N-sulfated regions of the C-5 epimerase that effects the postpolymerization epimerization of glucuronic to iduronic acid, polymeric areas of high iduronic acid content are associated with areas of high N-sulfated glucosamine content. Conversely, areas of high glucuronic acid content are associated with areas of high N-acetylated glucosamine content (Lindhal *et al.*, 1977; Jacobsson *et al.*, 1979).

It is clear that two populations of heparan polymers exist in the SAM of these cell lines. Figure 4 shows the molecular sieve chromatographic profile of $^{35}SO_4$/[^{3}H] glucosamine doubly radiolabeled SAM polysaccharides after chondroitinase ABC digestion. Although both areas Ia and Ib have been identified as heparan sulfate (Rollins and Culp, 1979a), area Ib is routinely much less sensitive to nitrous acid depolymerization than is area Ia. The difference between Ia and Ib relates not only to their hydrodynamic sizes, but also to sulfate content, with area Ia having a higher $^{35}SO_4$/[^{3}H]glucosamine ratio than area Ib. In addition, when GAG from SAM is passed through a column of CIg immobilized on Sepharose 4B beads, essentially the only material remaining bound to the column is the Ia class of heparan sulfate along with much smaller amounts of Ib material and some dermatan sulfate (Laterra *et al.*, 1980). These data are consistent with area Ia representing the highly N-sulfated, iduronic acid-rich subpopulation of heparan sulfate and area Ib representing the N-acetylated, glucuronic acid-rich material.

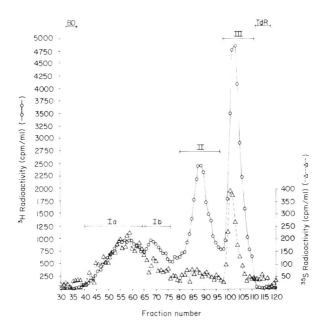

Figure 4 Gel filtration of SVT2 SAM polysaccharide after chondroitinase ABC digestion. SVT2 cells were grown for 72 hours in the presence of 5 μCi/ml of [^{3}H]glucosamine and 50 μCi/ml of Na$_2$^{35}SO$_4$. The cells were detached with EGTA treatment. SAM was then harvested with SDS and exhaustively digested with Pronase. The polysaccaride was digested with chontroitinase ABC and chromatographed on a column of Sepharose CL6B (1 × 120 cm, eluted with 15 mM Tris-HCl, pH 7.4, 0.2% SDS). Tritium H (O——O) and ^{35}S (△– –△) radioactivities were determined for each fraction, along with the markers blue dextran (BD) or thymidine (TdR). The following areas of the profile have been characterized: Ia, highly N-sulfated heparan sulfate; Ib, undersulfated heparan sulfate; II, glycoprotein-derived glycopeptide; and III, disaccharides liberated by chondroitinase digestion of hyaluronate, the various chondroitins, and dermatan sulfate.

The preferential binding of iduronic acid-rich heparan sulfate to CIg fits well into the accepted mechanism of the heparin–antithrombin III interaction in which the protein-binding sites of heparin are contained within N-sulfated, iduronic acid-rich segments (Rosenberg and Lam, 1979; Lindahl *et al.*, 1979). This may explain why some dermatan sulfate also binds to CIg–Sepharose columns (Laterra *et al.*, 1980), since this is the only other GAG that contains iduronic acid. The binding determinants for heparin on antithrombin III seem to involve lysine (Rosenberg and Damus, 1973) and a crucial tyrosine residue (Blackburn and Sibley, 1980), and such a dependence may, by analogy, eventually be found for the binding determinants of fibronectin or CIg for heparan sulfate. At any rate, all the evidence to date points to a central role for heparan sulfate binding to substratum-bound CIg in a well-accepted paradigm as an initial step in cell–

substratum adhesion. Further work on characterizing the heparan sulfate subfractions and their functions awaits the general availability of specific and well-purified heparanases.

C. Adhesion Sites of Cell Lines with Differing Adhesive Properties

The GAG content in the adhesion sites of some other cell lines has also been analyzed in much the same manner as described above for Balb/c 3T3 cells. In all cases, the GAG content in SAM is strikingly different from the GAG content in the rest of the cell. The GAG patterns show some interesting correlations with the phenotypes of these cells. Table III lists the distributions in the long-term-radiolabeled and reattaching SAM's from Balb/c 3T3, Kirsten murine sarcoma virus-transformed (KiMSV) Balb/c 3T3, C6 rat glioma, and B104 rat neuroblastoma cells (with and without neurites) (Culp *et al.*, 1980). In all cell lines tested

TABLE III
Comparative Analyses of GAG Distribution in Adhesion Sites of Various Cell Types[a]

SAM preparation	Polysaccharide[b]	Radioactivity (%)[c]				
		Balb/c 3T3	KiMSV	C6	B104[−]	B104[+]
Long-term	HS	26.3	39.0	19.8	44.4	64.7
	C6S	2.3	5.2	2.8	13.6	5.8
	C4S	22.4	15.2	28.0	14.8	7.3
	DS	2.7	10.0	0.0[d]	8.0	5.0
	COS	23.6	6.3	1.1	5.4	2.0
	HA	22.7	24.3	48.3	13.8	15.2
		100.0	100.0	100.0	100.0	100.0
Reattaching	HS	80.2	25.6	51.8	82.9	96.7
	C6S	1.6	10.7	0.0[d]	0.0[d]	0.0[d]
	C4S	8.0	18.2	10.7	0.0[d]	3.3
	DS	0.9	16.5	0.0[d]	0.0[d]	0.0[d]
	COS	4.8	8.2	0.0[d]	0.0[d]	0.0[d]
	HA	4.5	20.8	37.5	17.1	0.0[d]
		100.0	100.0	100.0	100.0	100.0

[a] Cells were grown in medium containing [³H]glucosamine for 72 hours, EGTA-detached, rinsed twice with PBS, and permitted to reattach to fresh dishes in serum-containing medium for 1 hour before EGTA-mediated detachment. The long-term and reattaching SAM's were harvested with SDS, and GAG determinations were performed as described by Rollins and Culp (1979a). KiMSV are Kirsten murine sarcoma virus-transformed Balb/c 3T3 cells, C6 are rat glioma cells, B104[−] are rat neuroblastoma cells without neurites, and B104[+] are rat neuroblastoma cells with neurites.

[b] Abbreviations for each GAG class are defined in footnote *b*, Table II.

[c] Percentage of each GAG type to the total GAG content of the SAM fractions.

[d] Below the sensitivity of detection for the amounts being analyzed.

except for the MSV-transformed cells, there is a concomitant enrichment of heparan sulfate in the newly formed adhesion sites and a progressive accumulation of the chondroitins and hyaluronic acid in the adhesion site as the cell spreads and initiates motility. Since the neural cells also spread effectively on CIg-coated dishes (Culp *et al.*, 1980), adhesion to the substratum by means of heparan sulfate–CIg interaction may be a more general phenomenon and not restricted to fibroblast cell types.

KiMSV-transformed cells are highly rounded in culture; i.e., they never form extensive adhesions that allow the cells to spread, and, in fact, the cells can be removed from the culture substratum by lightly pipetting medium over the cell layer. With this characteristic in mind, it is interesting to note that, from the moment they are plated in culture, MSV-transformed cells have large amounts of the chondroitins and hyaluronate in their adhesion sites and a low concentration of heparan sulfate compared to other related cell lines (such as the 3T3 or SVT2 cell types that do spread well) (Table III). That is, their adhesion sites do not go through the maturation process involving early high concentrations of heparan sulfate followed by accumulation of the other GAG's. This suggests that the chondroitins and hyaluronate may have a negative modulating effect on the strength and formation of cell–substratum adhesions and thus facilitate detachment of cells. Interestingly, KiMSV cells also display a much lower concentration of fibronectin in their SAM than the other cell types (Culp, 1976a). The C6 glioma cells are highly motile on the substratum, are easily detached, and have large quantities of hyaluronate in their long-term and newly formed adhesion sites (Table III) (Culp *et al.*, 1980). A number of glioma tumor lines have been examined and have been found to have much more hyaluronate and chondroitin in their SAM than a number of ''normal'' glial lines (Glimelius *et al.*, 1978).

On the other hand, the neuroblastoma cells are much less motile than the other cells in Table III. These cells have a high proportion of heparan sulfate in their new adhesion sites and a low concentration of hyaluronate. In fact, in the neurite-producing cells, which are perforce highly spread, there is essentially no hyaluronate and chondroitin in the newly formed adhesion site. The amounts of the chondroitins and hyaluronate do rise in the maturing adhesion sites, but to levels lower than those seen in the other cell types.

V. SUPRAMOLECULAR INTERACTIONS IN ADHESION SITES

The finding of selected and specific GAG's in adhesion sites has led to experiments that have tried to define the functionality of the GAG's in the adhesion process. These attempts have centered around demonstrating the interactions between GAG-containing molecules and other components of adhesion sites and generally coincide with the growing consensus over the last decade that GAG's function primarily as binding molecules (Lindahl and Höök, 1978). One major focus of this interest has been the interaction between GAG's and collagen (Mathews and Decker, 1968; Toole and Lowther, 1968a,b; Toole, 1976), an interaction that, interestingly, is enhanced by the presence of iduronic acid in the GAG polymer (Toole and Lowther, 1968a,b). The absence of any detectable collagen in the adhesion sites of 3T3 or SVT2 cells (Culp and Bensusan, 1978), however, suggests that GAG–collagen interactions are not relevant in the *in vitro* cell–substratum adhesion systems under consideration here but may be very relevant to collagen-dependent adhesion processes (Linsenmayer *et al.*, 1978).

A. Proteoglycans in Adhesion Sites

Before functional studies on interactions among adhesion site components can be undertaken, it is necessary to know the physical state in which they exist in the adhesion site. This caveat is particularly relevant to GAG's, since all of the GAG's except hyaluronic acid have been shown, in nearly all other systems, to be covalently linked to protein as proteoglycans (Lindahl and Höök, 1978). Heparan sulfate proteoglycans are present on the surface of hepatocytes (Oldberg *et al.*, 1979) and fibroblasts (Kraemer, 1971a,b, 1977; Kraemer and Smith, 1974; Kraemer and Barnhart, 1978; Johnston *et al.*, 1979), and chondroitin and dermatan sulfate proteoglycans are present on the surface of hamster fibroblasts (Dunham and Hynes, 1978; Rollins and Culp, 1979b; Perkins *et al.*, 1979). The presence of cell surface receptors for heparan sulfate proteoglycan was shown by the ability of exogenously added heparin to displace a portion of the endogenous heparan sulfate (Kraemer, 1977; Kjellen *et al.*, 1977; Oldberg *et al.*, 1979). Similar arguments have been made for the ability of exogenously added heparin to displace lipoprotein lipase from the surface of hepatic capillary endothelial cells (Olivercrona *et al.*, 1977). This, too, is presumptive evidence for a cell surface heparin-like molecule acting as a lipoprotein lipase receptor. It should be noted, however, that heparin cannot displace the heparan sulfate deposited in 3T3 substratum adhesion sites (L. A. Culp, unpublished data), emphasizing the firmness of binding of heparan proteoglycans in these sites.

Proteoglycans have also been shown to exist in the adhesion sites of fibroblasts (Rollins and Culp, 1979b). This was demonstrated by a Pronase-induced change in hydrodynamic size of the GAG species in solubilized SAM. When each of three areas of SAM proteoglycan separated on a Sepharose CL2B column in SDS-containing buffer is subjected to exhaustive Pronase digestion and re-chromatographed, the profiles of Figure 5 are seen. The areas designated 2BI and 2BII in Figure 5A were not altered in apparent hydrodynamic size by Pronase digestion. (This material is hyaluronic acid, and its resistance to proteolysis is consistent with its generally accepted protein-free structure). Biochemical

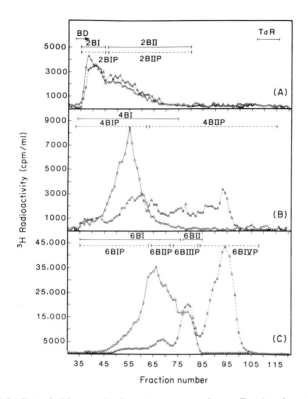

Figure 5 GAG's in SAM, except hyaluronate, are proteoglycans. Fractions from areas marked I, II, and III in Figure 1 of Rollins and Culp (1979b) after gel filtration on a Sepharose CL2B column of whole SAM in an SDS-containing buffer were combined and divided into two aliquots. One aliquot of each area was digested with Pronase, and the other was mock-digested. The sham-digested (○) and Pronase-digested (△) samples were then chromatographed on the same Sepharose columns (all 1 × 120 cm, eluted with 150 mM Tris-HCl, pH 7.4, 0.2% SDS). (A) Area I material chromatographed on Sepharose CL2B; the GAG is principally hyaluronic acid. (B) Area II material chromatographed on Sepharose CL4B; contains equivalent amounts of heparan sulfate and the various chondroitins, along with a small amount of hyaluronate. (C) Area III material chromatographed on Sepharose CL6B; contains small amounts of heparan sulfate and the chondroitins and >95% of the glycoprotein in SAM. The position of marker blue dextran (BD) and thymidine (TdR) elution was very similar for all three columns. Taken from Rollins and Culp (1979b) with permission.

analysis of area 4BI in Figure 5B shows it to be about half heparan sulfate and half chondroitin sulfate. After Pronase digestion, area 4BIP contains mostly heparan sulfate, whereas 4BIIP contains about half of the heparan sulfate and all of the chondroitins that were in 4BI. Similarly, area 6BI in Figure 5C consists of heparan sulfate, the chondroitins, and glycoprotein (area 6BII is entirely glyco-protein). Overall, approximately two-thirds of the heparan sulfate and all of the galactosaminoglycans in SAM were shown to chromatograph (in SDS-containing buffers) as much smaller entities after Pronase digestion, thus establishing their existence as proteoglycans (Mathews, 1971). In fact, initial analysis of the size distributions of these proteoglycans showed that the heparan sulfate behaved in a manner independent of the chondroitins (6-sulfated, 4-sulfated, and unsulfated), which themselves behaved as a single group. Thus, there are probably separate heparan sulfate and chondroitin proteoglycans in SAM. The dermatan sulfate proteoglycan in SAM showed a still different distribution, suggesting its independent existence as a third proteoglycan species (Toole and Lowther, 1968b).

The size of the native heparan sulfate proteoglycan is similar to that of the heparin proteoglycan derived from mast cells [approximately 200,000 MW (Yurt *et al.*, 1977)] and of the macromolecular heparan sulfate proteoglycan of human skin fibroblasts (Kleinman *et al.*, 1975). The smaller Pronase-digested pools of heparan sulfate are comparable to the single carbohydrate chains from these sources, although the persistence of Pronase-resistant ''mini''-proteoglycans cannot be ruled out by these experiments.

B. Associative Molecular Interactions

The presence of proteoglycans in SAM increases the possible complexity of the structural arrangements of various molecular entities in the adhesion site. As indicated above, a primary function of the GAG's as a class is that of multivalent binding molecules, and there are several models for proteoglycan binding in other systems. Considered below are the formation of proteoglycan complexes by protein–carbohydrate interactions and possibly by carbohydrate–carbohydrate interactions. Each of these supramolecular interactions is considered in the context of cell–substratum adhesion.

1. Proteoglycan Complexes

A decade of elegant experimentation has demonstrated that the proteoglycans in cartilage exist as a complex in which the core proteins of chondroitin–keratan sulfate proteoglycans bind to a ten-sugar region of hyaluronic acid (Hascall and Sajdera, 1969; Hardingham and Muir, 1972; Gregory, 1973; Hascall and Heinegard, 1974a,b; Heinegard and Hascall, 1974; Rosenberg *et al.*, 1975; Hascall, 1977). This interaction is further stabilized by "link" glycoproteins, which coordinately bind to the core protein and the hyaluronate (Tang *et al.*, 1979; Hardingham, 1979). These experiments have been repeated using normal chondrocyte cultures (Hascall *et al.*, 1976; DeLuca *et al.*, 1978; Kimura *et al.*, 1978), Swarm rat chondrosarcoma (Caterson and Baker, 1979; Kimura *et al.*, 1979), and glioma cells (Norling *et al.*, 1978), demonstrating the formation of similar proteoglycan complexes by cells cultured *in vitro*.

The same sort of analysis has been performed for the proteoglycans in the adhesion sites of Balb/c 3T3 and SVT2 cells (Rollins and Culp, 1979b), demon-

strating the *potential* for proteoglycans and GAG's from SAM to form supramolecular complexes. When [³H]glucosamine/³⁵SO₄²⁻ doubly radiolabeled
SAM is extracted with 4.0 M guanidine hydrochloride (dissociative extraction),
then dialyzed to associative conditions (0.4 M guanidine hydrochloride) and
subjected to equilibrium density gradient centrifugation, the profile of Figure 6A
results. Three peaks of radioactivity are seen. The densest (with a high ³⁵S:³H
ratio) consists of about 40% heparan sulfate, 40% chondroitins plus hyaluronate,
and 20% glycoprotein, as determined by biochemical analysis of Pronase-

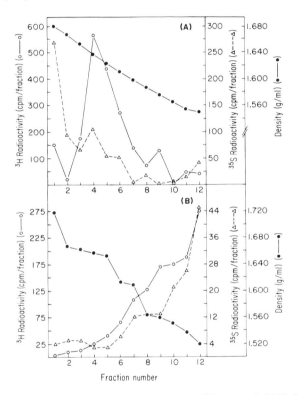

Figure 6 Associative and dissociative isopycnic centrifugation of SAM from long-term
[³H]glucosamine/³⁵SO₄ doubly radiolabeled 3T3 cells. SAM was extracted with 4.0 M guanidine
hydrochloride in buffer, concentrated, and dialyzed against 0.4 M guanidine hydrochloride in buffer
to permit reassociation of proteoglycans and glycoproteins into high molecular weight complexes, as
described by Rollins and Culp (1979b). Cesium chloride was added to this extract to a density of 1.63
gm/ml, and the mixture was centrifuged to equilibrium (A) as an associative gradient. Fractions 1–8
from (A) were combined and made 4.0 M in guanidine hydrochloride. Cesium chloride was added to
a density of 1.54 gm/ml, and the mixture was centrifuged to equilibrium (B) as a dissociative
gradient. Approximately 1-ml fractions were collected from the bottoms of the centrifuge tubes, and
the density (●), ³H (○), and ³⁵S (△) of each fraction were determined. Taken from Rollins and Culp
(1979b) with permission.

liberated polysaccharide classes. The mid-density fraction is principally hyaluronic acid (Garner and Culp, 1981, in press) with a small amount of associated glycoprotein, and the least dense fraction (with the least radioactivity) is composed of equal proportions of heparan sulfate, chondroitins, hyaluronate, and glycoprotein. All of the cytoskeletal protein extracted from SAM is observed in this least dense fraction. The highest-density and mid-density aggregates are very large and excluded from Sepharose CL2B columns eluted with associative buffers. When these two denser areas of radioactivity in the gradient are collected and taken to dissociative conditions, equilibrium density centrifugation shows the dissociation of this material into less dense components (Fig. 6B). The material at the top of these dissociative density gradients can be dialyzed back to associative conditions, and recentrifugation of this material then reproduces the two dense peaks of radioactivity shown in Figure 6A. Both aggregate classes are excluded from Sepharose CL2B columns eluted with associative buffers. Thus, the proteoglycans and hyaluronate in SAM can undergo a reversible association reminiscent of the associations in cartilage.

To examine the protein dependence of this aggregation in SAM, the proteoglycan extract was subjected to Pronase treatment and analyzed by associative CsCl gradient centrifugation (Rollins and Culp, 1979b). The densest gradient aggregate is no longer apparent, whereas the mid-density peak persists. Similarly, the addition of hyaluronate oligomers (HA-80, approximately 80 repeating units) displaces some of the material from the densest to the least dense fractions under associative conditions but has no effect on the mid-density aggregate. These observations are consistent with the presence of some macromolecular hyaluronate–chondroitin proteoglycan aggregates in SAM in the highest-density fraction. The correlation with the cartilage system is not complete, however, and would not be expected considering the enrichment of hyaluronate in the mid-density fractions and some heparan proteoglycan in the high-density fractions of the associative gradients.

Comparative analyses of long-term-radiolabeled SAM's from Balb/c or Swiss 3T3 cells reveal similar distributions of the high-density and mid-density aggregates (Garner and Culp, 1981) with very little sulfated GAG in the mid-density aggregate. Associative CsCl gradient analysis of materials from newly formed adhesion sites identifies a much higher proportion of aggregate at the bottom of the gradients. We have also shown that these aggregates can be only partially disaggregated with 4 M guanidine hydrochloride and require a detergent in order to achieve complete disaggregation. It appears likely, then, that much of this proteoglycan aggregation is biochemically more complex than the processes observed in the cartilage system (Hascall, 1977), and some of this proteoglycan may contain hydrophobic protein sequences, which result in a unique mode of aggregation (J. A. Garner, M. Lark, and L. A. Culp, unpublished data).

2. GAG and Proteoglycan Binding to Fibronectin

The fact that fibronectin is essentially a connective tissue glycoprotein makes it very likely that it interacts in some way with the other components of connective tissue. Early tissue studies established the presence of fibronectin in the extracellular fibrils coursing through connective tissue and basement membranes (Linder *et al.*, 1975). Fibronectin is also present in the tissue culture "connective tissue," i.e., in extracellular fibrils (Mautner and Hynes, 1977; Hedman *et al.*, 1978; Yamada and Olden, 1978; Murray *et al.*, 1980), and can be recovered in the glycocalyx fraction from certain plasma membrane preparations (Graham *et al.*, 1975, 1978). The fact that these plasma membrane fractions also stain heavily with ruthenium red is indirect evidence for a possible association between fibronectin and GAG-containing structures (Graham *et al.*, 1978).

There have been several direct demonstrations of interaction between fibronectin or CIg and GAG's or proteoglycans. Perkins *et al.* (1979) demonstrated the close proximity of fibronectin to chondroitinase-sensitive elements *in situ* on the cell surface by cross-linking fibronectin, derivatized with one end of a heterobifunctional agent, to proteoglycan material. Another demonstration of fibronectin–GAG interaction comes from treating SAM with testicular hyaluronidase *in situ*. The addition of this enzyme to SAM from long-term-radiolabeled 3T3 cells causes a small amount of fibronectin to be released (Culp *et al.*, 1978a). Similar treatment of SAM from newly reattaching cells leads to the release of essentially no fibronectin. On the other hand, much of the newly synthesized fibronectin deposited in 3T3 or SVT2 SAM turns over with a half-life of 1–2 hours, whereas the remainder becomes metabolically stabilized in this adhesive material. Hyaluronidase digestion of SAM from cells pulsed with radiolabel for 2 hours leads to *complete solubilization* of the metabolically labile fibronectin without loss of any cytoskeletal protein, indicating that this pool of labile fibronectin is associated with hyaluronate/chondroitin-containing structures. These data suggest the presence of two pools of fibronectin associated with the adhesion site: One is added to already formed adhesion sites by means of hyaluronidase-labile binding; this pool can then be chased into a second pool, either inaccessible to enzyme or bound by hyaluronidase-resistant ligands (Murray and Culp, 1981).

The first direct demonstration of GAG binding to CIg came from the studies of Stathakis and Mosesson (1977). They showed that human CIg forms complexes with several commercial preparations of heparin independent of the presence of fibrinogen and that these complexes form a precipitate at $2°$. In the precipitate, the molar ratio of heparin to CIg is $1.5–2:1$, and the ability of CIg to bind to heparin is independent of the anticoagulant activity of the heparin. Similarly, Jilek and Hörmann (1979) showed that the interaction between fibronectin and native collagen is greatly enhanced by the presence of heparin, and, interestingly, this heparin effect is inhibited by hyaluronic acid. In addition, the latter investigators were able to show a direct interaction between fibronectin and heparin (in the absence of collagen) in which heparin induced the fibronectin to assume a multimeric, elongated, filamentous form as seen by electron microscopy. It is tempting to consider that this may reflect the possible *in vivo* role of extracellular proteoglycans in inducing the characteristic filaments of fibronectin seen outside the cell (Singer, 1979). Cold-insoluble globulin bound noncovalently to gelatin–Sepharose columns subsequently binds heparin or heparan sulfate (Ruoslahti and Engvall, 1980).

Yamada *et al.* (1980) provided direct evidence for the ability of cellular fibronectin to bind specific cellular GAG's. Taking advantage of the insolubility

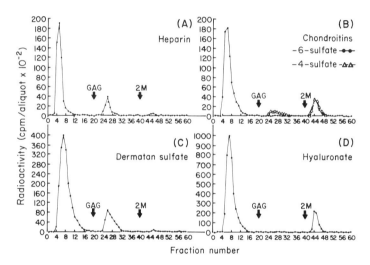

Figure 7 Specific displacement of SAM-derived radiolabeled GAG bound to a CIg–Sepharose column with reference standard GAG's. SVT2 cultures were grown for 72 hours in [³H]glucosamine-containing medium, long-term SAM was harvested with SDS and digested exhaustively with Pronase, and polysaccharides (GAG plus glycopeptide) were purified by ethanol precipitation as described by Rollins and Culp (1979a). The polysaccharide was passed through a 2-ml minicolumn of human CIg covalently linked to Sepharose 4B as described by Laterra *et al.* (1980) in a low ionic strength buffer that allows some of the GAG to bind. At the fraction number labeled "GAG," a 100μg/ml solution of a specific nonradioactive reference standard GAG (kindly provided by Dr. M. Matthews and Dr. A. Cifonelli of the University of Chicago) was passed through the column in a low ionic strength buffer to determine whether it would "chase" the radiolabeled GAG off the column. This shows that heparan sulfate or dermatan sulfate contains sequences similar to that of the SAM-derived, radiolabeled GAG bound to the column. Any persistent GAG bound to the column was then eluted with 2 *M* NaCl (2M in the figure) in buffer. (A) Heparin; (B) chondroitins 6-sulfate (●) and 4-sulfate (△); (C) dermatan sulfate; (D) hyaluronate. Taken from Laterra *et al.* (1980) with permission.

of fibronectin at neutral pH, they mixed radiolabeled GAG's with chick embryo fibroblast fibronectin at alkaline pH, then reduced the pH and collected the precipitate on nitrocellulose filters. In this way, they demonstrated the high affinity of fibronectin for heparin and hyaluronate at separate binding sites, a lesser affinity for a standard preparation of heparan sulfate, and a very low affinity for the chondroitins.

Of relevance to the cellular adhesion systems per se has been direct demonstration of an interaction between GAG isolated from SAM to human plasma CIg immobilized on Sepharose columns (Laterra *et al.,* 1980). In these experiments GAG's are purified from EGTA-detached fibroblasts or their SAM and applied to a column of human CIg-derivatized Sepharose. SAM is thereby shown to be considerably enriched in CIg-binding GAG's. After thorough washing of the column, the bound material (eluted with $2M$ salt) consists almost exclusively of the highly N-sulfated sequences of heparan sulfate. A very small percentage of dermatan sulfate binds to the CIg as well. Further, purified standard preparations of heparin, dermatan sulfate, and chondroitin 4-sulfate are able to dissociate 90%, 95%, and 20% respectively of the bound cell-derived GAG when added in excess, whereas hyaluronate and chondroitin 6-sulfate could displace none of the bound radiolabeled GAG (Fig. 7). This suggests that plasma CIg differs from cellular fibronectin in that it lacks the binding site for hyaluronic acid and/or that this affinity assay, which reflects unimolecular reactions, fails to detect weak binding of hyaluronate, whereas the aggregate-forming assay of Yamada *et al.* (1980) detects many fibronectin molecules in the aggregate bound to very long sequences of hyaluronate. Both of these possibilities are supported by the inability of cell surface fibronectin to bind hyaluronate as a dimer but its ability to do so after it aggregates into multimers (Laterra and Culp, submitted for publication). The aggregated form of CIg fails to bind hyaluronate. These results suggest very important structural and functional differences between plasma and cell surface fibronectins.

VI. MOLECULAR MODEL OF CELL SUBSTRATUM
   ADHESION

A. The Cell Surface Adhesion Complex—Its Formation
   and Fate

The data summarized in the foregoing discussion have led to the formulation of a model in which the supramolecular organization of the cell surface adhesion complex (Culp, 1976b) can effect binding of the cell to a serum-coated substratum (or to the extracellular matrix *in vivo*) (Fig. 8). Considering the well-established interaction between CIg and heparin (Stathakis and Mosesson, 1977)

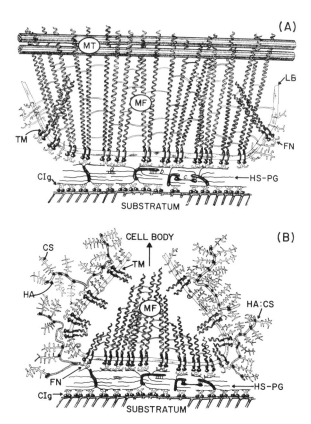

Figure 8 Prospective models of the attachment (A) and detachment (B) of fibroblasts. Identifiable components of a single footpad are shown in the newly formed adhesion site of reattaching cells (A) and in the fully matured adhesion site just before the cell breaks its connection with the footpad (B). These binding relationships are explained more fully in the text. These mechanisms may apply when the "substratum" is either a tissue culture dish *in vitro* or a collagen matrix *in vivo*. Key to abbreviations: CIg, cold-insoluble globulin in the adsorbed serum layer with bivalent sites for binding heparan sulfate chains; CS, chondroitin sulfate proteoglycan; FN, cell-surface fibronectin, which has distinguishable binding sites for heparan sulfate and hyaluronate; HA–CS, complexes composed of chondroitin proteoglycans bound to long sequences of hyaluronate; HS–PG, heparan sulfate proteoglycans, which have three potential (but unproved at this time) mechanisms of aggregation—by interaction of HS chains with each other (a), by binding of an HS chain to the core protein of another proteoglycan (b), or by cross-linkage of the core proteins (heparan sulfate chains) with a bivalent "link-type" glycoprotein (c); MF, microfilaments that are organized into bundles by cross-linkage proteins, which bind to the surface membrane by a transmembrane protein (TM) linking external fibronectin with internal MF and which may be laterally bound to the lipid bilayer (LB) membrane by some other mechanism and possibly to microtubules (MT). Not shown are the 10 nm diameter filaments and the possible binding of heparan sulfate proteoglycans to the membrane by a hydrophobic core protein sequence. It is also likely that hyaluronate chains do not bind to single dimers of fibronectin as shown in B but to aggregates of fibronectin as indicated by the data of Yamada *et al.* (1980). There also appear to be other cell surface molecules that bind hyaluronate to the cell surface (Underhill and Toole, 1979). Models adapted from Culp *et al.* (1979) with permission.

or heparan sulfate (Yamada *et al.*, 1980; Laterra *et al.*, 1980; Ruoslahti and Engvall, 1980) and considering the established role of substratum-bound CIg as an "adhesive factor" (Höök *et al.*, 1977; Grinnell and Hayes, 1978; Kleinman *et al.*, 1979; Murray *et al.*, 1980), the initial step in cell–substratum adhesion is envisaged as the binding of cell surface heparan sulfate proteoglycans to substratum-bound CIg (Fig. 8A). This proposal is also supported by (a) the striking enrichment of heparan sulfate in newly formed adhesion sites (Rollins and Culp, 1979a); (b) the aggregation of heparan proteoglycans into large complexes, which permits a high concentration of sites for binding to CIg and fibronectin in a positively cooperative manner (Rollins and Culp, 1979b; Garner and Culp, 1981); (c) the *coordinate resistance* to solubilization off the substratum of heparan proteoglycans, cellular fibronectin, and cytoskeletal proteins upon treatment of SAM with trypsin, testicular hyaluronidase, or guanidine hydrochloride (Culp *et al.*, 1978a; Cathcart and Culp, 1979b; Rollins and Culp, 1979a); (d) the enrichment in adhesion sites, particularly newly formed sites, of heparan sulfate sequences that bind CIg or cellular fibronectin (Laterra *et al.*, 1980); (e) the ability of cells to form primitive adhesion sites with the serum-coated substratum at 4°, which no doubt reflects the ability of heparan sulfate sequences in SAM to bind to CIg at this low temperature (Laterra *et al.*, 1980); and (f) the inability of large concentrations of CIg in the medium of attaching cells to inhibit initial attachment, because the unimolecular affinity between dimeric CIg and heparan sulfate chains is low and a "fixed" array of many CIg molecules bound to the extracellular matrix is required to form a stable bond between the cell surface and the matrix.

Other indications of the adhesive properties of heparan sulfate have also been described. It has recently been found (J. Laterra and L. A. Culp, unpublished data) that purified antithrombin III (AT-III) bound to tissue culture plastic does not promote physiologically compatible adherence and spreading of 3T3 and SVT2 cells; AT-III binds to heparin but *not* to the heparan sulfates or other GAG's produced by these cells. In contrast, hen egg white lysozyme which does bind to the heparan sulfates of these cells, as well as to other GAG's, promotes attachment and partial spreading of cells (Fig. 9). Also, a variant CHO cell line which detaches very slowly from its substratum was found to have three times the amount of heparan sulfate as its parental line (Barnhart *et al.*, 1979). In a completely different sort of system, heparan sulfate has been identified as the sole source of the high concentration of anionic sites in glomerular basement membranes (Kanwar and Farquhar, 1979a,b). This has been confirmed by immunochemical localization of heparan proteoglycans in basement membranes of skin, kidney, and cornea (Hassell *et al.*, 1980). Furthermore, the localization of heparan sulfate in discreet densities in the lamina rara externa suggests a role for heparan sulfate in the adhesion of glomerular epithelial footpads to the basement membrane (Caulfield and Farquhar, 1976). The presence of fibronectin in

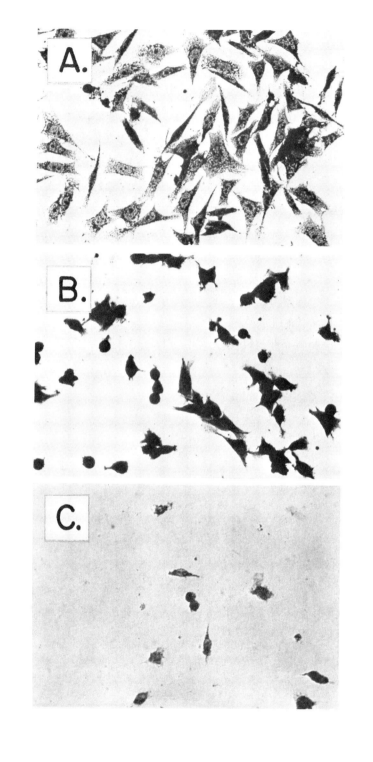

glomerular basement membrane (Linder *et al.*, 1975) is certainly consistent with this idea and provides an attractive *in vivo* reflection of the *in vitro* cell-substratum adhesion system described above. A somewhat similar mechanism of heparan sulfate–fibronectin–collagen interaction (Ruoslahti and Engvall, 1980) may also form the basis of basement membrane deposition (David and Bernfield, 1979) and subsequent organogenesis (Banerjee *et al.*, 1977) with its adhesion-directed morphogenic movements (Letourneau *et al.*, 1980).

The relationship between the cell surface heparan proteoglycans and the plasma membrane is unclear. The heparan sulfate proteoglycan might have its protein core embedded in the lipid matrix, as suggested by the detergent sensitivity to disaggregation of these components (J. A. Garner, M. Lark, and L. A. Culp, unpublished data), or the heparan sulfate could be bound to cell surface fibronectin (Fig. 8A) (Yamada *et al.*, 1980; Laterra *et al.*, 1980; Ruoslahti and Engvall, 1980). In the latter case, the heparan proteoglycans would be acting as a divalent ligand by linking the disulfide-cross-linked matrix of fibronectin on the cell surface (Hynes and Destree, 1977) to substratum-bound CIg (Fig. 8A). Consistent with this, Cohn *et al.* (1976) found cellular heparan sulfate to be resistant to *in situ* trypsin treatment, suggesting its presence as a protected and integral membrane constitutent. Culp *et al.* (1978a) showed that some 10–15% of the heparan sulfate in SAM is labile to *in situ* trypsinization, whereas the rest is highly resistant. Furthermore, some heparan sulfate is capable of being displaced from some regions of the cell surface by exogenously added heparan (Kraemer, 1977; Kjellen *et al.*, 1977) but not from SAM (L. A. Culp, unpublished data).

The issue of how the heparan sulfate is bound to the cell surface is not a trivial one, since the next phase of the adhesive process is cell spreading, and the mechanism of transmission of the signal to spread is unknown. The substratum adhesion sites are associated with a well-organized web of subcortical microfilaments (McNutt *et al.*, 1971, 1973), which are in turn transmembrane-linked to fibronectin "fibrils" (Singer, 1979). It is possible that either fibronectin or the heparan proteoglycan core protein is connected across the membrane (either directly or by an intermediate molecule) to the cytoskeleton (Fig. 8A). In fact, evidence is accumulating for transmembranous association between fibronectin

Figure 9 Adherence of murine fibroblasts to substrata coated with hen egg white lysozyme. Tissue culture dishes (35-ml) were incubated with 1 ml of serumless Eagle's medium plus 10 μg/ml CIg (A), 10 μg/ml hen egg white lysozyme free of any contaminating CIg (B), or 10 μg/ml bovine serum albumin (C). The dishes were rinsed well with PBS, fed 1 ml of protein-free Eagle's medium, and inoculated with 50,000 SVT2 cells which had been previously detached by EGTA treatment and rinsed twice with PBS. After 4 hours of attachment, the cultures were rinsed with PBS, fixed with cold 5% trichloroacetic acid, stained with Coomassie blue as described by Haas and Culp (1979), dried, and photographed on a Zeiss inverted microscope (original magnification 150X).

and actin-containing microfilaments, possibly via a transmembrane protein with binding sites for both actin and fibronectin (Ash and Singer, 1976; Mautner and Hynes, 1977; Hynes and Destree, 1978; Singer, 1979). However, there is no concrete evidence that a subset of fibronectin does not span the lipid bilayer and bind directly to actin (Keski-oja et al.. 1980). There is also evidence that the 10 nm filaments may be involved (possibly to a greater extent than actin) in the newly formed adhesion site since the ratio of the 10 nm filament subunit protein to actin is much higher in the newly formed site than in older sites (Culp, 1976a,b; Culp et al., 1978a). Thus, the heparan proteoglycan or fibronectin may be directly associated across the membrane with this cytoskeletal component. As the external molecules (heparan proteoglycans and fibronectin) become organized in the adhesion site, this organization would be reflected across the membrane by the ordering of cytoskeletal components. The cytoskeleton thus acts to reinforce the initial cell–substratum contact by inducing the cell to spread locally and increase the surface area of contact between the cell and the substratum (Rees et al., 1977). By lateral diffusion (either passive or directed by the cytoskeleton), the fibronectin and heparan proteoglycans would become focally concentrated to provide a strong adhesive interaction.

This focality of adhesion thus explains the phenomenon of footpad-mediated adhesion seen both in vitro (Harris, 1973; Revel and Wolken, 1973; Revel et al., 1974; Culp, 1975; Rosen and Culp, 1977; Vogel, 1978; Britch and Allen, 1980) and in vivo (Chernoff and Overton, 1977, 1979; Overton, 1977; Karfunkel, 1977). It is also reminscent of the focal collections of heparan sulfate in the glomerular basement membrane at sites of epithelial cell attachment (Caulfield and Farquhar, 1976). This phenomenon highlights the concept of cooperativity within the adhesion site. That is, the binding constant of a single molecule of heparan sulfate to a single molecule of CIg is minuscule compared to the binding constant achieved by a pair of two-dimensional arrays of these molecules in apposition. This would also explain why a high concentration of CIg in the culture medium fails to block binding sites for CIg in the substratum-bound serum layer, because the latter two-dimensional matrix provides a high concentration of fixed sites for binding to a weak surface receptor such as the heparan proteoglycans. Positive cooperativity and multivalent binding therefore become very important aspects of the biochemistry of adhesion. Thus, experiments designed to inhibit or enhance adhesion by the addition of exogenous heparan sulfate or univalent fragments of fibronectin or CIg will be ineffective. Not only are the aggregated heparan proteoglycans extremely multivalent (thus obviating any inhibitory action), but a single molecule binding at random must have a negligible effect on an organized "plate" of membrane literally bristling with similar or identical molecules.

The next step to consider in cell adhesion to a substratum is de-adhesion. How this may occur is suggested by the distribution of carbohydrates seen in the SAM from long-term-radiolabeled cells. During the 72-hour period of radiolabeling, cells move several cell diameters across the substratum and their SAM is enriched in abandoned footpads, i.e., footprints. The data described above (Section IV,A) show a steady accumulation of the various chondroitins and hyaluronic acid in these footprint preparations. Their accumulation in unvarying ratios may be a reflection of their presence as well-defined supramolecular complexes (see Section V,B,1), as further evidenced by the proteoglycan aggregation experiments (Rollins and Culp, 1979b; Garner and Culp, 1981). Yamada et al. (1980) have shown that hyaluronic acid can bind to fibronectin at a site on the molecule distinct from the heparin-binding site. Thus, the appearance of hyaluronate-proteoglycan complexes in the adhesion site could lead to individual complexes binding to the fibronectin and thereby destroying the highly ordered array of heparan sulfate–fibronectin complexes, as depicted in Figure 8B. The binding of chondroitin proteoglycans to localized regions of the hyaluronate chain may permit only fibronectin binding at widely spaced sites along the hyaluronate chain. Since the disulfide-cross-linked fibronectin matrix (Hynes and Destree, 1977) is functionally connected to the subcortical cytoskeleton, disruption of the external fibronectin organization will lead to disruption of the internal cytoskeletal organization (Fig. 8B). This leads to local rounding and stretching of the cell surface away from its footpad and the creation of a retraction fiber from the cell body to the footpad, the shearing of which finally leads to local separation of the cell from the substratum. The stimulus for the local accumulation of these components is uncertain, but is is clearly the focality of their accumulation which can lead to such an effect which is not reproducible by the random addition of GAG's.

There is extensive, but indirect evidence for hyaluronic acid production being associated with increased cell division, movement, and detachment (Tomida et al., 1975; Lembach, 1976; reviewed in Letourneau et al., 1980). Conversely, a cessation of growth and movement has been observed to be accompanied by decreased hyaluronic acid production (Morris, 1960; Tomida et al., 1974). As mentioned earlier, neural-derived glioma cell lines and MSV-transformed murine fibroblast cell lines show a much higher proportion of hyaluronate in their SAM, which correlates with their greater mobility and ease of detachment (Culp, 1980; Culp et al., 1980). There is a series of variant CHO cells selected for their differing ease of detachment from the tissue culture substratum. The variant most resistant to detachment synthesizes not only threefold more heparan sulfate than its parental line, but only one-half the amount of hyaluronate (Barnhart et al., 1979). Another variant, which is more easily detached, synthesizes threefold more hyaluronate than the parental line, again reflecting the negative modulatory effect of hyaluronate (Atherly et al., 1977). Finally, enzymatic digestion of hyaluronate and the chondroitins off the cell surface has only a marginal effect on the adherence and spreading of attaching fibroblasts (Bruns and Gross, 1980).

All of these data are consistent with the differing roles in cellular adhesion proposed here for heparan sulfate proteoglycans and the chondroitin proteoglycan–hyaluronate complexes.

B. Implications of the Model

There are several steps in this complex model for cellular adhesion in which adhesive dysfunction could occur. The most peripheral component of the model is actually the artificial substratum to which the cells adhere. *In vivo* correlates in which differing substrata play a role in adhesion might include the various cellular substrata over which migratory embyronic epithelia move. Although very little direct information on the chemistry of these *in vivo* substrata exists, there are a few *in vitro* correlates. Hyaluronate-derivatized substrata lead to increased cellular motility (Turley and Roth, 1979; reviewed in Letourneau *et al.*, 1980). Also, the quantity of glycosaminoglycan in SAM from cells attaching to pure CIg-coated substrata is greater than the quantity seen in SAM from cells adhering to serum-coated substrata (Murray *et al.*, 1980). This may reflect the direct heparan sulfate–CIg interaction, which would be somewhat diluted by adherence to whole serum-coated dishes.

The next organizational layer consists of the cell surface proteoglycans themselves. One would expect, according to this model, that if cells had similar GAG's they might adhere, initially, in the same way. This is certainly the case for the cell lines discussed earlier. The 3T3 and SVT2 cells have very similar proportions of GAG in their adhesion sites and do appear to adhere similarly [although subsequent cytoskeletal differences can change the strength of the connection between the cell body and the cell footpad (see below)]. In contrast, however, there is accumulating evidence that transformation of 3T3 cells by SV40 leads to increased sulfation of cell surface heparan sulfate (Underhill and Keller, 1975; Johnston *et al.*, 1979). Whether this has any effect on adhesion remains to be demonstrated, but considering the affinity of CIg or fibronectin for highly N-sulfated heparan sulfate (Laterra *et al.*, 1980) this is likely to be an important parameter.

The next structural level is that of cell surface fibronectin. Part of the initial excitement about this glycoprotein was its "apparent" absence (or presence in decreased amounts) on transformed cells (Hynes, 1976). However, most of the fibronectin in adhesion sites is inaccessible to antibody or to lactoperoxidase-catalyzed iodination (Cathcart and Culp, 1979b; Murray and Culp, 1981). If one argues along the transformation-decreased adhesion line, then decreased amounts of fibronectin would be correlated with decreased adhesiveness. Although the inverse correlation between fibronectin and transformation has proved not to be airtight, the role of fibronectin in adhesion is becoming well established

(Yamada and Olden, 1978), and as yet undefined structural alterations in fibronectin may account for alternations in adhesive behavior.

The final level of organization is the subcortical cytoskeleton with which the fibronectin is presumably associated across the membrane. According to the model described in Figure 8B, inherent instability of the cytoskeleton should lead to a weakened connection between the cell body and its adhesive footpads. This is apparently the case in many SV40 transformants of murine fibroblasts, such as the SVT2 cell line. It is clear that the membranes of transformants adhere well to the serum-coated substratum, consistent with the high levels of heparan sulfate in their footpad SAM (Rollins and Culp, 1979a). The cells themselves, however, are less adherent to the substratum than their normal counterparts. But when they are detached, for example, by a stream of medium, they leave their footpads behind still attached to the substratum. Thus, the weakened aspect of their substratum adhesion is localized at the connection between the cell body and the adhesive footpads, a function ascribed to the cytoskeleton. This functional inference is confirmed by the structural evidence that the subcortical microfilaments of SV40-transformed 3T3 cells are highly disorganized (McNutt *et al.*, 1971, 1973).

C. Adhesion in Embryogenesis and in Metastatic Disease

The adhesion model described above provides a different approach to discussing such adhesion-dependent phenomena as embryogenesis and metastasis. The movement of epithelia over the extracellular matrix separating the epithelium from mesenchyme appears morphologically to be similar to cell movement *in vitro*, i.e., the formation of stretched retraction fibers at the trailing edge of the epithelial cells, which, after shearing, leave behind footprints still adherent to the matrix (Chernoff and Overton, 1977, 1979; Overton, 1977; Karfunkel, 1977). The question arises as to how a given epithelial layer directs itself (or is directed) to a specific site. One possibility is that morphogenesis would be directed by the mesenchymal layer. Alternatively, the epithelium could direct its own movement by means of the epithelial-produced basal lamina. Bernfield's work, for example, has shown that mouse salivary gland and mammary epithelia produce a basal lamina rich in GAG's including heparan sulfate, hyaluronate, and the chondroitins (Bernfield and Banerjee, 1972; Banerjee *et al.*, 1977; Cohn *et al.*, 1976; Gordon and Bernfield, 1980). Furthermore, the epithelium appears to require this intact lamina in order to maintain its lobulated morphology (Bernfield *et al.*, 1972; Banerjee *et al.*, 1977; Gordon and Bernfield, 1980). It is interesting that the clefts in the glandular lobules, which are dependent on intact basal lamina GAG, are also dependent on intact, well-organized subcortical microfilaments (Spooner and Wessells, 1970). This is quite reminiscent of the microfilament-extracellular matrix relation described above for fibroblast adhesion in tissue culture. One wonders if the same sort of fibronectin–heparan sulfate proteoglycan interaction is occurring at established cleft sites to maintain this morphology. Subsequent hyaluronate and chondroitin accumulation at other sites would then lead to microfilament disruption, an altered morphology, and perhaps new cleft formation. Entirely consistent with this idea is the observation that the largest amount of new GAG synthesis occurs at new branching sites in the eipthelial nodule (Bernfield *et al.*, 1972).

Another area for which this model has relevance is metastatic disease. The adhesive defect in malignant cells could occur at any of the structural levels in the adhesion complex. These include the amounts of heparan proteoglycans, hyaluronate, and fibronectin made by the cell, as well as the myriad elements that control from within the cell its membrane-associated cytoskeleton. A defect in any one of these elements might generate a cell with a radically different phenotype. Moreover, an alteration of only one of these parameters might not be *necessary and sufficient* to generate tumorigenic and/or highly metastatic cells. The importance of an intact cytoskeletal apparatus and the possible generality of cytoskeletal dysfunction in metastatic cell lines are suggested by the morphological disorganization of the cytoskeleton in tumorigenic cells as a class (McNutt *et al.*, 1971, 1973; Pollack *et al.*, 1975).

The importance of adhesion in metastasis was further indicated by the fact that the more metastatic variants of B16 melanoma cells (Fidler, 1973; Fidler *et al.*, 1976) showed greater rates of adhesion to homotypic and heterotypic cell monolayers (Winkelhake and Nicolson, 1976) or cell aggregates (Raz *et al.*, 1980). This suggests again that the ability to consolidate adhesions is an important if not crucial early step in metastasis. Fidler's group has shown that the highly metastatic B16 variants possess well-organized clumps of anionic material on their surfaces (proteoglycan?) (Raz *et al.*, 1980). Anionic sites on the surface of the nonmetastatic variants are present only as diffuse, nonclumped material. These anionic clumps, stained with cationic ferritin, look similar to the anionic clumps in the renal glomerular basement membrane, which Farquhar's group stained with lysozyme and showed to be predominantly heparan sulfate (Caulfield and Farquhar, 1976, 1978).

The ability to assign functional roles to molecular species within the adhesion site opens new avenues for experimental approaches to the questions raised by adhesive specificity. The characterization GAG's and proteoglycans in the adhesion site and their association with other molecular entities in this site should shed further light on the molecular mechanisms controlling embryogenesis and metastatic disease. The binding of highly multivalent complexes of these molecules to each other in a positively cooperative manner will undoubtedly be a critical determinant in the adhesive process.

ACKNOWLEDGMENTS

The authors acknowledge support for these studies from NIH research grant CA13513, AM25646, CA27755, and NS17139; NIH Career Development Award (L.A.C.) CA70709; NIH training grants GM07225 (M.K.K.) and GM07250 (B.J.R.); and American Cancer Society grant BC217. We also appreciate the assistance of the members of our laboratory and Dr. Ben Murray (University of California, San Diego) in critically reviewing this manuscript.

REFERENCES

Albrecht-Buehler, G. (1976). *J. Cell Biol.* **69**, 275–286.
Ash, J. F., and Singer, S. J. (1976). *Proc. Natl. Acad. Sci. U.S.A.* **73**, 4575–4579.
Atherly, A. G., Barnhart, B. J., and Kraemer, P. M. (1977). *J. Cell. Physiol.* **89**, 375–385.
Bader, J. P. (1972). *J. Virol.* **10**, 267–276.
Badley, R. A., Lloyd, C. W., Woods, A., Carruthers, L., Alcock, C., and Rees, D. A. (1978). *Exp. Cell Res.* **117**, 231–244.
Badley, R. A., Woods, A., Carruthers, L., and Rees, D. A. (1980). *J. Cell Sci.* **43**, 379–390.
Banerjee, S. D., Cohn, R. H., and Bernfield, M. R. (1977). *J. Cell Biol.* **73**, 445–463.
Barnhart, B. J., Cox, S. H., and Kraemer, P. M. (1979). *Exp. Cell Res.* **119**, 327–332.
Bernfield, M. R., and Banerjee, S. D. (1972). *J. Cell Biol.* **52**, 664–673.
Bernfield, M. R., Banerjee, S. D., and Cohn, R. H. (1972). *J. Cell Biol.* **52**, 674–689.
Blackburn, M. N., and Sibley, C. C. (1980). *J. Biol. Chem.* **255**, 824–826.
Britch, M., and Allen, T. D. (1980). *Exp. Cell Res.* **125**, 221–231.
Brown, S., Levinson, W., and Spudich, J. A. (1976). *J. Supramol. Struct.* **5**, 119–130.
Bruns, R. R., and Gross, J. (1980). *Exp. Cell Res.* **128**, 1–7.
Caterson, B., and Baker, J. R. (1979). *J. Biol. Chem.* **254**, 2394–2399.
Cathcart, M. K., and Culp, L. A. (1979a). *Biochemistry* **18**, 1167–1176.
Cathcart, M. K., and Culp, L. A. (1979b). *Biochim. Biophys. Acta* **556**, 331–343.
Cauldwell, C. B., Henkart, P., and Humphreys, T. (1973). *Biochemistry* **12**, 3051–3055.
Caulfield, J. P., and Farquhar, M. G. (1976). *Proc. Natl. Acad. Sci. U.S.A.* **73**, 1646–1650.
Caulfield, J. P., and Farquhar, M. G. (1978). *Lab. Invest.* **39**, 505–512.
Chen, W. T. (1977). *J. Cell Biol.* **75**, 411a.
Chernoff, E. A. G., and Overton, J. (1977). *Dev. Biol.* **47**, 33–46.
Chernoff, E. A. G., and Overton, J. (1979). *Dev. Biol.* **72**, 291–307.
Chiarugi, V. P., Vannucchi, S., and Urbano, P. (1974). *Biochim. Biophys. Acta* **345**, 283–293.
Cohn, R. H., Cassiman, J. J., and Bernfield, M. R. (1976). *J. Cell Biol.* **71**, 280–294.
Collins, F. (1980). *Dev. Biol.* **79**, 247–252.
Culp, L. A. (1974). *J. Cell Biol.* **63**, 71–83.
Culp, L. A. (1975). *Exp. Cell Res.* **92**, 467–477.
Culp, L. A. (1976a). *Biochemistry* **15**, 4094–4104.
Culp, L. A. (1976b). *J. Supramol. Struct.* **5**, 239–255.
Culp, L. A. (1978). *Curr. Top. Membr. Transp.* **11**, 327–396.
Culp, L. A. (1980). *Nature (London)* **286**, 77–79.
Culp, L. A., and Bensusan, H. B. (1978). *Nature (London)* **273**, 680–682.
Culp, L. A., and Black, P. H. (1972a). *Biochemistry* **12**, 2161–2172.
Culp, L. A., and Black, P. H. (1972b). *J. Virol.* **9**, 611–620.
Culp, L. A., and Buniel, J. F. (1976). *J. Cell. Physiol.* **88**, 89–106.

Culp, L. A., Rollins, B. J., Buniel, J., and Hitri, S. (1978a). *J. Cell Biol.* **79**, 788–801.
Culp, L. A., Buniel, J. F., and Rosen, J. J. (1978b). *In* "Cell Surface Carbohydrate Chemistry" (R. Harmon, ed.), pp. 205–224. Academic Press, New York.
Culp, L. A., Murray, B. A., and Rollins, B. J. (1979). *J. Supramol. Struct.* **11**, 401–427.
Culp, L. A., Ansbacher, R., and Domen, C. (1980). *Biochemistry* **19**, 5899–5907.
David, G., and Bernfield, M. R. (1979). *Proc. Natl. Acad, Sci. U.S.A.* **76**, 786–790.
Defendi, V., and Gasić, G. (1963). *J. Cell. Comp. Physiol.* **62**, 23–32.
DeLuca, S., Caplan, A. I., and Hascall, V. G. (1978). *J. Biol. Chem.* **253**, 4713–4720.
Dermer, G. B., Lue, J., and Neustein, H. B. (1974). *Cancer Res.* **34**, 31–38.
Dietrich, C. P., Sampaio, L. O., Toledo, O. M. S., and Cassaro, C. M. F. (1977). *Biochem. Biophys. Res. Commun.* **75**, 329–336.
DiPasquale, A., and Bell, P. B. (1974). *J. Cell Biol.* **62**, 198–214.
Dorfman, A. (1963). *J. Histochem. Cytochem.* **11**, 2–13.
Dunham, J. S., and Hynes, R. O. (1978). *Biochim. Biophys. Acta* **506**, 242–255.
Ericksen, S., Eng, J., and Morgan, H. R. (1961). *J. Exp. Med.* **114**, 435–440.
Fidler, I. J. (1973). *Nature (London) New Biol.* **242**, 148–149.
Fidler, I. J., Gersten, D. M., and Budmen, M. B. (1976). *Cancer Res.* **36**, 3160–3165.
Folkman, J., and Moscona, A. (1978). *Nature (London)* **273**, 345–349.
Gail, M. H., and Boone, C. W. (1972). *Exp. Cell Res.* **70**, 33–40.
Garner, J. A., and Culp, L. A. (1981). *Biochemistry* **20**, in press.
Gasić, G., and Baydak, T. (1962). *In* "Biological Interactions in Normal and Neoplastic Growth" (M. J. Brennan and W. L. Simpson, eds.), Henry Ford Int. Sum., No. 12, pp. 1–14. Little, Brown, Boston, Massachusetts.
Glimelius, B., Norling, B., Westermark, B., and Wasteson, A. (1978). *Biochem. J.* **172**, 443–456.
Goggins, J. F., Johnson, G. S., and Pastan, I. (1972). *J. Biol. Chem.* **247**, 5759–5764.
Goldman, R. D., Lazarides, E., Pollack, R., and Weber, K. (1975). *Exp. Cell Res.* **90**, 333–344.
Gordon, J. R., and Bernfield, M. R. (1980). *Dev. Biol.* **74**, 118–135.
Graham, J. M., Hynes, R. O., Davidson, E. A., and Bainton, D. F. (1975). *Cell* **4**, 353–365.
Graham, J. M., Hynes, R. O., Rowlatt. C., and Sandall, J. K. (1978). *Ann. N.Y. Acad. Sci.* **312**, 221–239.
Gregory, J. D. (1973). *Biochem. J.* **133**. 383–386.
Grinnell, F. (1974). *Arch. Biochem. Biophys.* **165**, 524–530.
Grinnell, F. (1976a). *Exp. Cell Res.* **97**, 265–274.
Grinnell, F. (1976b). *Exp. Cell Res.* **102**, 51–62.
Grinnell, F. (1978). *Int. Rev. Cytol.* **53**, 63–144.
Grinnell, F., and Hays, D. G. (1978). *Exp. Cell Res.* **115**, 221–229.
Gruenstein, E., Rich, A., and Weihing, R. R. (1975). *J. Cell Biol.* **64**, 223–234.
Haas, R., and Culp, L. A. (1979). *J. Cell. Physiol.* **101**, 279–292.
Hamerman, D., Todaro, G. J., and Green, H. (1965). *Biochim. Biophys. Acta* **101**, 343–351.
Hardingham, T. E. (1979). *Biochem. J.* **77**, 237–247.
Hardingham, T. E., and Muir, H. (1972). *Biochim. Biophys. Acta* **279**, 401–405.
Harper, P. A., and Juliano, R. L. (1980). *J. Cell Biol.* **87**, 755–763.
Harris, A. (1973). *Dev. Biol.* **35**, 97–114.
Hascall, V. C. (1977). *J. Supramol. Struct.* **7**, 101–120.
Hascall, V. C., and Heinegard, D. (1974a). *J. Biol. Chem.* **249**, 4232–4241.
Hascall, V. C., and Heinegard, D. (1974b). *J. Biol. Chem.* **249**, 4242–4249.
Hascall, V. C., and Sajdera, S. W. (1969). *J. Biol. Chem.* **244**, 2384–2396.
Hascall, V. C., Oegema, T. R., Brown, J., and Caplan, A. I. (1976). *J. Biol. Chem.* **251**, 3511–3519.

Hassell, J. R., Robey, P. G., Barrach, H., Wilczek, J., Rennard, S. I., and Martin, G. R. (1980). *Proc. Natl. Acad. Sci. U.S.A.* **77**, 4494-4498.

Hedman, K., Vaheri, A., and Wartiovaara, A. (1978). *J. Cell Biol.* **76**, 748-760.

Heinegard, D., and Hascall, V. C. (1974). *J. Biol. Chem.* **249**, 4250-4256.

Henkart, P., Humphreys, S., and Humphreys, T. (1973). *Biochemistry* **12**, 3045-3050.

Höök, M., Rubin, K., Oldberg, A., Obrink, B., and Vaheri, A. (1977). *Biochem. Biophys. Res. Commun.* **79**, 726-733.

Humphreys, S., Humphreys, T., and Sano, J. (1977). *J. Supramol. Struct.* **7**, 339-351.

Hynes, R. O. (1976). *Biochim. Biophys. Acta* **458**, 73-107.

Hynes, R. O., and Destree, A. T. (1977). *Proc. Natl. Acad. Sci. U.S.A.* **74**, 2855-2859.

Hynes, R. O., and Destree, A. T. (1978). *Cell* **13**, 151-160.

Ishimato, N., Temin, H. M., and Strominger, J. L. (1966). *J. Biol. Chem.* **241**, 2052-2060.

Izzard, C. S., and Lochner, L. R. (1976). *J. Cell Sci.* **21**, 129-159.

Izzard, C. S., and Lochner, L. R. (1980). *J. Cell Sci.* **42**, 81-116.

Jacobsson, I., Backström, G., Höök, M., Lindahl, U., Feingold, D. S., Malmström, A., and Rodén, L. (1979). *J. Biol. Chem.* **254**, 2975-2982.

Jilek, F., and Hörmann, H. (1979). *Hopp-Seyler's Z Physiol. Chem.* **360**, 597-603.

Johnston, L. S., Keller, K. L., and Keller, J. M. (1979). *Biochim. Biophys. Acta* **583**, 81-94.

Juliano, R. L., and Gagalang, E. (1977). *J. Cell. Physiol.* **92**, 209-220.

Kanwar, Y. S., and Farquhar, M. G. (1979a). *Proc. Natl. Acad. Sci. U.S.A.* **76**, 1303-1307.

Kanwar, Y. S., and Farquhar, M. G. (1979b). *Proc. Natl. Acad. Sci. U.S.A.* **76**, 4493-4497.

Karfunkel, P. (1977). *Wilhelm Roux's Arch. Dev. Biol.* **181**, 31-40.

Keski-φja, J., Sen, A., and Todaro, G. J. (1980). *J. Cell Biol.* **85**, 527-533.

Kimura, J. H., Osdoby, P., Caplan, A. I., and Hascall, V. C. (1978). *J. Biol. Chem.* **253**, 4721-4729.

Kimura, J. H., Hardingham, T. E., Hascall, V. C., and Solursh, M. (1979). *J. Biol. Chem.* **254**, 2600-2609.

Kjellen, L., Oldberg, A., Rubin, C., and Höök, M. (1977). *Biochem. Biophys. Res. Commun.* **74**, 126-133.

Klebe, R. J. (1974). *Nature (London)* **250**, 248-251.

Kleinman, H. K., Silbert, J. E., and Silbert, C. K. (1975). *Connect. Tissue Res.* **4**, 17-23.

Kleinman, H. K., McGoodwin, E. B., Martin, G. R., Klebe, R. J., Fietzels, P. P., and Wooley, D. E. (1978). *J. Biol. Chem.* **253**, 5642-5646.

Kleinman, H. K., McGoodwin, E. B., Rennard, S. I., and Martin, G. R. (1979). *Anal. Biochem.* **94**, 308-312.

Knox, P., and Griffiths, S. (1979). *Exp. Cell Res.* **123**, 421-424.

Koyama, H., Tomida, M., and Ono, T. (1975). *J. Cell. Physiol.* **87**, 189-198.

Kraemer, P. M. (1971a). *Biochemistry* **10**, 1437-1444.

Kraemer, P. M. (1971b). *Biochemistry* **10**, 1445-1452.

Kraemer, P. M. (1977). *Biochem. Biophys. Res. Commun.* **78**, 1334-1340.

Kraemer, P. M., and Barnhart, B. J. (1978). *Exp. Cell Res.* **114**, 153-157.

Kraemer, P. M., and Smith, D. A. (1974). *Biochem. Biophys. Res. Commun.* **56**, 423-430.

Laterra, J., Ansbacher, R., and Culp, L. A. (1980). *Proc. Natl. Acad. Sci. U.S.A.* **77**, 6662-6666.

Lembach, K. J. (1976). *J. Cell. Physiol.* **89**, 277-288.

Letourneau, P. C., Ray, P. N., and Bernfield, M. R. (1980). *In* "Biological Regulation and Control" (R. Goldberger, ed.), pp. 339-376. Plenum, New York.

Lindahl, U., and Höök, M. (1978). *Annu. Rev. Biochem.* **47**, 385-417.

Lindahl, U., Backström, G., Jansson, L., and Hallen, A. (1973). *J. Biol. Chem.* **248**, 7234-7241.

Lindahl, U., Höök, M., Backström, G., Jacobsson, T., Riesenfeld, J., Malmström, A., Rodén L., and Feingold, D. S. (1977). *Fed. Proc., Fed. Am. Soc. Exp. Biol.* **36**, 19–24.

Lindahl, U., Backström, G., Höök, M., Therberg, L., Franssön, L. A., and Linker, A. (1979). *Proc. Natl. Acad. Sci. U.S.A.* **76**, 3198–3202.

Linder, E., Vaheri, A., Ruoslahti, E., and Wartiovaara, J. (1975). *J. Exp. Med.* **142**, 41–49.

Linker, A., and Hovingh, P. (1973). *Carbohydr. Res.* **29**, 41–62.

Linker, A., and Hovingh, P. (1975). *Biochim. Biophys. Acta* **385**, 324–333.

Linsenmayer, T. F., Gibney, E., Toole, B. P., and Gross, J. (1978). *Exp. Cell Res.* **116**, 470–474.

McClay, D. R., Gooding, L. R., and Fransen, M. E. (1977). *J. Cell Biol.* **75**, 56–66.

McNutt, N. S., Culp, L. A., and Black, P. H. (1971). *J. Cell Biol.* **50**, 691–708.

McNutt, N. S., Culp, L. A., and Black, P. H. (1973). *J. Cell Biol.* **56**, 412–428.

Makita, A., and Shimojo, H. (1973). *Biochim. Biophys. Acta* **304**, 571–574.

Mapstone, T. B., and Culp, L. A. (1976). *J. Cell Sci.* **20**, 479–495.

Mathews, M. B. (1971). *Biochem. J.* **126**, 257–273.

Mathews, M. B., and Decker, L. (1968). *Biochem. J.* **109**, 517–526.

Mautner, V., and Hynes, R. O. (1977). *J. Cell Biol.* **75**, 743–768.

Morris, C. C. (1960). *Ann. N.Y. Acad. Sci.* **86**. 179–187.

Murray, B. A., and Culp, L. A. (1981). *Exp. Cell Res.* **131**, 237–249.

Murray, B. A., Ansbacher, R., and Culp, L. A. (1980). *J. Cell. Physiol.* **104**, 335–348.

Nath, K., and Srere, P. A. (1977). *J. Cell. Physiol.* **92**, 33–42.

Norling, B., Glimelius, B., Westermark, B., and Wasteson, A. (1978). *Biochem. Biophys. Res. Commun.* **84**, 914.-921.

Oldberg, A., Kjellen, L., and Höök, M. (1979). *J. Biol. Chem.* **254**, 8505–8510.

Olivercrona, T., Bergtsson, G., Marklund, S., Lindahl, U., and Höök, M. (1977). *Fed. Proc., Fed. Am. Sco. Exp. Biol.* **36**, 60–65.

Overton, J. (1977). *Exp. Cell Res.* **105**, 313–323.

Overton, J. (1979). *Tissue Cell* **1**, 89–98.

Pearlstein, E. (1976). *Nature (London)* **262**, 497–500.

Perdue, J. F. (1973). *J. Cell Biol.* **58**, 265–283.

Perkins, M. E., Ji, T. H., and Hynes, R. O. (1979). *Cell* **16**, 941–952.

Pessac, B., and Defendi, V. (1972). *Science* **175**, 898–900.

Pollack, R., Osborn, M., and Weber, K. (1975). *Proc. Natl. Acad. Sci. U.S.A.* **72**, 994–998.

Poste, G., and Fidler, I. J. (1980). *Nature (London)* **283**, 139–146.

Pratt, R. M., Yamada, K. M., Olden, K., Ohaman, S. H., and Hascall, V. C. (1979). *Exp. Cell Res.* **118**, 245–252.

Raz, A., Bucana, C., McLellan, W., and Fidler, I. J. (1980). *Nature (London)* **284**, 363–364.

Rees, D. A., Lloyd, C. W., and Thom, D. (1977). *Nature (London)* **267**, 124–128.

Revel, J. P., and Wolken, K. (1973). *Exp. Cell Res.* **78**, 1–14.

Revel, J. P., Hoch, P., and Ho, D. (1974). *Exp. Cell Res.* **84**, 207–218.

Roblin, R., Albert, S. O., Gelb, N. A., and Black, P. H. (1975). *Biochemistry* **14**, 347–357.

Rollins, B. J., and Culp, L. A. (1979a). *Biochemistry* **18**, 141–148.

Rollins, B. J., and Culp, L. A. (1979b). *Biochemistry* **18**, 5621–5629.

Rosen, J. J., and Culp, L. A. (1977). *Exp. Cell Res.* **107**, 139–149.

Rosenberg, L., Hellmann, W., and Kleinschmidt, A. K. (1975). *J. Biol. Chem.* **250**, 1877–1883.

Rosenberg, R. D. (1978). *Annu. Rev. Med.* **29**, 367–378.

Rosenberg, R. D., and Damus, P. E. (1973). *J. Biol. Chem.* **245**, 6490–6505.

Rosenberg, R. D., and Lam, L. (1979). *Proc. Natl. Acad. Sci. U.S.A.* **76**, 1218–1222.

Ruoslahti, E., and Engvall, E. (1980). *Biochim. Biophys. Acta* **631**, 350–358.

Saito, H., and Uzman, B. (1971). *Biochem. Biophys. Res. Commun.* **43**, 723–728.

Saito, H., Yamagita, T., and Suzuki, S. (1968). *J. Biol. Chem.* **243**, 1536–1542.

Sato, C. S., and Gyorkey, F. (1974). *Anal. Biochem.* **61,** 305–310.

Satoh, C., Duff, R., Rapp, F., and Davidson, E. A. (1973). *Proc. Natl. Acad. Sci. U.S.A.* **70,** 54–56.

Shields, R., and Pollock, K. (1974). *Cell* **3,** 31–38.

Singer, I. I. (1979). *Cell* **16,** 675–685.

Spooner, B. S., and Wessells, N. K. (1970). *Proc. Natl. Acad. Sci. U.S.A.* **66,** 360–365.

Stathakis, N. E., and Mosesson, M. W. (1977). *J. Clin. Invest.* **60,** 855–865.

Takeichi, M. (1971). *Exp. Cell Res.* **68,** 88–96.

Tang, L. H., Rosenberg, L., Reiner, A., and Poole, A. R. (1979). *J. Biol. Chem.* **254,** 10523–10531.

Temin, H. M. (1965). *J. Natl. Cancer Inst.* **35,** 679–694.

Terry, A. H., and Culp, L. A. (1974). *Biochemistry* **13,** 414–425.

Thom, D., Powell, A. J., and Rees, D. A. (1979). *J. Cell Sci.* **35,** 281–305.

Tomida, M., Koyama, H., and Ono, T. (1974). *Biochim. Biophys. Acta* **338,** 352–363.

Tomida, M., Koyama, H., and Ono, T. (1975). *J. Cell Physiol.* **86,** 121–130.

Toole, B. P. (1976). *J. Biol. Chem.* **251,** 895–897.

Toole, B. P., and Lowther, D. A. (1968a). *Biochem. J.* **109,** 857–866.

Toole, B. P., and Lowther, D. A. (1968b). *Arch. Biochem. Biophys.* **128,** 567–576.

Turley, E. A., and Roth, S. (1979). *Cell* **17,** 109–115.

Turner, R. S., and Burger, M. M. (1973). *Nature (London)* **244,** 509–510.

Underhill, C. B., and Dorfman, A. (1978). *Exp. Cell Res.* **117,** 155–164.

Underhill, C. B., and Keller, J. M. (1975). *Biochem. Biophys. Res. Commun.* **63,** 448–454.

Underhill, C. B., and Toole, B. P. (1979). *J. Cell Biol.* **82,** 475–484.

Vessey, A. R., and Culp, L. A. (1978). *Virology* **86,** 556–561.

Vogel, K. G. (1978). *Exp. Cell Res.* **113,** 345–357.

Vogel, K. G., and Kelley, R. O. (1977). *J. Cell. Physiol.* **92,** 469–480.

Weber, M. J., Hale, A. H., and Losasso, L. (1977). *Cell* **10,** 45–51.

Weinbaum, G., and Burger, M. M. (1973). *Nature (London)* **244,** 510–512.

Weiss, L., Poste, G., Mackearnin, A., and Willett, K. (1975). *J. Cell Biol.* **64,** 135–145.

Willingham, M. C., Yamada, K. M., Yamada, S. S., Pouysségur, J., and Pastan, I. (1977). *Cell* **10,** 375–380.

Winkelhake, J. L., and Nicolson, G. L. (1976). *J. Natl. Cancer Inst.* **56,** 285–291.

Winterbourne, D. J., and Mora, P. T. (1977). *J. Supramol. Struct.* **7,** 91–100.

Winterbourne, D. J., and Mora, P. T. (1978). *J. Biol. Chem.* **253,** 5109–5120.

Yamada, K. M., and Olden, K. (1978). *Nature (London)* **275,** 179–184.

Yamada, K. M., Kennedy, D. W., Kimata K., and Pratt, R. M. (1980). *J. Biol. Chem.* **255,** 6055–6063.

Yamagata, T., Saito, H., Habucki, O., and Suzuki, S. (1968). *J. Biol. Chem.* **243,** 1523–1535.

Yurt, R. W., Leid, R. W., Jr., Rusten, K. F., and Silbert, J. E. (1977). *J. Biol. Chem.* **252,** 518–521.

SECTION 2

Biochemistry of Fibronectin

KENNETH M. YAMADA

I. INTRODUCTION

The glycoprotein fibronectin (*fibre* = fiber + *nectere* = to bind, tie) is thought to play important roles in a variety of cellular events, including adhesion, phagocytosis of colloids, embryonic development, and possibly malignancy. Each of these events involves the cell surface or the extracellular matrix, of which fibronectin is frequently a prominent constituent. Studies of purified fibronectin and its interactions with cells, as well as of its localization *in vivo*, have

331

THE GLYCOCONJUGATES, VOL. III

provided insights into the possible molecular mechanisms of these cellular and cell surface events.

Here, we will first review the properties of fibronectin and its biological activities and will conclude with an analysis of its mechanisms of action at the biochemical level. A number of other reviews on fibronectin and related subjects have been published recently (Hynes, 1976; Mosesson, 1977; Yamada and Olden, 1978; Grinnell, 1978; Vaheri and Mosher, 1978; Hynes *et al.*, 1979; K. M. Yamada *et al.*, 1980a; Ruoslahti *et al.*, 1980; Pearlstein *et al.*, 1980; Mosher, 1980; Mosesson and Amrani, 1980; Saba and Jaffe, 1980; Yamada, 1981).

A. Types of Fibronectin

Fibronectin exists in at least two forms: *plasma* fibronectin and *cellular* fibronectin (Yamada and Olden, 1978; Vaheri and Mosher, 1978; Hynes *et al.*, 1979; Mosher, 1980; Pearlstein *et al.*, 1980). These two types of fibronectin are very similar in molecular properties (Table I). A third form of fibronectin is found in amniotic fluid and is synthesized by amniotic cells. It is heavily glycosylated in comparison to cellular and plasma fibronectins, and its classification is currently uncertain (Crouch *et al.*, 1978; Ruoslahti *et al.*, 1981). Plasma and cellular fibronectins have nearly identical amino acid compositions and secon-

TABLE I
Molecular Properties of Fibronectin[a]

Property	Cellular fibronectin	Plasma fibronectin
Monomer molecular weight	220,000–240,000	215,000–220,000
Subunit structure	Disulfide-linked dimers and multimers	Disulfide-linked dimers
Sedimentation constant	8.5 (pH 8) 7.6 (pH 11)	12–14 (pH 7) 8.0 (pH 11)
Stokes radius	110 Å (pH 11)	—
Partial specific volume	0.72	0.72
Frictional ratio	2.8–2.9 (pH 11)	2.8 (pH 11); 1.7 (pH 7)
Isoelectric point	5.5–6.2(?)	5.5–6.2
Extinction coefficient at 280 nm ($E_{mg/ml}^{1\ cm}$)	1.2	1.28
Circular dicroism (far UV), minimum/maximum	212 nm/227 nm	213 nm/227 nm
Solubility	Insoluble (pH 7); soluble (pH 11)	Soluble (pH 7)

[a] Data from Mosesson and Umfleet (1970), Mosesson *et al.* (1975), Yamada *et al.* (1977a), Vuento *et al.* (1977), and Alexander *et al.* (1978, 1979).

TABLE II
Comparison of Biological Activities of
Cellular and Plasma Fibronectins

Biological activity	Relative specific activity[a]
Cell attachment to collagen[b]	1
Cell spreading on plastic[b,c]	1
Opsonic activity with macrophages[d]	1
Hemagglutination[b]	150–200
Restoration of alignment to transformed cells[b]	50
Attachment of transformed cells to plastic[e]	2–3

[a] Plasma fibronectin as 1.
[b] Yamada and Kennedy (1979).
[c] Pena and Hughes (1978).
[d] Marquette et al. (1981).
[e] Hynes et al. (1978).

dary and tertiary polypeptide structures as determined by spectrophotometric criteria (Yamada et al., 1977a; Vuento et al., 1977; Alexander et al., 1979).

Moreover, nearly all antibodies raised against either type of fibronectin cross-react completely with the other type within the same species, and each antigen completely absorbs out reactivity with the other (Ruoslahti and Vaheri, 1975). However, it is now possible to raise monoclonal antibodies that are specific only for cellular fibronectin. These antibodies are produced by hybridomas after immunization with cellular fibronectin; although many hybridoma clones react with both forms of fibronectin, some do not. This type of reagent should make it possible to determine the type of fibronectin that is present in various *in vivo* locations (Noonan et al., 1981; Atherton et al., 1981).

Cellular and plasma fibronectins also differ in certain biological activities (Table II), in apparent size, and in substructure of the molecules (see Section IV). These multiple differences suggest that there is more than one fibronectin gene or type of fibronectin messenger RNA.

Plasma fibronectin is a soluble glycoprotein present in blood at a concentration of approximately 0.3 mg/ml. Related molecules of similar sizes that bind to denatured collagen have been reported in mammals, birds, and fish (Mosesson et al., 1975; Engvall et al., 1978).

Cellular fibronectin is produced by many types of cells cultured *in vitro* (Table III). It is found on the cell surface, in the extracellular matrix surrounding the cell, and as a secreted protein in tissue culture medium (reviewed by Vaheri and Mosher, 1978; Yamada and Olden, 1978; Hynes et al., 1979). On early passage fibroblasts, cell surface fibronectin constitutes 1–3% of total cellular

TABLE III
Cells Synthesizing Fibronectin[a]

Fibroblasts
Endothelial cells
Macrophages
Hepatocytes
Epidermal cells
Amniotic cells
Myoblasts
Chondrocytes (*in vitro*)

[a] See reviews listed in Section I for details and possible additional examples.

protein and therefore probably constitutes over half of the protein associated with the plasma membrane (Yamada *et al.*, 1977b). Unlike plasma fibronectin, cellular fibronectin is relatively insoluble unless it is maintained at an alkaline pH, e.g., pH 11. The purified molecule readily self-aggregates at physiological pH, and it adheres readily to a variety of substrates (Hynes *et al.*, 1976; Yamada *et al.*, 1977a).

B. Location of Fibronectin

Fibronectin is found *in vivo* in a variety of mesenchymal and connective tissues, in extracellular spaces, and in association with many basement membranes (Linder *et al.*, 1975; Stenman and Vaheri, 1978). It is also present in blood, amniotic fluid, and cerebrospinal fluid (Mosesson and Umfleet, 1970; A. B. Chen *et al.*, 1976; Kuusela *et al.*, 1978). During embryonic development, the quantity of fibronectin can change in certain tissues; these developmentally regulated alterations and their possible significance in differentiation are discussed in Section II,G.

Fibronectin is often associated with basement membranes underlying a variety of epithelia. Although it is prominent in the basement membranes of blood vessels and the kidney mesangium, fibronectin is apparently present in only low amounts in the normal kidney glomerular basement membrane (Stenman and Vaheri, 1978; Quaroni *et al.*, 1978; Bray, 1978; Madri *et al.*, 1980; Courtoy *et al.*, 1980; Gospodarowicz and Tauber, 1980).

One important uncertainty in these *in vivo* immunological localization studies is that the form of fibronectin in each tissue or structure is not known, because the antibodies previously used to detect fibronectin could not distinguish between plasma and cellular fibronectin. A solution to this problem should be provided by the use of monoclonal antibodies from hybridomas that recognize cellular, but not plasma, fibronectin.

Studies of cells placed into tissue culture have identified interesting variations in the location of fibronectin on different cell types. The cellular fibronectin

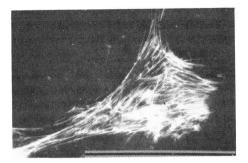

Figure 1 Cellular fibronectin on a fibroblast. Fibronectin was localized on this chick embryo fibroblast in tissue culture by immunofluorescence using goat anti-chicken fibronectin antibodies labeled with fluorescein. Note the fibrillar arrangement of fibronectin on the cell surface. Bar = 50 μm.

associated with fibroblasts is initially found underneath and in between cells (Fig. 1); this localization suggests a role in cell–substrate or cell–cell adhesion (L. B. Chen *et al.*, 1976; Mautner and Hynes, 1977; Yamada, 1978; Hedman *et al.*, 1978; Furcht *et al.*, 1978; and Chapter 3, Section 1, this volume.) As cells become confluent, the fibronectin becomes organized into an extracellular matrix

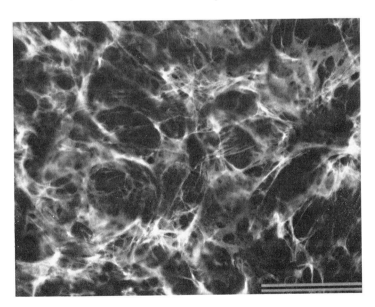

Figure 2 Fibronectin in the extracellular matrix. Fibronectin was localized by immunofluorescence using antifibronectin antibodies labeled with fluorescein. Cellular fibronectin is present as fibrils in the extracellular matrix of this densely confluent culture of chick embryo fibroblasts. Bar = 50 μm.

that completely surrounds the cells (Fig. 2). In chick fibroblast cultures, the formation of this matrix can reportedly occur without collagen, although these two proteins do codistribute in other cell types (Chen *et al.*, 1978; Vaheri *et al.*, 1978; Furcht *et al.*, 1980). In contrast, the fibronectin of cultured endothelial cells is strictly restricted to their basal surfaces, in a structure that resembles a basement membrane (Gospodarowicz and Tauber, 1980, and references therein).

In all of these locations *in vitro,* cellular fibronectin exists as aggregates and fibrils. This aggregated supramolecular organization might be expected from its known insolubility under physiological conditions. Fibronectin *in vivo* also exists in fibrillar structures.

It is of interest that minute amounts of the plasma fibronectin present in serum that is used in culture media can also be found in fibrillar arrangements on the cell surface (Hayman and Ruoslahti, 1979). It is not known whether such plasma fibronectin forms fibrils by means of binding interactions with other molecules such as collagen or because of the formation of complexes with preexisting cellular fibronectin. The latter mechanism of fibronectin binding to the cell surface has been demonstrated by adding cellular fibronectin labeled with the fluorescent dye fluorescein to cells. The labeled fibronectin binds to preexisting fibronectin fibrils rather than to new sites (Schlessinger *et al.*, 1977).

II. BIOLOGICAL ACTIVITIES OF FIBRONECTIN

The functions of fibronectin have been examined in a variety of *in vitro* biological assays. A central activity appears to be that of mediating cellular

TABLE IV
Biological Activities of Fibronectin

I. Cell–cell aggregation
   A. Erythrocytes (fixed sheep and trypsinized human)
   B. Living cells (BHK and chicken embryo)
II. Cell–substrate adhesion
   A. Cell attachment to collagen
   B. Cell attachment to fibrinogen
   C. Cell spreading on plastic or glass
III. Reconstitution of normal phenotype to transformed cells
   A. Cell alignment and overlapping
   B. Cell morphology
   C. Cell surface (microvilli, ruffles, and blebs)
   D. Cytoskeleton (microfilament bundles)
IV. Cell motility and chemotaxis
V. Nonimmune opsonic activity
VI. Inhibition or stimulation of differentiation
   A. Inhibition of chondrogenesis and myoblast fusion
   B. Stimulation of neural crest adrenergic phenotype

TABLE V
Materials Bound by Fibronectin

Collagen/gelatin
Fibrinogen/fibrin
Cells and gangliosides
Heparin and heparan sulfate
Hyaluronic acid
Transglutaminase substrates
Bacteria (*Staphylococcus aureus*)
Actin
DNA

adhesion, as well as other effects that may be related to its binding activity to the cell surface or to a variety of carbohydrate-containing macromolecules such as collagen or heparin (Tables IV and V).

A. Cell–Cell Adhesion

Cellular fibronectin can promote the adhesiveness of cells to other cells. Living cells that have been dissociated mechanically or by trypsin can be placed into assays that measure aggregation or attachment to a monolayer of cells. Purified cellular fibronectin increases adhesion in these assays (Yamada *et al.*, 1978).

A simpler model system for cell–cell adhesion is the aggregation of fixed erythrocytes. Formalin-fixed sheep erythrocytes or glutaraldehyde-fixed, trypsinized human erythrocytes are readily agglutinated by cellular fibronectin but only poorly by plasma fibronectin (Yamada *et al.*, 1975; Vuento, 1979; Yamada and Kennedy, 1979). This erythrocyte-agglutinating activity is lectin-like, with half-maximal agglutination observable at 1 μg/ml fibronectin. Although monosaccharides and disaccharides are relatively ineffective in inhibiting agglutination (e.g., only at concentrations of 25 mM), the more complex oligosaccharides of gangliosides, e.g., GTlb, are effective at 20 μM, and intact gangliosides are effective at as low as 1 μM. Fibronectin might therefore be considered to be a type of vertebrate lectin, although possessing an unusual and probably not absolute hapten specificity (Yamada *et al.*, 1981).

The mechanisms of the effects of fibronectin on cell–cell adhesion are not yet known. Fibronectin does not appear to be present in highly specialized cell–cell contact regions according to immunoelectron microscopic localization studies (Chen and Singer, 1981). Although it is in cell–cell contact areas of epithelial cells *in vitro*, it is absent *in vivo* (Quaroni *et al.*, 1978). Since fibronectin is also decreased or absent in the specialized focal contact sites at which actin-containing microfilament bundles of fibroblasts attach to substrates (Birchmeier *et al.*, 1980; Chen and Singer, 1980; Fox *et al.*, 1980), yet is present in other "close"-contact regions (Fox *et al.*, 1980), the molecule may be acting over intermediate distances to cause a general increase in adhesion. One interpretation is that it acts via cell–extracellular matrix–cell interactions in aggregating cells, and by cell–extracellular matrix–substrate interactions in adhesion to substrates.

B. Cell–Substrate Adhesion

The activity of fibronectin in mediating cell–substrate adhesion has been documented extensively. Two general areas of research have concerned cell–collagen adhesion and the adhesion and spreading of cells on plastic tissue culture materials.

1. Cell Attachment to Collagen

Fibronectin stimulates the attachment of fibroblastic cells to collagen types I, II, III, and IV (Klebe, 1974; Pearlstein, 1976; Kleinman *et al.*, 1978b). In these assays, purified collagen is adsorbed as a thin film on dishes, and the requirement for exogenous factors in cell adhesion to the collagen is determined. Cellular and plasma fibronectins are equally active for cell attachment to type I collagen (Yamada and Kennedy, 1979). Although a variety of cell lines can utilize fibronectin for attachment, chondrocytes and at least several epithelial cell types do not appear to respond to fibronectin. Instead, the latter cells can utilize the glycoproteins chondronectin and laminin, respectively (Hewitt *et al.*, 1980; Terranova *et al.*, 1980; Vlodavsky and Gospodarowicz, 1980).

Fibronectin can mediate cell attachment to collagen by directly binding to specific sites, e.g., between residues 757 and 791 on the $\alpha 1(I)$ collagen chain. This site is also the site recognized by mammalian collagenases, and it contains a unique amino acid sequence that may destabilize the collagen triple helix (Kleinman *et al.*, 1978a; Dessau *et al.*, 1978a).

Fibronectin appears to bind first to the collagen and then to cells. Fibronectin can be incubated with collagen and the excess unbound material washed away; the bound fibronectin alone readily mediates attachment of cells (Klebe, 1974). Reversing the order of plasma fibronectin addition, i.e., preincubating with cells and then washing, results in little or no cell attachment. These findings suggest either that the binding of fibronectin to collagen causes some activation of its capacity to bind to cells (Pearlstein, 1978) or, perhaps more likely, that it immobilizes the fibronectin to form a complex with multiple cell-binding sites that is much more effective in binding cells than a single site. Support for this notion comes from experiments with cellular fibronectin, which readily self-aggregates and can interact with cells in the absence of substrate, although probably with much lower efficiency (Yamada, 1980, and unpublished data). Cellular and plasma fibronectins act by means of specific binding sites for collagen and for cells (see Section III,C).

Although fibronectin can bind to native collagen (Furcht *et al.*, 1980; Kleinman *et al.*, 1981), it binds most effectively to denatured collagen, in which the triple helix has been disrupted (Engvall and Ruoslahti, 1977; Jilek and Hörmann, 1978; Engvall *et al.*, 1978). In addition, when care is taken to avoid denaturation of collagen substrates, cells show a decreased dependence on fibronectin for attachment. However, fibronectin still increases the extent or strength of attachment to these more native substrates (Linsenmayer *et al.*, 1978; F. Grinnell, personal communication; cf. Grinnell and Minter, 1978).

The possible role of fibronectin in platelet adhesion is currently controversial. Some laboratories have found that fibronectin is utilized in some cases in platelet spreading on collagen but may not be required for attachment or degranulation, whereas others have demonstrated significantly increased platelet attachment to collagen via exogenously added fibronectin (Bensusan *et al.*, 1978; Hynes *et al.*, 1978; Santoro and Cunningham, 1979; Grinnell *et al.*, 1980; Koteliansky *et al.*, 1981). A possible reason for these discrepancies is the difference in conditions, e.g., whether the collagen utilized was completely native or variably denatured, and whether calcium was present in the incubation mixtures.

An interesting report claims the identification of a human mutation in fibronectin–platelet interaction that can be corrected by adding normal fibronectin to the otherwise normal platelets (Arneson *et al.*, 1980). A further uncertainty in this area concerns the role of endogenous platelet fibronectin, which is stored in α granules and is released and bound onto the platelet surface after activation (Ginsberg *et al.*, 1980). Clearly, there is a need for many more studies of the role of fibronectin in platelet function.

2. Cell Spreading on Plastic

The attachment and spreading of a variety of cells on substrates in tissue culture often involves fibronectin, although it is likely that other factors are also involved (see Chapter 3, Section 1, this volume). For example, a number of cell types utilize plasma fibronectin derived from serum in culture media for normal spreading of their cytoplasm during cell attachment. As described for cell attachment to collagen, the fibronectin can be pre-attached to plastic substrates to mediate cell spreading. Cell spreading does not require the collagen-binding site of fibronectin, and it can be mediated only by the cell-binding portion of the molecule (see Section III,C,5).

Quantitative estimates suggest that approximately 45,000 fibronectin molecules are needed to mediate the spreading of a single cell (Hughes *et al.*, 1979). Some fibroblastic cells do not require fibronectin in order to attach and spread normally in serum-free medium. These cells appear to do so by synthesizing large amounts of their own fibronectin, and they can even promote the spreading of adjacent cells lacking fibronectin (Grinnell, 1978; Grinnell and Feld, 1979). Comparisons of the capacity of cellular and plasma fibronectins to mediate this biological activity show that they have identical activity (Pena and Hughes, 1978; Yamada and Kennedy, 1979).

The cell attachment and spreading activities of fibronectin have proved to be useful for the cultivation of certain cells in the complete absence of serum. Plasma fibronectin is precoated onto the plastic substrate before the addition of cells and serum-free medium, and it promotes the spreading and subsequent growth of cell lines such as rat follicular cells and others (Orly and Sato, 1979; Rizzino and Crowley, 1980).

C. Reconstitution and Inhibition Experiments

The roles of fibronectin in cell behavior have been examined in experiments in which fibronectin is added back to cells from which it is missing or in which monospecific antibodies are added to cells already containing fibronectin in an attempt to inactivate it. A number of transformed or tumor cells lack fibronectin (reviewed in Hynes, 1976; Yamada and Olden, 1978; Vaheri and Mosher, 1978). The addition of purified fibronectin to the culture medium of these cells is followed by a spontaneous reattachment of fibronectin to the cells (in one study, 15–20% of added fibronectin was reattached). The fibronectin organizes into normal-appearing cell-surface fibrils that can even be redistributed and capped normally by the addition of antifibronectin antibodies. The binding of fibronectin results in striking alterations in a number of aspects of cell behavior, as described below (Yamada and Weston, 1975; Yamada et al., 1976a,b; Willingham et al., 1977; Ali et al., 1977; Yamada, 1978; Chen et al., 1978).

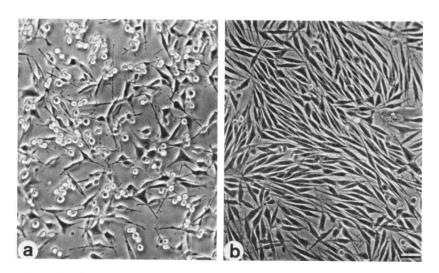

Figure 3 Effect of reconstituting fibronectin on transformed cells. Normal rat kidney cells transformed by Kirsten sarcoma virus (a) were treated for 48 hours with 50 μg/ml purified cellular fibronectin. Note the return to a more normal fibroblastic morphology, including a flatter, more bipolar shape with parallel alignment of adjacent cells (b). Bar = 50 μm.

Fibronectin-reconstituted cells tend to flatten and to become more normal and fibroblastic in appearance (Fig. 3; Table VI). In approximately half of the cell lines examined, one aspect of cell–cell social behavior is also restored to normal. Both normal and successfully reconstituted cells display a parallel alignment of cell bodies when cultures become dense; this process is thought to be due to a combination of side-to-side cell adhesion, adequate cell–substratum adhesion, and cellular response to the presence of another cell (Abercrombie, 1979; Bell, 1977; summarized in Yamada *et al.*, 1978). In contrast, many cells transformed by tumor viruses or carcinogens lose this ability to align and instead lie randomly or on top of one another *in vitro* (Abercrombie, 1979; Bell, 1977).

Besides altering cell shape, fibronectin reconstitution affects cell-surface architecture. The number of plasma membrane projections, such as microvilli, ruffles, or blebs, is substantially reduced to levels characteristic of "normal" cells (Yamada *et al.*, 1976b).

Finally, fibronectin treatment also affects the cytoskeleton. Transformed or protease-treated cells lacking fibronectin often display an altered organization of

TABLE VI
Effects of Fibronectin Reconstitution on Transformed Cells *in Vitro*[a]

Cell type[b]	Alignment	Morphology change
Chicken		
Schmidt–Ruppin RSV chick fibroblasts	+	+ +
Bryan high-titer RSV chick fibroblasts	−	+
Mouse		
L929	−	+ +
SV40 Balb/c 3T3 (SVT2)	+ +	+ +
SV40 Balb/c 3T3 tumor (SV1)	+ +	+ +
Harvey sarcoma virus Balb/c 3T3	−	+ +
Kirsten sarcoma virus Balb/c 3T3	+	+
Methylcholanthrene Balb/c 3T3	+	+ +
Rat		
Kirsten sarcoma virus NRK	+ +	+ +
Harvey sarcoma virus NRK	−	+ +
Moloney sarcoma virus NRK	−	+ +
Schmidt–Ruppin RSV NRK	−	+
Hamster		
Hamster sarcoma virus NIL	+ +	+ +
Chinese hamster ovary (CHO)	+	+ +
Human		
SV40 WI38 (VA13)	−	+

[a] + +, substantial effect, similar to that shown in Figure 3; +, moderate effect; −, questionable or undetectable effect. For documentation, see Yamada *et al.* (1976a,b, 1978) and Ali *et al.* (1977).
[b] RSV, Rous sarcoma virus; NRK, normal rat kidney.

microfilament bundles. There are fewer bundles visible by immunofluorescence microscopy and, in some cases, a loss of membrane-associated microfilaments (Goldman *et al.*, 1976). In other cases, the number of microfilaments is not detectably changed, although the presence of immunofluorescence changes implies some change in organization. In four transformed cell lines to which fibronectin had been added, there was a return of microfilament bundle organization as detected by immunofluorescence microscopy (Ali *et al.*, 1977; Willingham *et al.*, 1977). This increase was attributed to a general increase in cell–substrate adhesion (Willingham *et al.*, 1977). Since fibronectin is difficult to find in focal adhesion plaques, it seems unlikely that it acts directly to attach the ends of microfilaments to substrates (but also see Singer, 1979).

Fibronectin reconstitution on transformed cells does not restore normal growth rates, saturation density, or nutrient transport rates, strongly suggesting that it is not involved in these events (Yamada *et al.*, 1976a; Yamada and Pastan, 1976; Ali *et al.*, 1977).

The converse experiment involves treating cells that possess cell-surface fibronectin with antifibronectin antibodies or Fab fragments. In both cases, there is a rapid, reversible change in the cells such that their appearance is similar to that of transformed cells lacking fibronectin (Yamada, 1978).

This experiment and the preceding studies indicate that this glycoprotein plays a major role in a number of aspects of normal cell behavior and morphology. However, it should also be emphasized that other fibroblast adhesive mechanisms exist (e.g., Wylie *et al.*, 1979; Urushihara and Takeichi, 1980) and that, within at least a tenfold range of concentration, fibronectin can be decreased on the surface of various cell lines without major effects on cell behavior (Yamada *et al.*, 1977b).

D. Fibronectin and Cancer

The amount of fibronectin is often decreased on the surfaces of transformed cells, i.e., cells that have been altered in behavior by tumor viruses or other agents and that display abnormal *in vitro* growth rates. The amount of fibronectin is also decreased on most tumorigenic cells, i.e., transformed cells that can cause tumors when injected into compatible host animals. However, there are well-documented exceptions to these generalizations. In addition, fibronectin reconstitution cannot restore normal growth control to these cells (reviewed by Yamada and Olden, 1978; Vaheri and Mosher, 1978; Hynes *et al.*, 1979). It is therefore highly unlikely that a loss of fibronectin is a direct cause of all facets of the transformed phenotype.

Fibronectin is also often absent from certain tumors and from metastatic cells (Chen *et al.*, 1979; Smith *et al.*, 1979). One hypothesis currently under investigation is that, for some cells, the decrease in fibronectin levels may promote metastasis. However, an alternative viewpoint is that these decreases in fibronectin relate to transformation or malignancy only secondarily and that they are a result of the diversion of the metabolic or receptor capacity of the cell from synthesis of this extracellular molecule to other activities. This area needs further work, since little effort has been made to compare the effects of experimental changes in fibronectin metabolism on tumorigenicity or invasiveness, e.g., by mutants or by recombinant DNA experiments. In addition, there is a need for careful studies of human materials using monoclonal antibodies specific for cellular fibronectin.

E. Motility and Chemotaxis

Fibronectin increases the rate or directionality of fibroblastic cell migration, whether the motility of cells with a mutation affecting adhesion (Pouysségur et al., 1977), migration of single cells (Ali and Hynes, 1978), or migration away from an aggregate of cells (Yamada et al., 1978). These effects may be due to an increase in cell–substrate traction, although other effects such as increased polarity of the cytoskeleton may also contribute.

Fibronectin is also chemotactic for fibroblasts. Fibronectin placed in one well of a Boyden chamber promotes the directional migration of cells across a nitrocellulose filter (Gauss-Müller et al., 1980). Both cellular and plasma fibronectins have this chemotactic activity. Studies with fragments of fibronectin revealed that the chemotactic activity is localized to the cell-binding part of the molecule and that the collagen-binding portion is devoid of activity (Postlethwaite et al., 1981; Seppa et al., 1981). The mechanism of this "chemotactic" activity requires further study.

It is possible that the effects of fibronectin on cell migration and its chemotactic activity could be involved in wound healing, since fibronectin accumulates in clots and would presumably be attacked by the proteases released in these sites (Seppa et al., 1981). Fibronectin may also promote directional migration of cells along certain pathways in the embryo (Critchley et al., 1979; Newgreen and Thiery, 1980).

F. Opsonic Activity

Plasma fibronectin also promotes the binding and ingestion of certain materials by macrophages and other reticuloendothelial cells. In investigations of the uptake of gelatinized colloidal test particles, it was found that a factor in serum was required for nonantibody "opsonization" of these particles during uptake. This factor was purified and found to be plasma fibronectin (Saba, 1970; Allen et al., 1973; Molnar et al., 1977; Blumenstock et al., 1977, 1978; Saba et al., 1978a). Cellular fibronectin displays equal amounts of this "opsonin" activity (Marquette et al., 1981).

Levels of plasma fibronectin and of this opsonizing activity are substantially reduced in shock and severe septicemia. A current area of clinical research involves the administration of fibronectin to such severely ill patients in hopes of restoring normal functions, e.g., possibly the clearance of debris from tissue trauma or even of bacteria (Kaplan and Saba, 1976; Saba *et al.,* 1978b; Saba and Jaffe, 1980; Molnar *et al.,* 1981).

These opsonic roles of fibronectin are as yet somewhat speculative. Fibronectin plus heparin can mediate binding of colloids to liver slices (Saba, 1970; Molnar *et al.,* 1977; Blumenstock *et al.,* 1977, 1978), but there is little subsequent ingestion of test materials *in vitro* (Molnar *et al.,* 1979). Fibronectin mediates the ingestion of gelatinized particles by isolated macrophages, but it also requires the concomitant addition of heparin to act (Gudewicz *et al.,* 1980). This heparin requirement is puzzling, since such levels may not exist even locally *in vivo*. Perhaps heparan sulfate might be utilized for this function *in vivo*.

As for binding to bacteria, fibronectin does bind to the gram-positive organism *Staphylococcus aureus* (Kuusela, 1978). However, it is not known which other bacteria are bound, or whether fibronectin affects their ingestion by host defense cells. Although there are therefore many uncertainties in this area of investigation, it currently provides the greatest promise for the clinical application of this glycoprotein.

G. Developmental Activities

Fibronectin is lost from cells during cartilage and muscle formation in developing embryos (Linder *et al.,* 1975; Stenman and Vaheri, 1978). Similar differences in levels of cell surface fibronectin have been reported *in vitro* in mesenchyme cells compared to chondrocytes (although the latter can still secrete soluble fibronectin) and in cell lines of myoblasts compared to myotubes (Lewis *et al.,* 1978; Dessau *et al.,* 1978b; Chen, 1977; Furcht *et al.,* 1978).

An inhibitory role in development for fibronectin has been suggested by experiments in which purified cellular fibronectin is added back to these differentiating systems in tissue culture. Artificially added fibronectin causes a dedifferentiation of chondrocytes to a more mesenchymal phenotype in terms of cell morphology, synthesis of sulfated proteoglycans, and the type of collagen synthesized (type II becomes type I). It has been suggested that fibronectin must be removed from these tissues in order for development to proceed normally (Pennypacker *et al.,* 1979; West *et al.,* 1979).

A further correlation between differentiation and fibronectin is seen when cartilage cells are treated with vitamin A. The development of chondrocytes from mesenchyme cells is inhibited, and fibronectin levels remain high. Conversely, treatment of chondrocytes with vitamin A alters the capacity of cells to bind the fibronectin they synthesize. Treated cells now bind fibronectin in cell surface fibrils, and the cells dedifferentiate (Hassell *et al.,* 1978, 1979).

A similar inhibitory effect is seen *in vitro* on the fusion of plasma membranes that occurs during muscle development. Myoblasts fusion therefore seems to be sensitive to high levels of fibronectin (Podleski *et al.*, 1979). However, it is also clear that myoblasts show an *in vitro* requirement for fibronectin at earlier times of culture before the fusion event in order for effective cell attachment to take place (Chiquet *et al.*, 1979).

A possible positive role for fibronectin in differentiation has been identified in neural crest cells. These cells arise from the edge of the neural tube, and they migrate throughout the body to form a multitude of tissue types at distant locations. These cells do not possess fibronectin (Loring *et al.*, 1977; Sieber-Blum *et al.*, 1981). However, they do migrate through regions of the body that contain fibronectin, suggesting the possibility that the exogenous fibronectin might affect their migration or their fate in differentiation (Newgreen and Thiery, 1980).

The addition of purified fibronectin to neural crest cells *in vitro* stimulates the adrenergic cell phenotype (Sieber-Blum *et al.*, 1981; Loring *et al.*, 1981). These results suggest that this glycoprotein might promote the formation of sympathetic ganglion cells from neural crest cells *in vivo*. This concept may be testable by antibody injection experiments.

In the future, it will be of interest to determine whether fibronectin can promote any other differentiative event, as well as its mechanisms of action in development. An important implication of these experiments involving the effect of exogenous fibronectin on differentiating systems is that an extracellular structural molecule can apparently help to regulate intracellular metabolic processes.

III. STRUCTURE OF FIBRONECTIN

A. Subunit Organization

The structure of fibronectin has been examined by a number of physical techniques. Its molecular properties are listed in Table I. Both cellular and plasma fibronectins are composed of disulfide-linked subunits. Plasma fibronectin consists of dimers, whereas cellular fibronectin exists as a mixture of dimers, higher polymers, and small amounts of monomer. The subunits of fibronectin in dimers may not be identical, since the subunits of plasma fibronectin differ in apparent size as determined by polyacrylamide gel electrophoresis (reviewed by Mosher, 1980). However, peptide maps comparing these subunits show no convincing differences in primary structure (Birdwell *et al.*, 1980; Kurkinen *et al.*, 1980).

The interchain disulfide bonds that link the fibronectin subunits are unusual in that they are confined to one end of the molecule (Fig. 4). Both cellular and plasma fibronectins possess a protease-sensitive region close to these disulfide

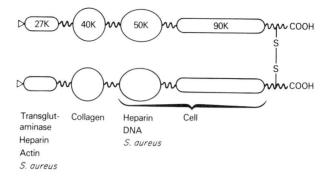

Figure 4 A current model of the specific structural and functional domains of fibronectin. A disulfide-linked dimer of identical subunits of 220,000 daltons is shown. The molecular weights of protease-resistant domains that have been purified are indicated for the upper fibronectin subunit (K = 1000). The functional or binding activities localized at these specific regions are indicated below the lower subunit. The shapes shown are current estimates of the shapes of the domains; the shape of the 27,000 fragment and the absolute length of the flexible polypeptide connecting regions are still unknown. The triangles to the left represent the blocked amino termini; for example, human plasma fibronectin contains an amino terminal pyrrolidonecarboxylic acid. Sites such as the binding region for hyaluronic acid are not shown; the relative order of the 50K and 90K domains are still uncertain.

bonds, which is located near the carboxyl terminus; cleavage at this point separates the two or more subunits into monomeric fragments (see Mosher, 1980, for references). It may be speculated that the asymmetric location of these disulfide bonds increases the ability of the molecule to stretch between distant points in mediating adhesive interactions.

B. Secondary and Tertiary Structure

In addition, the polypeptide subunits of fibronectin are themselves highly elongated or unfolded (Fig. 4). Circular dicroism studies of cellular and plasma fibronectins show little evidence for regular secondary structure, such as α helix or β sheet, even though the molecule is known to have an elongated shape with a large Stokes radius and an unusually low sedimentation coefficient for a protein of its size. These findings suggest that fibronectin is elongated because the polypeptide backbone is unfolded, rather than because it is helical and rigid as in collagen (Yamada *et al.*, 1977a; Alexander *et al.*, 1978, 1979).

Although there are no major regions of ordered secondary polypeptide structure, the molecule does appear to possess tertiary structure. Fluorescence studies indicate that the tryptophan residues are buried within the protein. They become exposed to water only after denaturation of the molecule by heating to >60° or by treatment with strong denaturing agents. The simplest interpretation of these results is that the parts of the molecule containing tryptophan residues are folded into regions of tertiary structure that exclude water and can be unfolded by strong denaturants (Alexander *et al.*, 1978; Colonna *et al.*, 1978).

In addition, when ultracentrifugation experiments are conducted at an elevated pH, there is a reduction in the sedimentation constant of fibronectin; this finding suggests that the molecule is flexible and can expand or stretch in an altered environment. Neutralization of pH results in a return to a more compact configuration that sediments more rapidly (Alexander *et al.*, 1978, 1979).

A general model explaining all of these results is one in which the fibronectin molecule is composed of structured, folded polypeptide domains that are separated by regions of flexible polypeptide chain, which can expand or shorten depending on ionic conditions (Fig. 4). This model is most clearly supported by proteolytic digestion experiments to be described in Section III,C.

C. Specific Structural and Functional Domains

In mediating certain biological effects, fibronectin can bind directly to a number of important targets, which are listed in Table V. It is thought that, by means of these specific binding interactions, fibronectin can mediate a variety of adhesive or binding events, e.g., cell adhesion to collagen via separate binding sites for cells and for collagen.

A major, recent advance has been the realization that fibronectin contains a series of functional as well as structural domains (Fig. 4). Fibronectin should be viewed as more than a simple, nonspecifically adhesive or gluelike molecule; instead, it appears to act by means of specialized domains that are specific for particular biological activities. For example, there appear to be separate domains for binding to collagen or to cells.

A valuable way to dissect these functional domains is to use proteases such as chymotrypsin or Pronase. Proteases tend to cleave proteins preferentially in regions where the polypeptide chain is flexible, and they tend to spare folded, globular regions in which cleavage sites are hidden. These facts make it possible to isolate intact binding sites of protease-digested molecules if these sites are located in globular regions in which the polypeptide backbone is folded (Fig. 5).

Two useful strategies are indicated in Figure 5, which describes a general approach to isolating any type of active site region that binds to a protease-resistant molecule or ligand. The approach might also prove useful in isolating binding sites of hormone receptors or of other glycoproteins that bind specific ligands. In one strategy, the molecule of interest (e.g., fibronectin) is permitted to bind to a ligand such as a collagen which has been covalently attached to agarose beads. The complex is digested with a protease, which cleaves unfolded regions and spares structured domains. After extensive washing, the only part of the fibronectin molecule that remains bound to the beads is the active binding site. A second strategy involves first digesting the fibronectin with the protease,

I   II

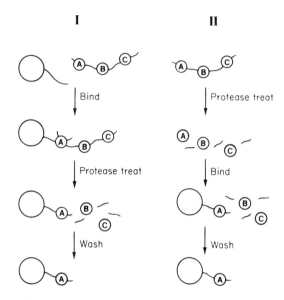

Figure 5 General methods for purifying biologically active proteolytic domains by affinity chromatography. The binding molecule, e.g., fibronectin, contains structural domains labeled A, B, and C. The ligand, e.g., collagen, is first coupled covalently to agarose beads. In strategy I, the molecule is permitted to bind to the ligand, which occurs at specific binding site A. The protease-resistant binding site domain remains bound after protease treatment of the complex and washing of the beads and can be recovered in relatively pure form. In strategy II, the isolated binding molecule is first cleaved by a protease, which is then inhibited by a protease inhibitor. Site A retains its binding capacity and can subsequently be bound to the ligand-coupled beads. These procedures may prove valuable in dissecting the binding interactions of other glycoconjugates.

then treating with protease inhibitors and purifying the binding domain by affinity chromatography with the ligand bound to beads (Fig. 5).

1. Collagen-Binding Domain

Both cellular and plasma fibronectins can mediate the attachment of cells to collagen; in addition, both molecules bind directly to collagen even in the absence of cells. Several laboratories have isolated this important collagen-binding portion of fibronectin. By the use of proteases such as chymotrypsin or trypsin to cleave the molecule into fragments, and affinity columns containing denatured collagen (to which fibronectin binds especially well), the collagen-binding domain of fibronectin was recovered in fragments of 30,000–40,000 daltons (Balian *et al.*, 1979; Ruoslahti *et al.*, 1979a; Hahn and Yamada, 1979a,b; Gold *et al.*, 1979). For example, chymotrypson cleavage yields a fragment of 40,000 daltons that is, as expected, relatively globular in shape (Fig. 4).

These collagen-binding domains contain an unusually large number of intra-

Fragments pre-incubated with:

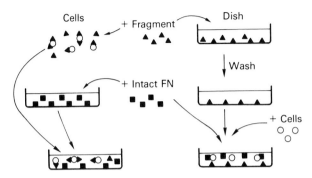

Figure 6 Tests for competitive inhibition by purified fibronectin fragments. Chinese hamster ovary cells require fibronectin to attach to dishes coated with type I collagen. To test which fragments interact with the cell surface or to collagen, each fragment is preincubated with either cells or dishes. Their inhibitory effects on attachment mediated by intact fibronectin are quantitated. The cell-binding fragment inhibits attachment only if preincubated with cells, and the collagen-binding fragment inhibits attachment only if preincubated with the collagen substrate (Hahn and Yamada, 1979b).

molecular disulfide bonds, which appear to be necessary for maintaining their collagen-binding activity (Balian *et al.*, 1979; Wagner and Hynes, 1979; cf. Pearlstein *et al.*, 1980). In contrast to intact fibronectin, this purified protein domain cannot by itself mediate the attachment of cells to collagen substrates. However, this fragment can competitively inhibit cell attachment to collagen mediated by intact fibronectin (Fig. 6). This experiment indicates that this domain is the functionally important site for fibronectin binding to collagen (Hahn and Yamada, 1979b).

Fibronectin also binds to fibrin or fibrinogen (Ruoslahti and Vaheri, 1975; Stemberger and Hörmann, 1976; Mosesson and Amrani, 1980). Although fibrin and collagen have been reported to compete with each other for binding to fibronectin (Engvall *et al.*, 1978), these molecules may not bind to the same domain. Fibrin-binding domains on fibronectin have been identified that are distinct from the collagen-binding domain (Sekiguchi and Hakomori, 1980b).

2. Heparin-Binding Domains

Fibronectin binds to the glycosaminoglycan heparin, and heparin can strengthen the binding of fibronectin to collagen (Stathakis and Mosesson, 1977; Jilek and Hörmann, 1979; Johansson and Höök, 1980; K. M. Yamada *et al.*, 1980a; Ruoslahti *et al.*, 1980). This binding is relatively strong, as indicated by a moderately high binding affinity ($K_D = 10^{-7}-10^{-9}M$; K. M. Yamada *et al.*, 1980a). A heparin-binding fragment of 50,000 daltons can be isolated after digestion of fibronectin with Pronase. It resembles the collagen-binding domain in possessing a relatively globular shape compared to the highly elongated shape of the fibronectin molecule itself. However, in striking contrast to the collagen-binding domain, it is nearly devoid of cysteine residues and carbohydrate (Hayashi et al., 1980).

The resistance of this fragment to degradation by moderate concentrations of Pronase, a very broad spectrum protease, is strong evidence for the existence of a tightly folded polypeptide domain containing no regions of exposed polypeptide that are accessible to proteolysis. As expected, the heparin- and collagen-binding domains can be separated readily, since only this heparin-binding domain binds to heparin affinity columns and vice versa (Hayashi et al., 1980). Other heparin-binding regions can be found on fibronectin that differ in affinity or divalent cation requirements (Sekiguchi and Hakomori, 1980a; Hayashi and Yamada, 1981).

Two molecules that are related to heparin are heparan sulfate (or its proteoglycan) and dextran sulfate. Both of these molecules are also bound by both plasma and cellular fibronectins (Perkins et al., 1979; Ruoslahti et al., 1979b; Yamada et al., 1980; Ruoslahti and Engvall, 1980; Laterra et al., 1980). It is not yet known whether the heparin-binding domain(s) mediate this binding.

3. Hyaluronic Acid-Binding Site

Hyaluronic acid may have important regulatory effects on cells during embryonic development in promoting migration and inhibiting differentiation (Toole, 1981). One of the molecules with which this glycosaminoglycan interacts is fibronectin (Jilek and Hörmann, 1979; K. M. Yamada et al., 1980a; Ruoslahti and Engvall, 1980). Purified preparations of fibronectin bind to hyaluronic acid, and the binding site is kinetically separate from the site for binding to heparin. Binding of one molecule does not affect the binding of the other, even though each site is readily saturated by its appropriate ligand (K. M. Yamada et al., 1980a). Hyaluronic acid appears to increase the strength of the binding of fibronectin to collagen (Jilek and Hörmann, 1979; Ruoslahti and Engvall, 1980).

A structural domain on fibronectin for binding hyaluronic acid has not been identified as yet. However, the site has considerable specificity regarding the size of repeating disaccharide of hyaluronic acid that it recognizes. The hyaluronic acid recognition site is found to be between 8 and 10 monosaccharide units in size in competitive inhibition experiments. The site has a moderately high binding affinity of 10^{-7} M (K. M. Yamada et al., 1980a). It is possible that this binding site might provide one of the means by which hyaluronic acid interacts with cells, or it might play a role in organizing the arrangement of hyaluronic acid chains in extracellular spaces. Interestingly, plasma fibronectin may be less effective in binding hyaluronic acid (Laterra *et al.*, 1980; Ruoslahti and Engvall, 1980; Jilek and Hörmann, 1979).

The existence of separate binding sites on fibronectin for collagen, heparin, and hyaluronic acid suggests that this molecule could play a central role in the organization and the binding to cells of a variety of extracellular or secreted molecules (see Chapter 3, Section I, this volume).

4. Binding to Actin and DNA

Other macromolecules are known to interact with fibronectin, but the specificity or physiological significance of these interactions is not yet clear. Actin and DNA have been shown to bind to fibronectin (Keski-Oja *et al.*, 1980; Zardi *et al.*, 1979). A prominent actin-binding domain of fibronectin is approximately 27,000 daltons in size and is separate from the collagen- and heparin-binding sites; it is adjacent to the collagen-binding domain (Keski-Oja and Yamada, 1981). A second actin-binding region is located on the other side of the collagen-binding domain (Hayashi and Yamada, 1981). The functions of these sites, which can bind important *intracellular* molecules, are not yet known. One possibility is that they assist the reticuloendothelial system in removing from blood the actin and DNA liberated by trauma from cells by means of an "opsonic" activity of fibronectin (Saba *et al.*, 1978a; Molnar *et al.*, 1977, 1981) (see Section II,F).

5. Cell-Binding Domain

A centrally important interaction of fibronectin is its binding to the cell surface in order to mediate its biological effects on cells, e.g., in mediating cell adhesion. A proteolytic fragment of 160,000 daltons that has been purified from fibronectin retains the ability to bind to cells (Hahn and Yamada, 1979b; see also Ruoslahti and Hayman, 1979; Sekiguchi and Hakomori, 1980a; McDonald and Kelley, 1980). This fragment lacks the collagen-binding site of fibronectin and is therefore unable to mediate the attachment of cells to collagen. However, it is still capable of binding to cells and competitively inhibiting the binding of intact fibronectin; it therefore competitively inhibits fibronectin-mediated cell attachment to collagen. This fragment can also directly mediate other adhesive events that do not involve collagen, such as mediating attachment and spreading of cells on tissue culture plastic (Hahn and Yamada, 1979b; McDonald and Kelley, 1980). This 160,000 dalton fragment is relatively elongated, even though part of it contains a globular heparin-binding site (Fig. 4).

Another, larger fragment of fibronectin with a molecular weight of 205,000 contains the same heparin- and cell-binding sites but also the collagen-binding site. It can therefore readily mediate cell attachment to collagen (Hahn and Yamada, 1979b; Wagner and Hynes, 1979; K. M. Yamada *et al.*, 1980a; Sekiguchi and Hakomori, 1980a).

D. Oligosaccharides

Certain structural features of fibronectin besides its binding sites are localized at specific regions of the molecule. The carbohydrate on fibronectin is localized at the collagen-binding and cell-binding parts of the molecule (Wagner and Hynes, 1979; Sekiguchi and Hakomori, 1980a; McDonald and Kelley, 1980; K. M. Yamada and D. W. Kennedy, unpublished results). The functional significance of this localization of sugars is not yet known.

Both cellular and plasma fibronectins contain only one type of oligosaccharide unit (Yamada et al., 1977a; Fukuda and Hakomori, 1979b; Wrann, 1978; Takasaki et al., 1979). There is an average of five of these "complex" asparagine-linked oligosaccharides per subunit, each containing the sugars N-acetylglucosamine, mannose, and galactose, with additional sialic acid or fucose residues; there appears to be less fucose in the plasma form of fibronectin.

The presence of only one class of oligosaccharide on fibronectin has proved useful. The glycoprotein glycosylation inhibitor tunicamycin has been shown to be a specific inhibitor of the formation of this class of oligosaccharide linkage (see Chapter 1, Section 3, this volume). It is therefore possible to synthesize fibronectin that is free of carbohydrate residues in order to determine the role of carbohydrates in its structure and function.

From these studies, the carbohydrate moiety of fibronectin is found to play no apparent functional role, since unglycosylated fibronectin is as active as the native molecule in the biological assays of hemagglutination, cell adhesion to collagen, cell spreading on plastic substrates, and reversion of the morphological phenotype of transformed cells. Instead, the carbohydrate appears to stabilize the molecule against proteolytic attack and abnormal rates of turnover. Carbohydrate-free fibronectin is substantially more susceptible to in vitro degradation by Pronase and has a severalfold higher rate of protein turnover on living cells (Olden et al., 1978, 1979).

E. Sulfhydryls and Disulfides

One or two free sulfhydryl groups are located at specific regions of fibronectin and it has been suggested that these groups could permit the molecule to form higher-order polymers on the cell surface (Wagner and Hynes, 1979, 1980; Fukuda and Hakomori, 1979a; McDonald and Kelley, 1980). However, there is no information as yet as to whether multimers of fibronectin actually make significant use of these groups when they form cross-linked complexes. As noted earlier, the intermolecular disulfide bonds are localized at the carboxyl terminus of fibronectin, and there is no convincing evidence as yet for functional disulfide bonding using the free sulfhydryl.

Intramolecular disulfide bonds are concentrated in a region of the molecule 40,000–70,000 daltons in size, that contains the collagen-binding site (Balian *et al.*, 1979; Ruoslahti *et al.*, 1979a; McDonald and Kelley, 1980). Since intact intrachain disulfide bonds are required in order for fibronectin to bind to collagen, these bonds may help to maintain a specific conformation of this domain that is necessary for binding to collagen. In striking contrast, the heparin-binding domain appears to lack enough cysteine for even one disulfide bond (Hayashi *et al.*, 1980).

F. Transglutaminase Reaction Site

Fibronectin can be covalently cross-linked by the enzyme transglutaminase into complexes with itself, fibrinogen, or collagen. This cross-linking occurs at a specific site to glutamine residues of a 27,000 dalton region of the molecule (Jilek and Hörmann, 1977; Mosher *et al.*, 1980) (Fig. 4). This domain for transglutaminase-mediated cross-linking is distinct from the site for fibronectin binding to collagen and is reported to be the domain closest to the amino terminus of the molecule (Furie *et al.*, 1980; Furie and Rifkin, 1980; Wagner and Hynes, 1980; McDonald and Kelley, 1980). This domain is also reported to be the site at which fibronectin binds to the bacterium *Staphylococcus aureus* (Kuusela, 1978; Mosher and Proctor, 1980), as well as to heparin and actin (Sekiguchi and Hakomori, 1980a; Keski-Oja and Yamada, 1981; Hayashi and Yamada, 1981).

IV. STRUCTURES OF CELLULAR VERSUS PLASMA FIBRONECTINS

As discussed in Section I, cellular and plasma fibronectins are very similar in composition and structure. However, they differ in some biological activities (Table II). In addition, they differ slightly in apparent molecular weight (e.g., see Yamada and Kennedy, 1979).

The structures of these fibronectins have recently been compared in detail (Hayashi and Yamada, 1981). Chicken cellular and plasma fibronectins have homologous functional, protease-resistant domains for binding to cells, collagen, heparin, actin, and *S. aureus*. However, they have three major regions of difference that appear to be deletions or insertions. The collagen-binding domain of cellular fibronectin is 1000 daltons smaller in apparent size than that of plasma fibronectin. A calcium-sensitive heparin-binding region of the molecule contains a similarly smaller cellular fibronectin region. However, the largest region of difference is an apparent deletion of 11,000 daltons from the center of the plasma fibronectin polypeptide.

These differences do not appear to result from differences in carbohydrate, since plasma and cellular fibronectins have virtually identical carbohydrate compositions and structures. Moreover, there is no detectable difference in carbohydrate locations by gel electrophoresis and periodic acid–Schiff staining of all proteolytic fragments (Hayashi and Yamada, 1981).

This identification of several differences at the interior of the molecule strongly suggests that the fibronectins are not related by simple processing events. Specifically, it suggests that plasma fibronectin is not produced from cellular fibronectin by a proteolytic cleavage from an end, as had been suggested previously; instead, plasma and cellular fibronectin are probably the products of differently spliced mRNA molecules or of two or more tissue-specific genes for fibronectin. Important future work will involve determining the exact nature of these apparent deletions, their cause, and whether one or more of these differences can account for the differences in biological activity between these two types of fibronectin.

V. CELLULAR RECEPTOR FOR FIBRONECTIN

What molecule in the plasma membrane binds the cell-binding domain of fibronectin? An experimental approach to this problem is to test for competitive inhibitors of fibronectin in biological assays. If a certain plasma membrane constituent is the natural receptor for fibronectin, then adding an excess amount of this molecule to assays should result in a competition for binding to fibronectin between the receptor on cells and this exogenously added free receptor. The result should be an increasing inhibition of fibronectin activity as its cell-binding domain becomes saturated with an increasing amount of added free ligand.

In experiments of this type, the best current candidates for the fibronectin receptor are negatively charged lipids such as glycolipids containing several sialic acid residues, especially the ganglioside GT1b (Kleinman et al., 1979). These lipids are bound by fibronectin, and they effectively block many of the fibronectin-mediated biological events described earlier, including cell attachment to collagen, cell spreading, hemagglutination, and restoration of a normal morphological phenotype to transformed cells (Kleinman et al., 1979; K. M. Yamada et al., 1980b, 1981). As might be predicted, the cell-spreading activity of the 160,000 dalton cell-binding domain itself is also inhibited by exogenously added gangliosides.

The portion of gangliosides that is recognized by fibronectin appears to be the carbohydrate moiety, since the purified oligosaccharide portion of gangliosides also inhibits cell attachment to collagen, cell spreading, and hemagglutination; in contrast, the ceramide portion is without effect (Kleinman et al., 1979; Yamada et al., 1981). Phospholipids that have been tested to date for inhibitory activity are at least an order of magnitude less effective, and the glycopeptide portion of glycoproteins has no detectable activity (Yamada et al., 1981). A current hypothesis regarding the receptor for fibronectin is therefore that fibronectin binds to a specific cell-surface receptor consisting of specific charged molecules such as gangliosides. It has also been suggested that the fibronectin receptor is a protein, since the binding and phagocytosis of fibronectin-coated latex beads are prevented from taking place by treatment of cells with protease (Grinnell, 1980). Recent cross-linking studies suggest that a 47,000 dalton glycoprotein could be a major fibronectin receptor (Aplin *et al.*, 1981). It will therefore be important to elucidate the role of protein in the "receptor" for fibronectin; perhaps the receptor contains both lipid and protein components.

VI. SYNTHESIS AND REGULATION

A. Biosynthesis, Organization, and Turnover

Fibronectin is synthesized in the rough endoplasmic reticulum of cells. It can be visualized in this site as well as in the Golgi apparatus by immunofluorescence and immunoelectron microscopy procedures (Yamada, 1978; S. S. Yamada *et al.*, 1980; Hedman, 1980). Its rate of biosynthesis can be regulated by the amount of fibronectin messenger RNA, and its characteristic decrease after neoplastic transformation of cells can be partially explained by a major decrease in levels of fibronectin mRNA (Adams *et al.*, 1977). The mRNA for fibronectin has been isolated, and recently DNA complementary to fibronectin mRNA has been cloned in bacteria. Recombinant DNA technology should prove to be useful for analyzing the regulation of this glycoprotein in malignancy and development, as well as for determining the structure of its gene(s) (Fagan *et al.*, 1979, 1981).

During biosynthesis, the carbohydrate moiety is added to fibronectin as an immature oligosaccharide unit similar to that described for other asparagine-linked oligosaccharides. As the protein proceeds to the cell surface, two-thirds of the original mannose residues are enzymatically removed, and the N-acetylglucosamine, galactose, sialic acid, and fucose residues characteristic of the oligosaccharides of the mature protein are added (Choi and Hynes, 1979; Olden *et al.*, 1980). It also becomes disulfide-linked into multimers near the time of secretion (Choi and Hynes, 1979). Surprisingly, the process of fibronectin secretion does not appear to require the presence of carbohydrate, since secretion itself occurs normally even when glycosylation is blocked by tunicamycin (Olden *et al.*, 1978). The fibronectin can be secreted either directly onto the cell surface or into culture media.

How secreted fibronectin becomes organized into its distinctive fibrillar patterns is not yet clear at the molecular level. Purified cellular fibronectin readily self-aggregates, although not into fibrils under conditions tested thus far. A fraction of plasma fibronectin has been reported to form fibrillar structures under certain *in vitro* conditions (Iwanaga *et al.*, 1978). Whether these fibrils are identical to those found *in vivo* is not yet known. It may also be pertinent that heparin can induce plasma fibronectin to form striking fibrillar complexes, suggesting that other extracellular components might be involved in normal fibronectin fibril formation (Iwanaga *et al.*, 1978; Jilek and Hörmann, 1979; Vuento *et al.*, 1980).

Once fibronectin has been organized on the cell surface, it can be lost by sloughing from cells or by degradation by cellular proteases. The turnover rate of fibronectin on the cell surface is relatively slow, with a half-life similar to that of total protein turnover (30–36 hours) (Olden and Yamada, 1977). This rate of turnover is accelerated in transformed cells or if fibronectin is synthesized lacking carbohydrate (Hynes and Wyke, 1975; Olden and Yamada, 1977; Olden *et al.*, 1978).

B. Hormonal Regulation

Fibronectin levels on the cell surface can be modulated by hormones or other small molecules. For example, the growth of certain mouse 3T3 cells in medium lacking serum results in a loss of fibronectin, which can be restored by treatment with the hormone epidermal growth factor (Chen *et al.*, 1977). Although transformed cells contain very low levels of cell-surface fibronectin, more normal quantities are seen on the surface after treatment with glucocorticoids such as dexamethasone or with butyrate (Furcht *et al.*, 1979a,b; Milhaud *et al.*, 1980; Hayman *et al.*, 1980; see also Marceau *et al.*, 1980). Treatment of chondrocytes with vitamin A causes a shift in cell-surface binding, so that the cells accumulate significantly more fibronectin (Hassell *et al.*, 1979). In all cases, the mechanism of action of the hormone or drug is not known. It may be acting via effects on quantities of receptor(s) for fibronectin, which may help to identify such receptors. Clearly, many more studies are needed to determine the patterns of hormonal regulation of fibronectin and to ascertain whether hormones and vitamins are involved in clinically important changes in the amounts of this glycoprotein.

VII. PROSPECTS

The glycoprotein fibronectin has provided a convenient model system for the analysis of a number of cell–substratum, cell–cell, and cell–extracellular matrix adhesive events. Additional detailed proteolytic dissection studies, domain-specific immunological inhibition experiments, and protein sequencing of important sites should eventually provide a detailed understanding of how this glycoprotein mediates adhesion and alters cell morphological and social behavior.

Moreover, this glycoprotein has been implicated in a number of complex cellular events, including cell motility, chemotaxis, phagocytosis, and embryonic differentiation of certain cell types. By determining the mechanism of action of fibronectin in these more complex events it should be possible to define its physiological roles in more detail, as well as gain insight into these vital cellular processes.

Finally, future research must provide the answers to a number of pressing questions: Where is plasma fibronectin synthesized? Are the different forms of fibronectin actually part of a developmentally regulated multigene family that regulates certain adhesive interactions? What is the role of fibronectin in metastasis? Does fibronectin provide clinically important host defense via the reticuloendothelial system against blood-borne cellular debris and bacteria? And, finally, what are the precise mechanisms by which fibronectin binds to the cell surface and alters the organization of the actin-containing cytoskeleton? The answers to these and many other questions about this cell surface and extracellular matrix glycoprotein should provide insight into a number of important cellular events.

REFERENCES

Abercrombie, M. (1979). *Nature (London)* **281**, 259–262.
Adams, S. L., Sobel, M. E., Howard, B. H., Olden, K., Yamada, K. M., de Crombrugghe, B., and Pastan, I. (1977). *Proc. Natl. Acad. Sci. U.S.A.* **74**, 3399–3403.
Alexander, S. S., Colonna, G., Yamada, K. M., Pastan, I., and Edelhoch, H. (1978). *J. Biol. Chem.* **253**, 5820–5824.
Alexander, S. S., Colonna, G., and Edelhoch, H. (1979). *J. Biol. Chem.* **254**, 1501–1505.
Ali, I. U., and Hynes, R. O. (1978). *Cell* **14**, 439–446.
Ali, I. U., Mautner, V., Lanza, R., and Hynes, R. O. (1977). *Cell* **11**, 115–126.
Allen, C., Saba, T. M., and Molnar, J. (1973). *RES, J. Reticuloendothel. Soc.* **13**, 410–423.
Aplin, J. D., Hughes, R. C., Jaffe, C. L., and Sharon, N. (1981). *Exp. Cell Res.* **134**, 488–494.
Arneson, M. A., Hammerschmidt, D. E., Furcht, L. T., and King, R. A. (1980). *JAMA, J. Am. Med. Assoc.* **244**, 144–147.
Atherton, B. T., Taylor, D., and Hynes, R. O. (1981). *J. Supramolec. Struct., Suppl.* **5**, 305.
Balian, G., Click, E. M., Crouch, E., Davidson, J., and Bornstein, P. (1979). *J. Biol. Chem.* **254**, 1429–1432.
Bell, P. B. (1977). *J. Cell Biol.* **74**, 963–982.
Bensusan, H. B., Koh, T. L., Henry, K. G., Murray, B. A., and Culp, L. A. (1978). *Proc. Natl. Acad. Sci. U.S.A.* **75**, 5864–5868.
Birchmeier, C., Kreis, T. E., Eppenberger, H. M., Winterhalter, K. H., and Birchmeier, W. (1980). *Proc. Natl. Acad. Sci. U.S.A.* **77**, 4108–4112.
Birdwell, C. R., Brasier, A. R., and Taylor, L. A. (1980). *Biochem. Biophys. Res. Commun.* **97**, 574–581.
Blumenstock, F. A., Weber, P., and Saba, T. M. (1977). *J. Biol. Chem.* **252**, 7156–7162.
Blumenstock, F. A., Saba, T. M., Weber, P., and Laffin, R. (1978). *J. Biol. Chem.* **253**, 4287–4291.
Bray, B. A. (1978) *J. Clin. Invest.* **62**, 745–752.
Chen, A. B., Mosesson, M. W., and Solish, G. I. (1976). *Am. J. Obstet. Gynecol.* **7**, 958–961.
Chen, L. B. (1977). *Cell* **10**, 393–400.

Chen, L. B., Gallimore, P. H., and McDougall, J. K. (1976). *Proc. Natl. Acad. Sci. U.S.A.* **73,** 3570–3574.

Chen, L. B., Gudor, R. C., Sun, T. -T., Chen, A. B., and Mosesson, M. W., (1977). *Science* **197,** 776–778.

Chen, L. B., Murray, A., Segal, R. A., Bushnell, E., and Walsh, M. L. (1978). *Cell* **14,** 377–391.

Chen, L. B., Summerhayes, I., Hsieh, P., and Gallimore, P. H. (1979). *J. Supramol. Struct.* **12,** 139–150.

Chen, W.-T., and Singer, S. J. (1980). *Proc. Natl. Acad. Sci. U.S.A.* **77,** 7318–7322.

Chen, W.-T., and Singer, S. J. (1981). *J. Cell Biol.* **91,** 258a.

Chiquet, M., Puri, E. C., and Turner, D. C. (1979). *J. Biol. Chem.* **254,** 5475–5482.

Choi, M. G., and Hynes, R. O. (1979). *J. Biol. Chem.* **254,** 12050–12055.

Colonna, G., Alexander, S. S., Yamada, K. M., Pastan, I., and Edelhoch, H. (1978). *J. Biol. Chem.* **253,** 7787–7790.

Courtoy, P. J., Kanwar, Y. S., Hynes, R. O., and Farquhar, M. G. (1980). *J. Cell Biol.* **87,** 691–696.

Critchley, D. R., England, M. A., Wakely, J., and Hynes, R. O. (1979). *Nature (London)* **280,** 498–500.

Crouch, E., Balian, G., Holbrook, K., Duksin, D., and Bornstein, P. (1978). *J. Cell Biol.* **78,** 701–715.

Dessau, W., Adelmann, B. C., Timpl, R., and Martin, G. R. (1978a). *Biochem. J.* **169,** 55–59.

Dessau, W., Sasse, J., Timpl, R., Jilek, F., and von der Mark, K. (1978b). *J. Cell Biol.* **79,** 342–355.

Engvall, E., and Ruoslahti, E. (1977). *Int. J. Cancer* **20,** 1–5.

Engvall, E., Ruoslahti, E., and Miller, E. J. (1978). *J. Exp. Med.* **147,** 1584–1595.

Fagan, J. B., Yamada, K. M., de Crombrugghe, B., and Pastan, I. (1979). *Nucleic Acids Res.* **6,** 3471–3480.

Fagan, J. B., Sobel, M. E., Yamada, K. M., de Crombrugghe, B., and Pastan, I. (1981). *J. Biol. Chem.* **256,** 520–525.

Fox, C. H., Cottler-Fox, M. H., and Yamada, K. M. (1980). *Exp. Cell Res.* **130,** 477–480.

Fukuda, M., and Hakomori, S. (1979a). *J. Biol. Chem.* **254,** 5442–5450.

Fukuda, M., and Hakomori, S. (1979b). *J. Biol. Chem.* **254,** 5451–5457.

Furcht, L. T., Mosher, D. F., and Wendelschafer-Crabb, G. (1978). *Cell* **13,** 263–271.

Furcht, L. T., Mosher, D. F., Wendelschafer-Crabb, G., Woodbridge, P. A., and Foidart, J. M. (1979a). *Nature (London)* **277,** 393–395.

Furcht, L. T., Mosher, D. F., Wendelschafer-Crabb, G., and Foidart, J.-M. (1979b). *Cancer Res.* **39,** 2077–2083.

Furcht, L. T., Smith, D., Wendelschafer-Crabb, G., Mosher, D. F., and Foidart, J. M. (1980). *J. Histochem. Cytochem.* **28,** 1319–1333.

Furie, M. B., and Rifkin, D. B. (1980). *J. Biol. Chem.* **255,** 3134–3140.

Furie, M. B., Frey, A. B., and Rifkin, D. B. (1980). *J. Biol. Chem.* **255,** 4391–4394.

Gauss-Müller, V., Kleinman, H. K., Martin, G. R., and Schiffman, E. (1980). *J. Lab. Clin. Med.* **96,** 1071–1080.

Ginsberg, M. H., Painter, R. G., Forsyth, J., Birdwell, C., and Plow, E. F. (1980). *Proc. Natl. Acad. Sci. U.S.A.* **77,** 1049–1053.

Gold, L. I., Garcia-Pardo, A., Frangione, B., Franklin, E. C., and Pearlstein, E. (1979). *Proc. Natl. Acad. Sci. U.S.A.* **76,** 4803–4807.

Goldman, R., Pollard, T., and Rosenbaum, J. (1976). "Cell Motility." Cold Spring Harbor Lab., Cold Spring Harbor, New York.

Gospodarowicz, D., and Tauber, J. P. (1980). *Endocr. Rev.* **1,** 201–227.

Grinnell, F. (1978). *Int. Rev. Cytol.* **53,** 65–144.

Grinnell, F. (1980). *J. Cell Biol.* **86,** 104–112.

Grinnell, F., and Feld, M. K. (1979). *Cell* **17,** 117–129.

Grinnell, F., and Minter, D. (1978). *Proc. Natl. Acad. Sci. U.S.A.* **75,** 4408–4412.

Grinnell, F., Feld, M., and Snell, W. (1980). *Cell Biol. Int. Rep.* (in press).

Gudewicz, P. W., Molnar, J., Lai, M. Z., Beezhold, D. W., Siefring, G. E., Credo, R. B., and Lorand, L. (1980). *J. Cell Biol.* **87,** 427–433.

Hahn, L.-H. E., and Yamada, K. M. (1979a). *Proc. Natl. Acad. Sci. U.S.A.* **76,** 1160–1163.

Hahn, L.-H. E., and Yamada, K. M. (1979b). *Cell* **18,** 1043–1051.

Hassell, J. R., Pennypacker, J. P., Yamada, K. M., and Pratt, R. M. (1978). *Ann. N.Y. Acad. Sci.* **312,** 406–409.

Hassell, J. R., Pennypacker, J. P., Kleinman, H. K., Pratt, R. M., and Yamada, K. M. (1979). *Cell* **17,** 821–826.

Hayashi, M., and Yamada, K. M. (1981). *J. Biol. Chem.* **256,** 11292–11300.

Hayashi, M., Schlesinger, D. H., Kennedy, D. W., and Yamada, K. M. (1980). *J. Biol. Chem.* **255,** 10017–10020.

Hayman, E. G., and Ruoslahti, E. (1979). *J. Cell Biol.* **83,** 255–259.

Hayman, E. G., Engvall, E., and Ruoslahti, E. (1980). *Exp. Cell Res.* **127,** 478–481.

Hedman, K. (1980). *J. Histochem. Cytochem.* **28,** 1233–1241.

Hedman, K., Vaheri, A., and Wartiovaara, J. (1978). *J. Cell Biol.* **76,** 748–760.

Hewitt, A. T., Kleinman, H. K., Pennypacker, J. P., and Martin, G. R. (1980). *Proc. Natl. Acad. Sci. U.S.A.* **77,** 385–388.

Hughes, R. C., Pena, S. D. J., Clark, J., and Dourmashkin, R. R. (1979). *Exp. Cell Res.* **121,** 307–314.

Hynes, R. O. (1976). *Biochim. Biophys. Acta* **458,** 73–107.

Hynes, R. O., and Wyke, J. A. (1975). *Virology* **34,** 492–504.

Hynes, R. O., Destree, A. T., and Mautner, V. (1976). *J. Supramol. Struct.* **9,** 189–202.

Hynes, R. O., Ali, I. U., Destree, A. T., Mautner, V., Perkins, M. E., Senger, D. R., Wagner, D. D., and Smith, K. K. (1978). *Ann. N.Y. Acad. Sci.* **312,** 317–342.

Hynes, R. O., Destree, A. T., Perkins, M. E., and Wagner, D. D. (1979). *J. Supramol. Struct.* **11,** 95–104.

Iwanaga, S., Suzuki, K., and Hashimoto, S. (1978). *Ann. N.Y. Acad. Sci.* **312,** 56–73.

Jilek, F., and Hörmann, H. (1977). *Hoppe-Seyler's Z. Physiol. Chem.* **358,** 1165–1168.

Jilek, F., and Hörmann, H. (1978). *Hoppe-Seyler's Z. Physiol. Chem.* **359,** 247–250.

Jilek, F., and Hörmann, H. (1979). *Hoppe-Seyler's Z. Physiol. Chem.* **360,** 597–603.

Johansson, S., and Höök, M. (1980). *Biochem. J.* **187,** 521–524.

Kaplan, J. E., and Saba, T. M. (1976). *Am. J. Physiol.* **230,** 7–14.

Keski-Oja, J., and Yamada, K. M. (1981). *Biochem. J.* **193,** 615–620.

Keski-Oja, J., Sen, A., and Todaro, G. J. (1980). *J. Cell Biol.* **85,** 527–533.

Klebe, R. J. (1974). *Nature (London)* **250,** 248–251.

Kleinman, H. K., McGoodwin, E. B., Martin, G. R., Klebe, R. J., Fietzek, P. P., and Woolley, D. E. (1978a). *J. Biol. Chem.* **253,** 5642–5646.

Kleinman, H. K., Murray, J. C., McGoodwin, B. S., and Martin, G. R. (1978b). *J. Invest. Dermatol.* **71,** 9–11.

Kleinman, H. K., Martin, G. R., and Fishman, P. H. (1979). *Proc. Natl. Acad. Sci. U.S.A.* **76,** 3367–3371.

Kleinman, H. K., Wilkes, C. M., and Martin, G. R. (1981). *Biochemistry* **20,** 2325–2330.

Koteliansky, V. E., Leytin, V. L., Sviridov, D. D., Repin, V. S., and Smirnov, V. N. (1981). *FEBS Lett.* **123,** 59–62.

Kurkinen, M., Vartio, T., and Vaheri, A. (1980). *Biochim. Biophys. Acta* **624,** 490–498.

Kuusela, P. (1978). *Nature (London)* **276,** 718–720.

Kuusela, P., Vaheri, A., Palo, J., and Ruoslahti, E. (1978). *J. Lab. Clin. Med.* **92,** 595.

Laterra, J., Ansbacher, R., and Culp, L. A. (1980). *Proc. Natl. Acad. Sci. U.S.A.* **77,** 6662–6666.

Lewis, C. A., Pratt, R. M., Pennypacker, J. P., and Hassell, J. R. (1978). *Dev. Biol.* **64,** 31–47.

Linder, E., Vaheri, A., Ruoslahti, E., and Wartiovaara, J. (1975). *J. Exp. Med.* **142,** 41–49.

Linsenmayer, T. F., Gibney, E., Toole, B. P., and Gross, J. (1978). *Exp. Cell Res.* **116,** 470–474.

Loring, J., Erickson, C., and Weston, J. (1977). *J. Cell Biol.* **75,** 71a (abstr.).

Loring, J. *et al.* (1981). Submitted for publication.

McDonald, J. A., and Kelley, D. G. (1980). *J. Biol. Chem.* **255,** 8848–8858.

Madri, J. A., Roll, F. J., Furthmayr, H., and Foidart, J.-M. (1980). *J. Cell Biol.* **86,** 682–687.

Marceau, N., Goyette, R., Valet, J. P., and Deschenes, J. (1980). *Exp. Cell Res.* **125,** 497–502.

Marquette, D., Molnar, J., Yamada, K. M., Schlesinger, D., Darby, S., and Van Alten, P. (1981). *Mol. Cell. Biochem.* **36,** 147–155.

Mautner, V., and Hynes, R. O. (1977). *J. Cell Biol.* **75,** 743–766.

Milhaud, P., Yamada, K. M., and Gottesman, M. M. (1980). *J. Cell. Physiol.* **104,** 163–170.

Molnar, J., McLain, S., Allen, C., Laga, H., Gara, A., and Gelder, F. (1977). *Biochim. Biophys. Acta* **493,** 37–54.

Molnar, J., Gudewicz, P. W., Lai, M. W., Credo, R. B., Siefring, G. E., and Lorand, L. (1979). *Fed. Proc., Fed. Am. Soc. Exp. Biol.* **38,** 411 (abstr.).

Molnar, J., Froehlich, J., and Rovin, B. (1981). *J. Supramol. Struct., Suppl.* **5,** 310.

Mosesson, M. W. (1977). *Thromb. Haemostasis* **38,** 742–750.

Mosesson, M. W., and Amrani, D. L. (1980). *Blood* **56,** 145–148.

Mosesson, M. W., and Umfleet, R. A. (1970). *J. Biol. Chem.* **245,** 5728–5736.

Mosesson, M. W., Chen, A. B., and Huseby, R. M. (1975). *Biochim. Biophys. Acta* **386,** 509–524.

Mosher, D. F. (1980). *Prog. Hemostasis Thromb.* **5,** 111–155.

Mosher, D. F., and Proctor, R. A. (1980). *Science* **209,** 927–929.

Mosher, D. F., Schad, P. E., and Vann, J. M. (1980). *J. Biol. Chem.* **255,** 1181–1188.

Newgreen, D., and Thiery, J. P. (1980). *Cell Tissue Res.* **211,** 269–291.

Noonan, K. D., Noonan, N. E., and Yamada, K. (1981). *J. Supramol. Struct., Suppl.* **5,** 302.

Olden, K., and Yamada, K. M. (1977). *Cell* **11,** 957–969.

Olden, K., Pratt, R. M., and Yamada, K. M. (1978). *Cell* **13,** 461–473.

Olden, K., Pratt, R. M., and Yamada, K. M. (1979). *Proc. Natl. Acad. Sci. U.S.A.* **76,** 3343–3347.

Olden, K., Hunter, V. A., and Yamada, K. M. (1980). *Biochim. Biophys. Acta* **632,** 408–416.

Orly, J., and Sato, G. (1979). *Cell* **17,** 295–305.

Pearlstein, E. (1976). *Nature (London)* **262,** 497–500.

Pearlstein, E. (1978). *Int. J. Cancer* **22,** 32–35.

Pearlstein, E., Gold, L., and Garcia-Pardo, A. (1980). *Mol. Cell. Biochem.* **29,** 103–127.

Pena, S. D. J., and Hughes, R. C. (1978). *Cell Biol. Int. Rep.* **2,** 339–344.

Pennypacker, J. P., Hassell, J. R., Yamada, K. M., and Pratt, R. M. (1979). *Exp. Cell Res.* **121,** 411–415.

Perkins, M. E., Ji, T. H., and Hynes, R. O. (1979). *Cell* **16,** 941–952.

Podleski, T. R., Greenberg, I., Schlessinger, J., and Yamada, K. M. (1979). *Exp. Cell Res.* **122,** 317–326.

Postlethwaite, A. E., Keski-Oja, J., Balian, G., and Kang, A. H. (1981). *J. Exp. Med.* **153,** 494–499.

Pouysségur, J., Willingham, M., and Pastan, I. (1977). *Proc. Natl. Acad. Sci. U.S.A.* **74,** 243–247.

Quaroni, A., Isselbacher, K. J., and Ruoslahti, E. (1978). *Proc. Natl. Acad. Sci. U.S.A.* **75,** 5548–5552.

Rizzino, A., and Crowley, C. (1980). *Proc. Natl. Acad. Sci. U.S.A.* **77,** 457–461.

Ruoslahti, E., and Engvall, E. (1980). *Biochim. Biophys. Acta* **631,** 350–358.

Ruoslahti, E., and Hayman, E. G. (1979). *FEBS Lett.* **97**, 221–224.

Ruoslahti, E., and Vaheri, A. (1975). *J. Exp. Med.* **141**, 497–501.

Ruoslahti, E., Hayman, E. G., Kuusela, P., Shively, J. E., and Engvall, E. (1979a). *J. Biol. Chem.* **254**, 6054–6059.

Ruoslahti, E., Pekkala, A., and Engvall, E. (1979b). *FEBS Lett.* **107**, 51–54.

Ruoslahti, E., Hayman, E. G., and Engvall, E. (1980). *In* "Cancer Markers: Developmental and Diagnostic Significance" (S. Sell, ed.), pp. 485–505. Human Press, Clifton, New Jersey.

Ruoslahti, E., Engvall, E., Hayman, E. G., and Spiro, R. G. (1981). *Biochem. J.* **193**, 295–299.

Saba, T. M. (1970). *Arch. Intern. Med.* **126**, 1031–1052.

Saba, T. M., and Jaffe, E. (1980). *Am. J. Med.* **68**, 577–594.

Saba, T. M., Blumenstock, F. A., Weber, P., and Kaplan, J. E. (1978a). *Ann. N.Y. Acad. Sci.* **312**, 43–55.

Saba, T. M., Blumenstock, F. A., Scovill, W. A., and Bernard, H. (1978b). *Science* **201**, 622–624.

Santoro, S. A., and Cunningham, L. W. (1979). *Proc. Natl. Acad. Sci. U.S.A.* **76**, 2644–2648.

Schlessinger, J., Barak, L. S., Hammes, G. G., Yamada, K. M., Pastan, I., Webb, W. W., and Elson, E. L. (1977). *Proc. Natl. Acad. Sci. U.S.A.* **74**, 2909–2913.

Sekiguchi, K., and Hakomori, S. (1980a). *Proc. Natl. Acad. Sci. U.S.A.* **77**, 2661–2665.

Sekiguchi, K., and Hakomori, S. (1980b). *Biochem. Biophys. Res. Commun.* **97**, 709–715.

Seppä, H. E. J., Yamada, K. M., Seppä, S. T., Silver, M. H., Kleinman, H. K., and Schiffmann, E. (1981). *Cell Biol. Int. Rep.* **5**, 813–819.

Sieber-Blum, M., Sieber, F., and Yamada, K. M. (1981). *Exp. Cell Res.* **133**, 285–295.

Singer, I. I. (1979). *Cell* **16**, 675–685.

Smith, H. S., Riggs, J. L., and Mosesson, M. W. (1979). *Cancer Res.* **39**, 4138–4144.

Stathakis, N. E., and Mosesson, M. W. (1977). *J. Clin. Invest.* **60**, 855–865.

Stemberger, A., and Hörmann, H. (1976). *Hoppe-Seyler's Z. Physiol. Chem.* **357**, 1003–1005.

Stenman, S., and Vaheri, A. (1978). *J. Exp. Med.* **147**, 1054–1064.

Takasaki, S., Yamashita, K., Suzuki, K., Iwanaga, S., and Kobata, A. (1979). *J. Biol. Chem.* **254**, 8548–8553.

Terranova, V. P., Rohrbach, D. H., and Martin, G. R. (1980). *Cell* **22**, 719–726.

Toole, B. P. (1981). *In* "Cell Biology of Extracellular Matrix" (E. D. Hay, ed.). Plenum, New York (in press).

Urushihara, H., and Takeichi, M. (1980). *Cell* **20**, 363–371.

Vaheri, A., and Mosher, D. F. (1978). *Biochim. Biophys. Acta* **516**, 1–25.

Vaheri, A., Kurkinen, M., Lehto, V.-P., Linder, E., and Timpl, R. (1978). *Proc. Natl. Acad. Sci. U.S.A.* **75**, 4944–4948.

Vlodavsky, I., and Gospodarowicz, D. (1980). *Nature* (*London*) **289**, 304–306.

Vuento, M. (1979). *Hoppe-Seyler's Z. Physiol. Chem.* **360**, 1327–1333.

Vuento, M., Wrann, M., and Ruoslahti, E. (1977). *FEBS Lett.* **82**, 227–231.

Vuento, M., Vartio, T., Saraste, M., von Bonsdorff, C.-H., and Vaheri, A. (1980). *Eur. J. Biochem.* **105**, 33–42.

Wagner, D. D., and Hynes, R. O. (1979). *J. Biol. Chem.* **254**, 6746–6754.

Wagner, D. D., and Hynes, R. O. (1980). *J. Biol. Chem.* **255**, 4304.

West, C. M., Lanza, R., Rosenbloom, J., Lowe, M., Holtzer, H., and Avdalovic, N. (1979). *Cell* **17**, 491–501.

Willingham, M. C., Yamada, K. M., Yamada, S. S., Pouysségur, J., and Pastan, I. (1977). *Cell* **10**, 375–380.

Wrann, M. (1978). *Biochem. Biophys. Res. Commun.* **84**, 269–274.

Wylie, D. E., Damsky, C. H., and Buck, C. A. (1979). *J. Cell Biol.* **80**, 385–402.

Yamada, K. M. (1978). *J. Cell Biol.* **78**, 520–541.

Yamada, K. M. (1980). *Lymphokine Rep.* **1**, 231–254.

Yamada, K. M. (1981). *In* "Cell Biology of Extracellular Matrix" (E. D. Hay, ed.). Plenum, New York (in press).

Yamada, K. M., and Kennedy, D. W. (1979). *J. Cell Biol.* **80,** 492–498.

Yamada, K. M., and Olden, K. (1978). *Nature (London)* **275,** 179–184.

Yamada, K. M., and Pastan, I. (1976). *J. Cell. Physiol.* **89,** 827–830.

Yamada, K. M., and Weston, J. A. (1975). *Cell* **5,** 75–81.

Yamada, K. M., Yamada, S. S., and Pastan, I. (1975). *Proc. Natl. Acad, Sci. U.S.A.* **72,** 3158–3162.

Yamada, K. M., Yamada, S. S., and Pastan, I. (1976a). *Proc. Natl. Acad. Sci. U.S.A.* **73,** 1217–1221.

Yamada, K. M., Ohanian, S. H., and Pastan, I. (1976b). *Cell* **9,** 241–245.

Yamada, K. M., Schlesinger, D. H., Kennedy, D. W., and Pastan, I. (1977a). *Biochemistry* **16,** 5552–5559.

Yamada, K. M., Yamada, S. S., and Pastan, I. (1977b). *J. Cell Biol.* **74,** 649–654.

Yamada, K. M., Olden, K., and Pastan, I. (1978). *Ann. N.Y. Acad. Sci.* **312,** 256–277.

Yamada, K. M., Kennedy, D. W., Kimata, K., and Pratt, R. M. (1980a). *J. Biol. Chem.* **255,** 6055–6063.

Yamada, K. M., Olden, K., and Hahn, L. E. (1980b). *In* "The Cell Surface: Mediator of Developmental Processes" (S. Subtelny and N. K. Wessells, eds.), pp. 43–77. Academic Press, New York.

Yamada, K. M., Kennedy, D. W., Grotendorst, G. R., and Kleinman, H. (1981). *J. Cell. Physiol.* **109,** 343–351.

Yamada, S. S., Yamada, K. M., and Willingham, M. C. (1980). *J. Histochem. Cytochem.* **28,** 953–960.

Zardi, L., Siri, A., Carnemolla, B., Santi, L., Gardner, W. D., and Hoch, S. O. (1979). *Cell* **18,** 649–657.

Index